Proceedings of the Institution of Mechanical Engineers

International Conference

Engineering Design

Volume II

22 – 25 August 1989
Harrogate International Centre

Sponsored by the Manufacturing Technology and Design Group of the Institution of Mechanical Engineers

In association with
Fédération Européenne d'Associations Nationales d'Ingénieurs
Commission of the European Communities Sprint Programme
American Society of Mechanical Engineers
Design Council
Engineering Council
Institution of Electrical Engineers
Schweizerischer Ingenieur und Architekten Verein
Société des Ingénieurs et Scientifiques de France
Verein Deutscher Ingenieure

Published for the Institution of Mechanical Engineers by
Mechanical Engineering Publications Limited

The Publishers are not responsible for any statement made in this publication. Data, discussion and conclusions developed by authors are for information only and are not intended for use without independent substantiating investigation on the part of potential users.

Printed by St Edmundsbury Press, Bury St Edmunds, Suffolk

Contents

RELIABILITY

INFORMATION

LATE PAPER

C377/128

A computer aided system for production design

H MEERKAMM, Dr-Ing, K FINKENWIRTH, Dipl-Ing and U RÄSE, Dipl-Ing
Universitat Erlangen-Nurnberg, Erlangen, West Germany

SUMMARY

Design for production demands processing a great quantity of information.
Therefore information modules and design modules, as the main components of a
computer aided system for production design, shall support the designing
engineer. Hierarchical structured elements for designing parts and a
knowledge based check of their production-feasibility enable him to realize
solutions designed for production which can be manufactured at lowest costs.

1   INTRODUCTION

The engineer's task is to design products and their components which fulfil
the function demanded on the one hand, and can be manufactured at lowest cost
on the other hand.

Therefore the designing engineer has to consider the details of manufacturing during all stages of his design work. For that he has to deal with a
great quantity of information and data of production. Requirements for this
are: either he himself has a large production knowledge (this should be general as well as specialized in the factory he works at), or this is available
to him in some way or other. Depth and breadth of this knowledge necessarily
depend on the actual stage of design.

This shows that the consideration of all manufacturing aspects puts
great demands on the designing engineer. Sometimes he is able to meet them
because of his personal production knowledge or after consulting a manufacturing specialist. Very often in practice, however, he does not (or not
enough) design for production. Wrong decisions of the designer can often be
revised very late only, or not at all. The consequences are: a big loss of
time and unnecessary cost in design and manufacturing.

The use of CAD exclusively cannot compensate this deficiency. What we
therefore need is a computer aided system for production design, based on
commercial CAD systems.

2 COMPONENTS OF DESIGN SYSTEM

The actual CAD systems are essentially concentrated on processing geometrical data and functions based on them. Information processing which is an important part of the designer's activity, is not at all or not enough supported.

In view of this situation a conception of a computer aided design system was worked out. This fulfils the following requirements:

o Suitable possibilities of information processing shall ensure a design for ease of manufacturing. For that the system must:

- support the designing engineer in all stages and types of design without restriction on his creativity

- meet the designer's working method by relevant modelling functions

o It shall be based on commercial CAD systems. Existing modellers should only be used for visualization.

In order to realize such a design system it is necessary to integrate into CAD systems all possibilities of rule processing and management of a great lot of data. Existing attempts and techniques such as data bases and expert-systems should be used.

2.1 Overall Concept

The main components (fig. 1) of the system are: information modules and construction modules. According to the groups of production techniques they have access to the production knowledge base. To attain an efficient coordination between design process and knowledge processing it is necessary to work with defined constructive elements for the description of parts.

A useful interaction between construction modules and information modules ensures an efficient support of the designing engineer. The designer works directly with the construction module which gives him access to the constructive elements having a hierarchical structure. Each finished design will be checked on manufacturing criteria by the information modules which activate the knowledge basis. Consequently the design engineer is able to design for ease of production, supported by the computer aided design system.

2.2 Construction Module

The construction modules according to the different groups of production techniques are a fundamental part of the design system. Each module has access to elements being either general or specific for a definite manufacturing process. These elements are hierarchically structured in:

- fundamental construction elements (KGE)

- construction elements (KE)

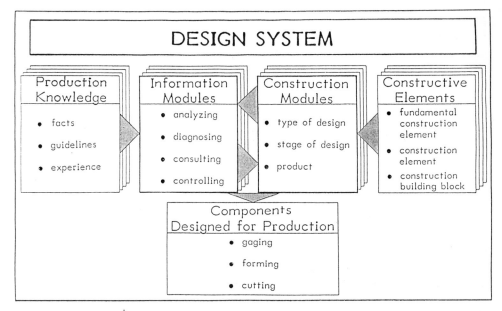

Fig. 1: Main components of Design System

- construction building block (KB)

In addition there are elements of technology (TE), function (FE) and organisation (OE). By combination of these elements complete parts can be described with regard to their shape, technology, function and organisation. The connection to individual manufacturing processes is ensured by specific modelling instructions.

By these constructive elements, a suitable design orientated interface and an object orientated data-structure the design engineer can be supported in all design stages. The design system proper can be adjusted to individual types of problems and product spectra by defining user-specific construction elements and construction building blocks, both based on the fundamental construction elements.

It is a philosophy of the system that parts whose production procedure will be expensive should demand much effort by the designer. This could lead him to low-cost design. Therefore plant-specific standards (e.g. construction elements and construction building blocks) should be defined only by the standards engineer, and not by the designer.

Figure 2 shows some examples of elements which are used by the construction module "design for bent parts". Most parts require different production procedures before they are really finished (e.g. casting, turning, milling). Therefore it is important for the designer that the construction modules, each based on different manufacturing processes, are compatible with each other. Because of this a dynamic change from one construction module to another will be possible. This is very important for the practical use of the

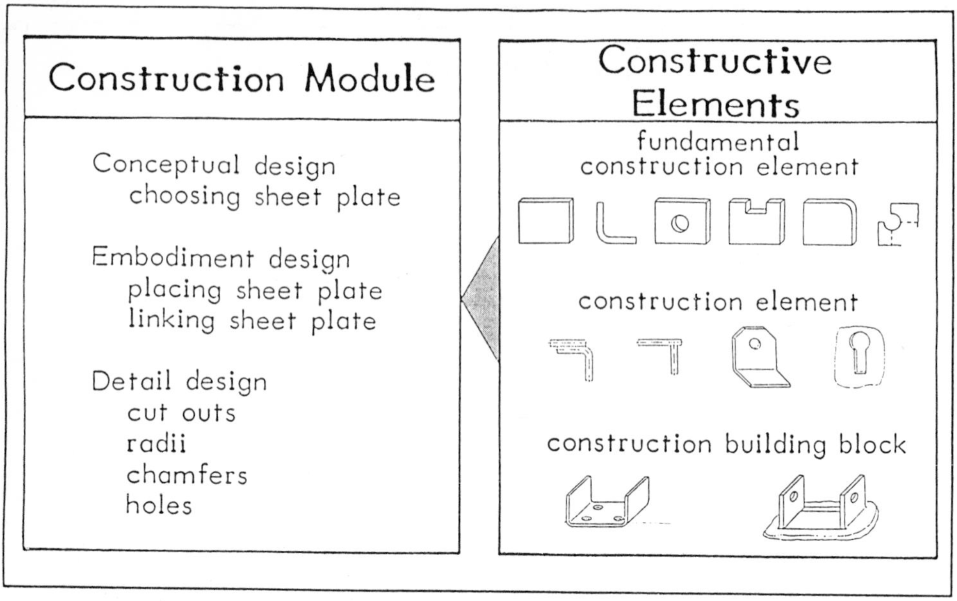

┌─────────────────────────────┬─────────────────────────────┐
│ **Construction Module** │ **Constructive Elements** │
├─────────────────────────────┼─────────────────────────────┤
│ Conceptual design │ fundamental │
│   choosing sheet plate │ construction element │
│ │ │
│ Embodiment design │ construction element │
│   placing sheet plate │ │
│   linking sheet plate │ construction building block │
│ │ │
│ Detail design │ │
│   cut outs │ │
│   radii │ │
│   chamfers │ │
│   holes │ │
└─────────────────────────────┴─────────────────────────────┘

Fig. 2: Construction Module and Constructive Elements

design system.

2.3   Information Module

All the manufacturing information necessary for the designing engineer is
available in an extensive knowledge basis. This could primarily be classified
according to the different manufacturing processes. All the guidelines and
limits of these processes, together with the inherent tools and machines form
the substance of this knowledge basis. Of course, it is necessary to differentiate between general production knowledge and a knowledge that is
specific for the workshop.

   During the design process the construction modules initiate the information modules enabling an access to the knowledge basis. It will hereby be
possible to analyse the components as to whether they fulfil the guidelines
for the production design or not. (fig. 3)

   The result of this component's analysis can be multiple. First of all it
may only be a diagnosis by stating that the designed component offends
against the guidelines of production design. (i.e. there is a difference
between the design and the rules put into the knowledge basis). Furthermore
the analysis can be a consultant. Then the system does not only point out the
mistakes made by the designer, but makes proposals for remedial actions. At
last the analysis may have a controlling function by having a restrictive
influence on the system.

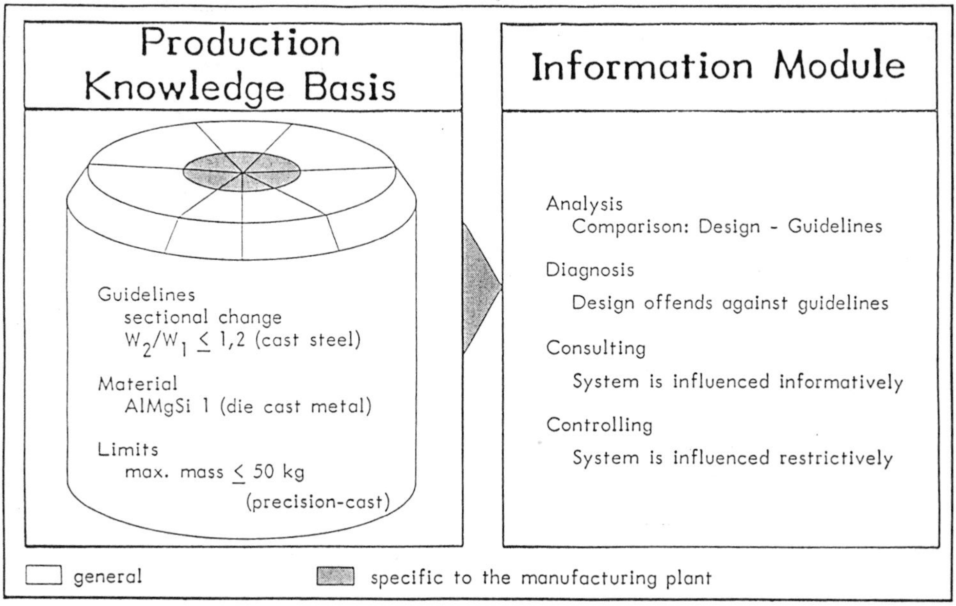

Production Knowledge Basis	Information Module

Guidelines
sectional change
$W_2/W_1 \leq 1,2$ (cast steel)
 	
Material
AlMgSi 1 (die cast metal)
 	
Limits
max. mass $\leq$ 50 kg
(precision-cast)
	Analysis
	  Comparison: Design - Guidelines
	
	Diagnosis
	  Design offends against guidelines
	
	Consulting
	  System is influenced informatively
	
	Controlling
	  System is influenced restrictively

☐ general   ▨ specific to the manufacturing plant

Fig. 3: Information Module and knowledge basis

This analysis should not only accompany the design process, but should rather follow when the design is finished. That is a condition for checking the accordance with the guidelines of production design at an existing component when manufacturing conditions have changed or are completely new.

2.4 Component Model - Data Structure

An efficient interaction between the different information modules and construction modules can only be guaranteed by a suitable data structure shown in figure 4.

It is important that constructive elements, i.e. slots, radii, chamfers etc. and constructional interconnections within the component must be identified correctly. Only in this manner it is possible to design with this computer aided constructional system and to carry out intelligent control operations.

Therefore the data structure has to be orientated to design, i.e. it must enable a hierarchical structure corresponding to that of the constructive elements. Orientation to objects and a hierarchical structure are not only fundamental conditions for the operation of the system, but present additional advantages:

- they enable simple modifications of the components without destroying the constructive connections

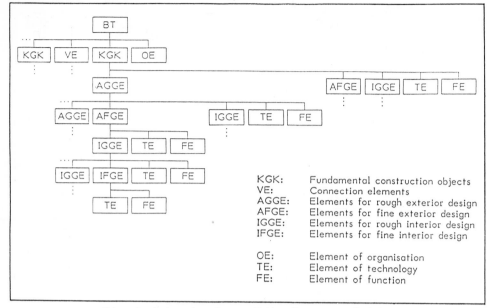

Fig. 4: Data structure of the Component Model

- they facilitate the operational standardization and the search of repeat components by structural analysis of already existing parts

- they enable an administration of design configurations free from redundancy by recording the actual change of the component model

3   CONSTRUCTION SYSTEM "DESIGN FOR PRODUCTION"

The preceding chapter presented the construction system "Design for Production" in its overall architecture, with its main components and its data structure. In the following we would like to deal with the function of its modules. With its plurality of construction modules and information modules which are specific to the manufacturing processes, the construction system "Design for Production" represents an extensive overall system. Designing for production of say cast components, sheet metal parts, turned parts means to use only shares of the complete volume of the system. These can be understood as subsystems for special fields, e.g. as a construction system for cast components, sheet metal parts or components machined by turning.

3.1   Construction System "Cast Components"

As explained in chapter 2.1 each manufacturing process has its own construction module and information module. Designing for production means - depending on the complexity of the components to be designed - either the use of one module or, more frequently, modules of different manufacturing processes. This procedure will be shown by some examples in the following.

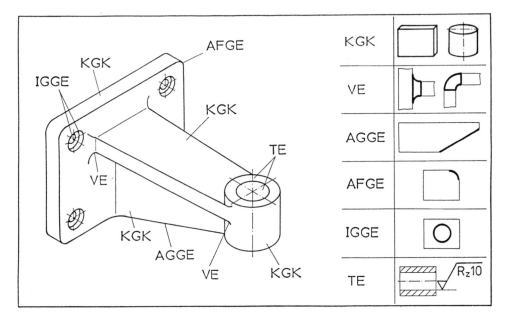

Fig. 5: Cast component and its Constructive Elements

Figure 5 shows a cast component which can be designed by means of the construction system. On the right hand sight there is a collection of elements which can be used for describing cast components. These elements have got a hierarchical structure with fundamental construction elements, construction elements and construction building blocks. The cast component shown is assembled by fundamental construction shapes (KGK). These are joined together by connection elements (VE), which are specific to the casting process. The fundamental construction elements can be modified by elements for rough exterior (AGGE) and interior (IGGE) design. Corner radius and inclined cast surfaces can be inserted by fine design elements (AFGE, IFGE).

The necessary reference to other manufacturing processes shall be made by elements of technology (TE), e.g. those describing surfaces (quality) or fundamental construction elements which are not specific for casting. This shows again that the different construction modules must be compatible with each other and handled flexibly.

The interaction of the construction module "Casting" with other construction modules and information modules finally guarantees that the component is really designed for production, including all processes necessary for its manufacture. Casting-specific design guidelines to be controlled by the system are e.g.: aim at uniform wall thicknesses and cross sections and gradual changes of the cross section. Aim at a corner radius of sufficient size.

3.2 Construction System "Sheet Metal Parts"

The functional sequence of the construction system "Sheet Metal Part" is analogous to the above mentioned system. The bent part with cutouts shown in figure 6 can be designed with the constructive elements shown on the right. First of all the three fundamental construction elements are positioned to each other and joined together by connection elements. The further rough design for exterior and interior shapes is effected by the relevant rough design elements. Detailing will be done with fine exterior and interior design elements.

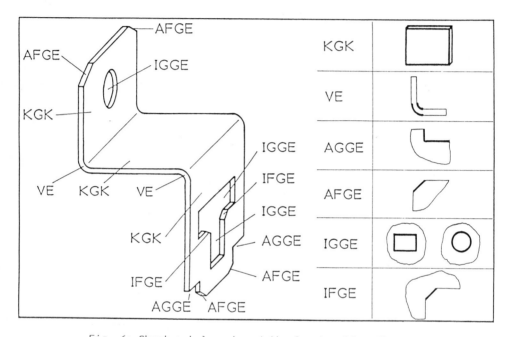

Fig. 6: Sheet metal part and its Constructive Elements

The schedule of the single steps of modelling is free for the designing engineer, it has no influence on the data structure of the component. It is obligatory, however, to follow the hierarchical structure of elements.

The information modules here guarantee also the consideration of guidelines for production design. They ensure, for example, that the part can be projected into a plane or bent one, that there are the right tools for defined radii or that suitable machines are available for the sheet size chosen.

3.3 Construction System "Turned Parts"

The procedure explained above is, in principle, valid for lathe-cut-parts, too (fig. 7).

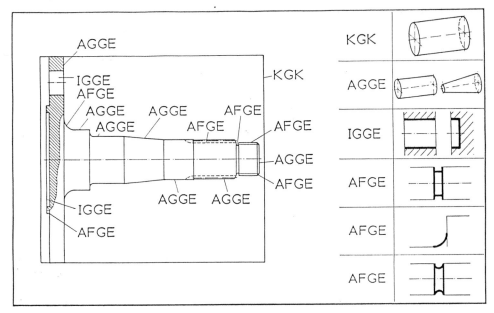

Fig. 7: Turned part and its Constructive Elements

Their basis is a cylindrical fundamental construction element which can be compared with semifinished material as a blank. Then the rough outline of the part will be defined with the rough exterior design elements: cylinder and cone. The shaft can be designed in detail with the fine exterior design elements, e.g. radii, recesses, threads, spline shafts, chamfers etc. It is true that the surfaces to be machined are described with design elements, but this does not include set operations.

The information module here controls the possibilities of manufacturing, too. Furthermore it checks whether the design elements (e.g. threads, pinhole images etc.) correspond to the actual standards or the guidelines of NC-technique.

4   EXAMPLE OF APPLICATION

The use of the computer aided construction system will be shown exemplarily by means of the turned part already discussed above.

Figure 8 shows some alternative blanks for the component to be machined by turning. Starting point for manufacturing of the part can be either semi-finished round steel (DIN), or a blank being forged, cast, or welded. These four alternatives for one simple turned part demonstrate that - especially in the beginning stage of the design work - one must be able to change the kind of the manufacturing process without any problems. Therefore it is necessary to have a random access to the different construction modules. That under-lines again that the modules must be compatible with each other.

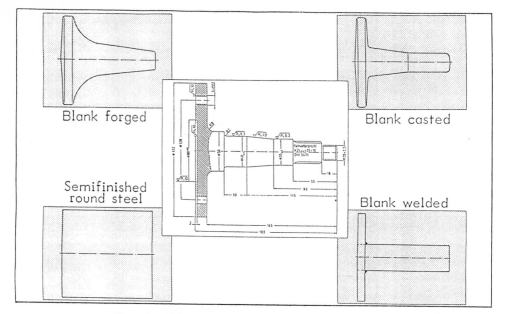

Fig. 8: Alternative blanks for a turned part

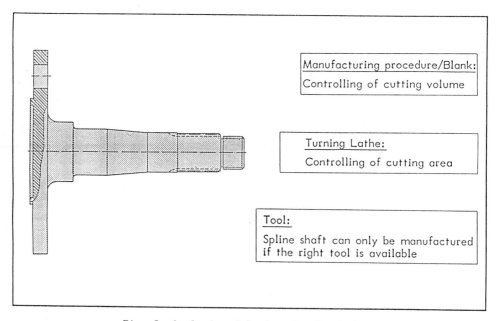

Fig. 9: Analysis of Design for Production

Figure 9 shows in what kind and to what extent the construction system "Turned Parts" supports the designer by suitable analysis of the component. Already during the rough design of the shaft the control of the cutting volume can give information as to whether the blank chosen is suitable and economic. You can check automatically whether the necessary semifinished material is available or not.

By controlling the dimensions of the part the system finds out whether the component can be machined on the turning lathes of a specific workshop. Spline shaft profiles, for example, will be checked on their feasibility. If there are priority profiles, including the necessary tools, they are directly offered to the designer.

This ensures that the designer will get the relevant information just in time. Accompanying the complete design process, the computer aided construction system takes care of the necessary analysis and controls: From the first idea relative to a component via the decision concerning a definite blank and the first layout, up to the stage of elaboration.

5   CONCLUSIONS

With the construction system "Design for Production", as proposed here, the designing engineer will be able to design for ease of manufacturing better than before. This is possible based on commercial CAD systems, as our first attempts show. The main components of this system are: construction modules and information modules being compatible with each other, though they are specialized on different manufacturing processes.

The efficient interaction of these modules is based on a suitable part model and relevant constructive elements for describing the parts. Combined with a quick access to an extensive manufacturing knowledge base this can support the designer efficiently by a consulting specialized on the workshop and the task. Solutions not suitable for production can be detected earlier than before. This saves a great lot of money and time. Therefore the construction system "Design for Production" can be an essential step into the correct direction, i.e. to provide the existing CAD systems with more intelligence.

First components of this design system, based on a commercial CAD system have been realized for turned parts. Our system will be available on main frames as well as on UNIX-workstations. First the implementation was effected on a mainframe, now we are preparing an installation on the workstation.

The design system in its actual stage enables us to generate and change the geometry of parts. During the design process the system checks the availability of machines and tools necessary for production.

Additionally the design system is used for training students of production engineering in the principal subject of design science.

6 REFERENCE

/1/ Meerkamm H., Finkenwirth K.
Fertigungsgerecht konstruieren mit CAD -
Anforderungen und Lösungsansätze
Proceedings of ICED 88 S. 259-266
Schriftenreihe WDK 16 Volume 2

C377/278

An information system about CAD software for a large industrial corporation

S HOSNEDL, Ing, PhD
Skoda Concern Enterprise, Plzen, Czechoslovakia

The high development costs of computer-aided design software created solely by teams of highly qualified experts impose evergrowing demands on the effectiveness of its creation and utilization. One of the most effective means leading to this end is the repeatable use of the existing software. But the main obstacle to this simple way is usually wrong information about the solved problems and unsuitable form for their repeatable use. The present contribution summerizes the basic principles of the classification, record-keeping and documentation system for CAD software that has been developed and put to use.

1 INTRODUCTION

One of the indispensable tools for computer-aided design (CAD) in practice is software solving of ´CAD problems´ of a certain organizational unit (individual, group, department, section, etc.). This software can be either bought, bought and adapted, or created by one´s own means. The first option is the most effective one, the second one is also suitable. But in this way, mostly only software of a universal character can be acquired, above all system and function software.

Only seldom do we succeed in acquiring application software dealing with a particular production programme of a corporation. Therefore, it is usually inevitable to create such software by one's own means using proper 'know-how' and available system and function software as much as possible. The high development costs of software, however, impose evergrowing demands on the effectiveness of its creation and utilization.

For an effective creation and utilization of software a number of generally known ways are valid. It is obvious that substantial benefits may be brought about especially by repeated application of already worked out problems. One of the principal problems is the inconsistence existing between the extensive, and, therefore, also repeatable use, and the simple applicability. The more extensively a certain problem is used (e.g. a mathematical problem in comparison with the problem of a construction part), the more difficult it is to use it in current engineering design practice and vice versa. But good CAD software must work with maximally concrete representation of the product to be designed, in order to be simply applicable for design engineers. A succesful solution to this problem is a necessary basis for the repeatable use of developed software.

Another principal problem of the repeated application of the existing software follows from a prominent feature of the design activities i. e. from their large variety and number. The spectrum of such problems is the broader, the more varied and demanding is the production programme of a given organizational unit.

Only a good information system about CAD software can, under such conditions, solve effective repeatable use of available program products. Among the users of such information system there must belong not only the practicing engineering designers but also all workers who take care of the acquisition, creation and maintenance of the CAD software.

The key problem of this information system is, therefore, a classification subsystem for a whole spectrum of CAD problems. Another problem is the record-keeping information subsystem which must be readily accessible, user-friendly and which, in the first step, can offer any user a suitable program product or products with the most important set of corresponding information. In case of satisfactory information, it is necesary, in the next step, to give the user a well arranged and more detailed set of information as a user's manual and in special cases also as a programmer's manual.

2 BASIS OF SOLUTION

The basic design principle of a good program architecture is a modular structure in which its program modules provide the procedures necessary to ensure the functionality of the program as a whole. Thus, the program module can be defined as a closed subsystem of methods and parameters designed to solve a particular problem within the program in such a way that the individual modules in their functionality:
- do not overlap in the program
- are not undersized, so that no 'gaps' appear in the program.

It is further desirable that the individual program modules have an immovable, unambiguous and complete interface defined toward the program and as flexible an interface toward the computer as possible.

The basic inconsistence between the required concreteness of the representation of the product to be designed within the program on the one hand, and the call for a wide applicability of the CAD software already worked out on the other, can, therefore, be simply reconciled by decomposing the CAD software into partial (CAD) modules solving (CAD) problems at differentiated levels of abstractness:

- mathematical-logical level
- physical-mathematical level
- structural-physical level
- product-structural level.

These levels are implicitly contained in every engineering design problem as shown in Fig. 1.

But the two program modules, in view of their definition, solving even an identically orientated CAD problem, are dependent in their scope of function and interface upon a particular CAD program so that neither their scope nor interface or form need be identical. Because of this, the program modules from the existing CAD programs, having generally a different origin, cannot serve satisfactorily as a source of worked-out (CAD) modules meeting special requirements when developing other CAD programs for different purposes. Therefore, an information system about CAD software must cover not only software in the form of programs and program systems but also of partial software blocks such as program modules for the creation of new CAD program products.

Up to now, a program product has usually been used as a basic information unit in the present information systems about CAD software. Any of these program products can serve a number of different, individually accesible CAD programmed problems for the solution of engineering design problems as shown in Fig. 2.

Because of the absence of a one-to-one correspondence between program products and solvable problems, it has been found that an information system orientated toward the program products is not satisfactory. Instead, the newly developed information system on CAD software presented here uses the ´(CAD) problem´ as a basic information unit because it is much easier for users to find suitable program products for any problem to be solved.

Actually, this ´(CAD) problem´ is any programmed (or programmable) engineerig design activity which can be used for CAD
immediately or during the creation of new CAD software. Not only
the computer graphics activities but also the solutions of calculations, finding and processing of textual information, etc.,
belong under the heading ´(CAD) problem´.

3 THE CLASSIFICATION SUBSYSTEM

The developed classification subsystem has a four-layer architecture. The subject and objective of (CAD) problems have been
chosen as the basic recognition factors on all four layers. The
abstractness/concreteness or universality/specialty of these
factors is an auxiliary criterion. This consistent approach to
the decomposition on all four layers leads to a simple, well-
arranged matrix architecture in the whole classification system.

The first layer of this system is divided into two main
groups of (CAD) software (Fig. 3):
- system and function (´tool problem´ orientated) software
- application (´CAD problem´ orientated) software.

The remaining three layers divide each of these two main
groups into several smaller areas of (CAD) problems, then into a
great number of classes of (CAD) problems and, finally, into
individual (CAD) problems.

The extent of applicability of any individual ´system or
function program product´ exceeds the field of CAD and even of a
factory. In addition, the amount of different products of this
kind used in a factory, compared with CAD ´application program
products´, is relatively small. Therefore, a classification mode
(1) has been adopted for this group of software with slight
arrangements. At the present stage it contains 70 classes.

The classification of the application software appeared as the key problem because it was necessary, to meet the needs of a large industrial corporation, to ensure an effective applicability of a large number of more or less specifically orientated CAD problems. The three-layer decomposition of this extensive group was realized in the following way.

First, all (CAD) problems were decomposed into three areas denoted by generally used names: mathematics, physics and technology. These areas differ from each other according to the degree of abstractness of the representation models (Fig. 1) of the corresponding (CAD) problems:

- The mathematics area contains mathematical-logical (CAD) problems. Their common feature is the highest degree of applicability. The corresponding program products should be a standard part of the mathematical library on every computer. With regard to the high degree of applicability of the problems of this area their outstanding merit is the high effectiveness of their creation. On the other hand, their rather difficult immediate practical applicability is a disadvantage that is brought about by the high degree of abstractness of the problems solved.

- The physics area contains physical-mathematical (CAD) problems whose creation should be performed to the maximum degree by the program products taken from the mathematics area. Their common feature is a medium degree of applicability. In view of a lower applicability and a higher degree of concreteness, the program products corresponding to this area represent a partial compromise between the effectiveness of their creation and the simplicity of their immediate practical applicability.

- The technology area contains structural-physical and product-structural CAD problems. Their common characteristic is the lowest degree of universal applicability. A common area for problems at the structural-physical and product-structural level of representation has been used because of the uniform way of subsequent decomposition and due to frequent difficulties in distinguishing among them in current practice. These two levels of abstractness of problems can be separated within the technology area, if necessary. In view of the generally high specialty of the problems of this area, it is necessary to pay regard to effectiveness during their creation. This will be best done by using the program products, taken from the two preceding areas and as wide a scope of the solved problems as possible. However, the main advantage is the immediate practical applicability due to the highest degree of concreteness of the solved problems.

In the classification system presented here, the arrangement of the above mentioned three areas, according to the abstractness/concreteness of their subjects and objectives of the corresponding (CAD) problems, is shown schematically in Fig. 3.

As mentioned above, the next layer of the classification system is sectionalized into particular classes of (CAD) problems. A generally used classification system (2), (3) have served as a basis for defining hierarchically arranged subjects and objectives to which the above mentioned classes of (CAD) problems vere assigned. In spite of hierarchical structure of classes of subjects and objectives, their denominations are suitable arranged within the classification system at the same formal level. With regard to the great number of (CAD) problems, it is advantageous that the matrices of the classes of (CAD) problems within individual areas may be partially incomplete and that, as need may be, they can be gradually supplemented and brought to a desired degree of refinement while maintaining compatibility with the original arrangement.

At the present stage, the newly developed classification system contains 479 classes in the mathematics area, 2790 classes in the physics area, and 39060 classes in the technology area. In the classification system presented here, the arrangement of the individual classes of (CAD) problems, according to the universality/specialty of their subjects and objectives of the corresponding (CAD) problems, is shown schematically within the particular areas in Fig. 3.

The bottom layer of the classification system meets the requirements of the different kinds, scopes and theoretical levels of solution of (CAD) problems. Each class of (CAD) problems can, therefore, contain even more program products that solve an identically orientated (CAD) problem but at a different qualitative and quantitative level. The arrangement of the particular (CAD) problems in the classification system, according to their universality/specialty of their subjects and objectives, is shown schematically within a particular class of (CAD) problems in greater detail in Fig. 3.

4   RECORD-KEEPING AND DOCUMENTATION SUBSYSTEMS

Every program product solving one or more individually accessible (CAD) problems is documented by standardized rules. The documentation consists of two basic parts: record-keeping forms for (CAD) problems and user's manuals with programmer's manuals for the relevant program products.

The most important set of information concerning every (CAD) problem of any kind, type or extent is recorded in one type of the record-keeping form which is shown in Fig. 4. This form has two separate parts. The first part of the form contains a basic set of information on the (CAD) problem, the other part contains a set of information concerning the relevant program product that solves the appropriate (CAD) problem.

The set of information on the record-keeping forms creates a basis for the developed record-keeping (sub)system. The key role of this system consists in the classification of the subject and the objective of every (CAD) problem according to the above mentioned classification system. The set of information about every program product provides the user with the same extent of services as the traditional record-keeping systems do.

A useful, powerful and user-friendly program system ESOPP has been developed for the discussed record-keeping system. ESOPP operates with the help of the function program dBASE III plus on the personal computers IBM PC AT and compatible ones. ESOPP facilitates the completion of the record-keeping forms with the aid of extensive menus including the classification of each (CAD) problem. Further every (CAD) problem obtains automatically an identification number according to the International Decimal Classification etc.

But the main role of the program system ESOPP consists in a quick and user-convenient selection of recorded (CAD) problems and relevant program products according to user's requirements (identification number, name, autor, classification of subject and/or objective at any level, type of hardware, programming language, etc.). Completed and displayed are only those filed record-keeping forms which are relevant to the (CAD) problem asked. A record-keeping form is then printed for each selected problem.

Every individual, group, department, section, factory etc. can maintain a local record-keeping system. This possibility of maintaining a local record-keeping systems is advantageous both for an effective decentralization of locally important (CAD) problems and for an effective centralization of a selected important set of information about the (CAD) problems that have been solved already (or are in the process of solution, or planned) at 'lower' organizational levels.

For the cases in which this effective reduction of the amount of stored data will not be sufficient, a more power record-keeping system is projected. This system is based upon ESOPP which, in the batch mode, makes use of memory discs of a minicomputer. Selected parts of the centralized record-keeping database will be processed on PCs as has been customary up to now. However, it is no substantial problem to implement the whole record-keeping system on any suitable computer with an appropriate database software.

Compendious and well-arranged user´s and programmer´s manuals identified by name and identification number of the relevant program product, are the next necessary conditions for an effective use of (CAD) program products in current engineering design practice and, during the creation of new (CAD) program products, for their effective maintenance, development etc. The two above mentioned types of manuals have prescribed the following uniform structure for any kind, type, or extent of a program product:
- introductory part
- functional part
- informative part
- processing part
- technological part
- organisational part
- economic part.
The internal structure of these parts is also standardized.

In the case of simple (CAD) problems, these manuals, at the same time, serve as a project documentation. In the case of large projects, this is not possible because they cover problems of many individual (CAD) program products. Hovewer, due to the same standardized structure of such a project documentation and the structure of the proposed manuals, it is quite simple to derive user´s and programmer´s manuals for the individual (CAD) program products.

Due to the modular and uniform structure of these manuals, the effectiveness of their creation, including algorithmization and programming proper, is gradually increased with the aid of computers and copying machines.

The basic form of these manuals including program products is the textual form with algorithms and a listing of program products, with the source program products on discs. A higher form of this documentation are the enclosed texts of documents and binary programs on store discs. The highest form of the documentation makes it possible to call simply program products from a computer library when solving engineering design problems, developing new (CAD) program products, etc.

It is obvious that the higher the form of the documentation, the more effective the (CAD) program products may be used, but with more specialized demands on hardware and system software. However, for many (CAD) problems and relevant program products the basic ´paper´ form is practically independent of hardware and system software, which may be often very advantageous.

This documentation system makes possible not only an effective use of (CAD) program products but especially a repeatable application of existing program modules for a rational development of new program products ranging from the most simple modules to the most extensive program systems. This is shown simplified in Fig. 5.

5 CONCLUSION

The above mentioned information system on CAD software is an informational kernel of the devoloped system of rules for an effective utilization and creation of CAD software in a large industrial corporation (4). It can be stated that the hithertho

experiences with the use of this system prove considerable efficiency of the use and creation of (CAD) software over a wide
range of subjects, objectives, solution levels and extents of
CAD problems in the engineering design of machine tools, rolling
mills, electric locomotives, electric generators etc.

The information system presented here will enable in the
future the development of a library of CAD methods governed by a
control system of expert type i.e. development of a bank of CAD
methods. It is expected that this intelligent bank will contribute to a rapid increase in the potential of CAD methods as an
effective tool for the engineering design.

REFERENCES

(1) SAMMET J.E., RALSTON A.R. The new (1982) computing reviews
    classification system – final version. Communications of the
    ACM, 1982, <u>25</u>, 13 – 25.

(2) BOISVERT R.F., HOWE S.E., KAHANER D.K. GAMS: A frame work
    for the management of scientic software. ACM transactions on
    mathematical software, 1985, <u>11</u>, 313 – 355.

(3) International Decimal Classification, 1981, Matica sloven-
    ská.

(4) HOSNEDL S. Developing CAD software in a large industrial
    corporation. ICED´88, Budapest, 1988, 181 – 188, WDK.

MATHEMATICAL – LOGICAL LEVEL:

$$P = f_P(T\ldots)[MP_a]$$

$$L_A = f_A(|F_A|, C_A\ldots)[h]$$

$$L_B = f_B(|F_B|, C_B\ldots)[h]$$

$$F_D = f_F(T, z, m\ldots)[N]$$

$$u_i = f_u(F_0, F_R, I_{ui}\ldots)[mm]$$

$$\delta_i = f_\delta(F_0, F_R, W_i\ldots)[MP_a]$$

PHYSICAL – MATHEMATICAL LEVEL:

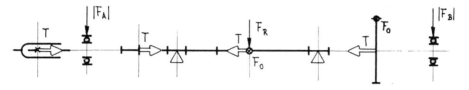

STRUCTURAL – PHYSICAL LEVEL:

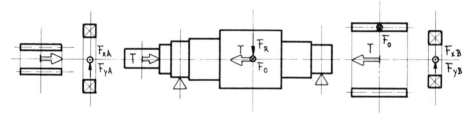

PRODUCT – STRUCTURAL LEVEL:

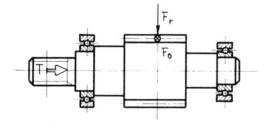

Fig 1   Decomposition of an engineering design problem into abstract levels

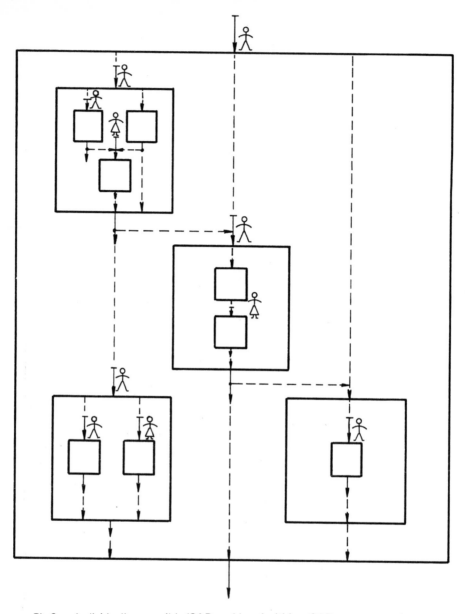

Fig 2   Individually accessible 'CAD problems' within a CAD program product

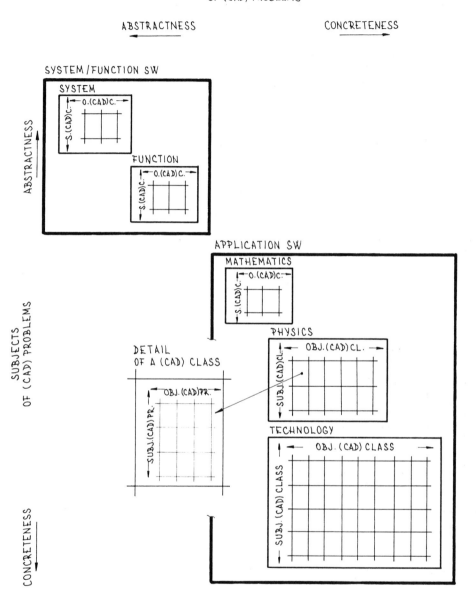

Fig 3   Scheme of the classification system for CAD problems

```
                  RECORD  CARD FOR (CAD) PROBLEM
Name,address and ident.number of organis.:!Owner of card:
ŠKODA k.p., 316 00 Plzeň                 !213201 ! AIP-AKP
Record num. of progr.product (PP):!CPsum.:!Acc. v.!Last v.
AIP/AKP - 88/018                 !   1   ! 2.1   ! 2.2
Name of PP:                               !PP designation:
Record-keeping system for (CAD) prog.prod.! ESOPP
Characteristic of PP:
Record-keeping and classification system for program pro-
ducts and problems applicable within CAD
Name of (CAD) problems (CP):      !CPnum.:!CP designation:
Record-keeping set of (CAD) PP    !  1    ! ESOPP
Aim of CP:
Record-keeping set of program products intended for re-
cord-keeping and classification of (CAD) problems. Inten-
ded for programmers and engineering designers.
Method of CP solving:
Utilization of database system dBASE III plus

Form of CP solving  :
Interactive mode

Subject of CP:                   !Objective of CP:
Program products                 !Record-keeping;classification
Area of CP:  !Field of CP:  !Sub.of CP cl.:!Obj.of CP cl.:
Mathematics  !Math.cybernet.!MATHEMATIC.S.  !INFORMATION
IDC number of CP:                !Id.num.of corr.industr.prod.:
519.72:681.3.06                  ! -
Built-in special mathematical SW:
-
Subsidiary PP:
dBASE III plus
Necessary accessories of computer:
Standard
Documentation for PP:
User´s and programmer´s manual
Sort of PP : !Lang. for PP: !Size of PP/CP:!Origin of PP :
SW SYST.     !dBASE III plus! 2500  /2500  !OWN
State of PP: !Computer:      !Oper. system :!Oper. memory.:
APPLICABLE   !IBM PC compat.!MS-DOS        !0.5 - 1 MB
Notes:
Coauthors: Ing. Hosnedl, Ing. Kvoch

Informant on PP     :        !Author´s org.:!Author:
AIP-AKP, Červený, 215-8274   !ŠKODA Plzeň   !Červený
Administrator of PP:         !Suppl. org.:  !Date of rec.:
AIP-AKP, [illegible], -8254  !ŠKODA Plzeň   !88/10/20
```

Fig 4   Record-keeping form with information on program product ESOPP

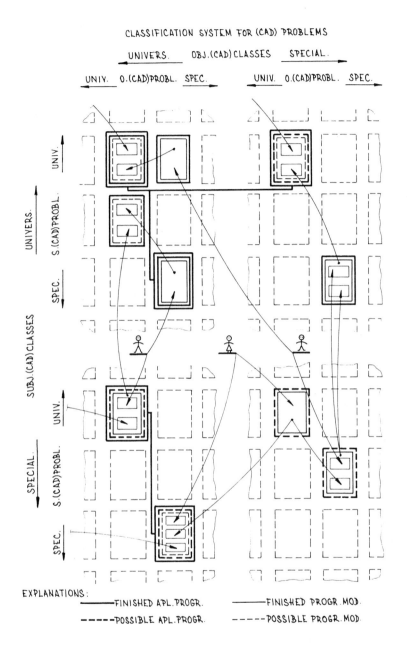

CLASSIFICATION SYSTEM FOR (CAD) PROBLEMS

EXPLANATIONS:
———— FINISHED APL.PROGR. ———— FINISHED PROGR.MOD.
————— POSSIBLE APL.PROGR. ————— POSSIBLE PROGR.MOD.

Fig 5   Modular creation of program products

C377/229

The application of computer aided design to large projects

R ELLIOTT, BSc, CEng, FIChemE
John Brown Engineers and Constructors Limited, London

The use of C.A.D. techniques is described as it is
applied by a major design contractor to a wide range
of projects, and details of the approach taken on
the control of multidiscipline design teams using
computer design techniques are given. The latest
computer techniques currently used are described,
and future trends are discussed.

1 INTRODUCTION AND BACKGROUND TO JOHN BROWN

JOHN BROWN in London are a large Engineering Design, Procurement and
Construction Company who have been established in London for the last 40
years, and are currently engaged in the fields of Oil and Gas, Nuclear and
Defence Projects. For the last 3 years we have been part of the Trafalgar
House Group.

Although we are an Engineering Design, Procurement and Construction
Company, we are interested in the C.A.D./CAM interface because of our
relationship with other Companies in the Trafalgar House Group, and to
improve the service we provide on projects.

In addition, we are increasingly approaching projects on an overall
design and supply basis which demands a close co-operation with our
sister Companies.

This has led us to look at ways to integrate Design with Fabrication
and Construction as it relates specifically to our business.

However, to date the main thrust of our application of C.A.D. has been
in the production of Design deliverables in the form of hard copy
drawings for issue to fabricators and constructors.

2 C.A.D. SYSTEM DESCRIPTION

Appendix I shows the computer hardware used by JOHN BROWN in London. The
bulk of the equipment is used for Computer Aided Design.

The C.A.D. facilities consist of two major systems:-

2.1 The Apollo Ring

This comprises of a DOMAIN ring network of APOLLO Graphic Workstations, which can operate on a stand alone basis or can be networked together to share file server units, printers and computing power.

The major advantages of this system is the portability of the individual units, its high level of computer power, and the ease with which it can be adapted to the changing market requirements. The APOLLO units are used for various Structural, Naval Architectural and Finite Element Analysis Routines, as well as for general 2-D draughting for Electrical, Instrument, Process, Building & Civil works for both Diagrams and 2D Layouts.

Because of our specialised application requirements many of the analysis programmes have been developed in-house.

The 2D Draughting programme we use is DOGS (Drawing Office Graphics System).

2.2 The Prime Facility

The second major C.A.D. facility is centred around a cluster of PRIME computers on which we have installed various programmes for 2-D and 3-D design, namely PDMS, DOGS, COMDACE and SCHEMA.

Attached to these computers are a series of graphic terminals, which enable us to design interactively the majority of our project deliverables.

2.3 Other Hardware Items

The other points of interest on our hardware are:

(a) The Silicon Graphics facility, which with software called "REVIEW", allows us to show '3-D' visualisation and step through of designs developed in the PRIME facility.

(b) The Prime UNIX 'box', which is a small but powerful multi-access facility used solely as a central database facility for procurement control and reporting.
By this I do **NOT** mean accounting records, but the status of each item from identification, enquiry, purchase, manufacture and delivery to site.
I mention this, even though it is not strictly a C.A.D. function, since its input is a material file from COMPIPE for piping materials, and interfaces are currently in hand to develop this auto-link for other materials, i.e. Structural, Electrical, Instruments, Mechanical, etc.

(c) The Versatec 7236 Plotters - We have invested heavily in high speed electrostatic plotters, as we found very early that a large plotting capacity was essential for efficient operation.

2.4 Software Used

Briefly, the software packages mentioned above are used as follows:-

PDMS (Plant Design Management Systems) is a 3-D space management program which enables us to prepare a fully defined spatial model of the project. Some examples of the 3D representations of typical projects are shown in Appendix II and III. These are in black and white, but I must emphasise that the visualisations from which they are derived are coloured, which considerably increases their impact and use. These electronic models are exact representations of the plants in each case.
From this model we then develop specialist layout drawings and detail drawings - Appendix IV and V (Cable Tray Routing Drawing and Typical Isometric Drawing).

DOGS (Drawing Office Graphics System) is used for the development of P&ID's and associated database information (line lists, equipment lists, etc), as well as other 2D drawings.

COMDACE is used to develop detailed piping isometrics and materials schedules from the PDMS data.

SCHEMA is a 2D draughting system used to develop views of 3D models, which we can convert into 2D DOGS Drawings for subsequent manipulation.

3 CONTROL OF MULTI-DISCIPLINE INTERFACE

In our business areas, the primary interface between the various disciplines of equipment, electrical, piping, structural and instruments occur at the dimensional level. That is to say the various design disciplines have to each identify their area of space, which must, as far as possible, fit into a minimum overall envelope. This is true not only for Offshore Platforms and Nuclear Design Projects but increasingly in all our areas of business.

The standard formula of space = weight = cost applies in almost every industry, but in the Offshore Industry, weight is a particularly major area of concern and in Nuclear Design, the limitations of designing a plant with an already defined biological shield enclosure size means space is always very limited.

In order to ensure that the conflicting space requirements are all satisfactorily achieved, we put the primary responsibility for the development, maintenance, and checking of the spatial model with our Layout Group.

Each discipline independently inputs its specific data into the overall 3D model, but the Layout Group has the responsibility for regularly identifying clashes between disciplines.

However, I would not want you to get the impression that all personnel could be working on one area at the same time. This in practice cannot happen, and in any case the sequence of design develops in a natural priority order.

In order to understand this, let me explain how, typically, a design develops. Appendix VI shows a typical offshore project plan.

Firstly, the equipment is sized, and the most optimum layout studied. A computer can assist at this stage, but often in practice this 'thinking time' phase is done using paper layouts, since it is more economical than tie-ing up a screen for relatively long periods on such work.

Once a basic layout has been agreed in principle, the equipment shapes are developed by the Layout designers and entered in the computer model. These are not accurate representations of the shape, but may be no more defined than a single rectangular block. All that is required is that the various connections are fully detailed, and sufficient volume is reserved for the equipment item.

The Piping & Layout Engineers then continue with the piping design development, and progressively enter the geometry of each pipeline into the system, usually starting with the largest diameter and working down to the smallest sizes.

In parallel to this, the Instrument, Electrical and Safety Engineering Groups are each developing their design detail, and some clashes therefore are inevitable at this stage.

However, at an early stage general services or utility corridor areas will have been allocated for each individual discipline, to reduce the incidence of clashes. See Appendix VII.

Where there are instances of one section needing space in another's area, e.g. for instrument panels and switchgear, or erecting instrument items at a sensible height for maintenance, the clash will be identified and a suitable alternate arrangement agreed by discussion and negotiation.

The Structural Design of the unit will also be carried out in parallel with Piping development, as secondary steelwork, ladders, platforms and pipe supports are added to the primary structure. The Structural data is then transferred at regular defined times into the overall model.

We have spent considerable development time recently in improving this interface. Firstly we have developed a computer routine in-house which takes the 3D space frame model from the analysis program and adds frame member dimensional data to it.

This then has sufficient dimensional data such that the primary structural framework information can be fed into the 3D spatial model.

The second stage we have now largely automated is in the Secondary Support Steelwork. This design is carried out in a 2D Graphics design package (DOGS). We have developed a computer routine such that the Structural Designer adds a third dimension to the essentially flat representation in order to provide a full 3D representation. This is then progressively input to the 3D spatial model.

Previously, the analysis program was used to develop the design, then the 3D structural detail was entered into the 3D model one section at a time. The automatic entry of dimensional data from the analysis program, together with the transfer of 3D data from 2D detail drawings has thus saved us considerable tedious input effort.

As the design proceeds, the clash routines are progressively run on a regular, structured basis, to ensure that each potential clash is identified and cleared as early as possible.

Once the detailed design for all individual disciplines - Piping, Structural, Electrical, etc has been fully developed, each area is then cleared completely, by running a total clash check by area.

These routines are necessarily computer intensive, so they are run overnight when the graphic design functions are not absorbing the majority of the computer power.

A detailed record is kept of the status of the computer model at any stage. This is usually presented in a tabular format for ease of reference.
See Appendix VIII - Tabular Status Report.

In addition to solid or "hard" volumes for pipes, beams etc we can also assign open space reserved for particular operations, e.g. bundle pulling area, escape routes, lifting area above removable items. This reserved but empty space is termed a soft volume.

Escape Routes are designed on offshore platforms for the safe evacuation of personnel, and this volume must therefore be kept free of all obstructions once the platforms are completed.

Maintenance access can be checked accurately either by visual inspection of the 3D visualisation, or by simulating the movement by repeated movement of the component electronically, and thus developing a required movement corridor. Appendix IX shows various views taken from a typical maintenance access corridor development. Sheet 3 of Appendix IX particularly illustrates the type of data which is easily developed in 3D CAD, but which would be prohibitively expensive to develop by manual design methods. Once developed this access corridor is also entered into the 3D model as a soft volume.

The regular clash checks give an indication of each clash between solid items ("hard to hard" clash) or between a reserved space ("soft" volume) and solid item (hard volume). These clash check reports are critically reviewed by the Designer, and progressively cleared, to ensure a workable design is developed.

The Clash checks may indicate certain apparent clashes which can in fact be accepted, e.g. "soft" area around Pipe for Construction access and/or emergency escape envelope, or pipe running above cable tray in soft area required for cable pulling and splicing. It is during the review of clashes that the expertise and experience of the designer comes into its own, as he often has to make sensible compromise decisions depending upon the degree of clash (See Appendix X). Often, the "soft" volumes are for different uses at different times, e.g. construction access and maintenance access on the completed platform. Sometimes, the designer may decide that even though "soft" volumes are involved, they cannot be allowed to overlap: for example, safety escape access and maintenance laydown areas. Clearly though, a "hard to hard" clash can never be accepted.

After inspection of each clash, it is either accepted, or cleared. The repeated printing of clashes accepted is then suppressed, but the information is still retained in the model file, so a full report can be generated if required at any time.

The following then summarise our approach to the Control of Interfaces within the design office:-

1. Always have a clearly defined single responsibility point for the spatial model. We usually vest this responsibility on our Piping and Layout Group, but for specialist layout areas, e.g. within a building or control room, other disciplines may be assigned this responsibility, for example the Instrument Group in the case of control rooms.
2. Always have a clearly defined programme, with parallel working as much as possible, but clearly planned and regulated checks being carried out at regular intervals.

Because the computer can do these checks as routine, a very high degree of accuracy can be achieved.

With regard to direct design to manufacturing interfaces, we currently generate data from our design model for direct input to Computer Numerically Controlled (CNC) bending machines for British Nuclear Fuels Ltd, to ensure the accurate production of seamless pipe spools.

For other projects, we are also setting up data links to our associated companies to enable direct data transfer from designer to fabricator to be carried out.

We have also successfully set up interactive data links to our construction sites, which have proved to be a major benefit during the construction phase, and have enabled documents and drawings to be developed at site, for use a construction aids. For example, we were able to develop 3D views of hydrotest pack systems which were of particular benefit during the execution of this work.

Whilst the same information could be produced by traditional manual means, the ease of production by use of an existing CAD data base made it economically possible and, therefore, acceptable (see Appendix XI).

4   MANAGING THE NEW TECHNOLOGY

The new technology places demands on all staff to face new challenges, but the present generation of Design Managers at all levels face the greatest challenge.

We have to be able to grasp the capabilities of the system, its pitfalls, problems and potentials, whilst not in most cases having had direct hands-on experience. We generally belong to the pre-computer generation, and have had to learn about computers at a relatively advanced age.

I personally do not advocate that direct, hands-on experience is absolutely necessary, although our Computer Systems people would not agree with me, but it does ensure that the basic computer language is understood by the managers. It is very easy to be misled into believing that the C.A.D. system can do more than it does, simply because we fail to ask the right questions.

In my Company, the major decisions on equipment and software purchase are made by the Computer Systems personnel. Their background in Engineering helps to ensure that these decisions are the right ones for the long term development of the C.A.D. systems.

We are particularly lucky in our Company, as our Engineering and Systems Support people work very closely together, and have a high degree of mutual respect and understanding of each others problems. I believe this is essential to avoid problems.

C.A.D. supervisors or toolmakers are also important. These people must have a clear grasp of the technology, and can thus objectively assess the time taken to do things, the capability of new equipment, and the validity of claims made by computer salespeople for new hardware or software.

As managers, it is each persons responsibility to apply some degree of cynicism to all you are told - my main computer buzzword is - "show me".....

The control of the technology needs the following:

1.   Sufficient personnel of adequate experience with clearly defined terms of reference and lines of communication.
2.   A clear structure with adequate feed back systems.
3.   An understanding, at all levels, of the systems potential, its particular areas of weakness, and the facility to assess potential or perceived bottlenecks in the system to overcome problems.

4. A close and ongoing Computer Systems and Engineering Department relationship, to ensure problems as they arise are tackled on a mutually beneficial basis, not with an "us and them" attitude. Any problem will usually occur at the least convenient time, when urgent deadlines have to be met, or high workloads are in progress.

5. Procedures to ensure that software systems and data are properly controlled, validated and checked, and that proper back up procedures are in place. There also needs to be some procedure to ensure these procedures are followed.

JOHN BROWN have always taken great care to ensure that computer software packages are properly controlled. Because some of our workload is Defence related, we have had our Quality Assurance and Control Procedures on the use of computer software assessed by the Ministry of Defence (MOD) and been issued with an AQAP 13 certificate, which is a NATO Standard relating to the control and use of computer software.

5 THE LATEST COMPUTER TECHNIQUES

5.1 3D Visualisation

We have recently acquired a high capacity, high resolution, stand alone Silicon Graphics, colour workstation which I previously referred to in the section on hardware.

This equipment, together with latest module of the PDMS software suite, denoted as 'REVIEW' gives us an ability to run a model review of the plant on a step through basis.

The increasing quality of the 3D C.A.D. model has resulted in our clients no longer demanding a physical model as part of the design. The traditional model review where all interested parties reviewed the plastic model, has been superseded by the the use of these electronic techniques, which enable us to show how a plant will appear from the inside; something it was always difficult to show with a plastic model (Appendix XII).

This equipment, despite its tremendous capacity, is truly portable, and is readily moved from office to office.

It is usually linked to the Ethernet for data transfer purposes. Like all computer equipment, it can be somewhat upset if moved, so we try not to move it too often.

An A3/A4 colour laser printer, is used to produce high quality colour copies on to film or paper.

I must emphasise that this 'REVIEW' facility does not allow change of the model in any way. However, for model reviews and presentations, this is exactly what is need.

5.2 Drawing Scanning

There are now Computer Systems which can scan a manually produced drawing and translate it into digital data. Our facility is centred in Holland and it is currently extensively used by several clients.

It is possible to scan even poor quality reproducible or microfilm copies and generate very high quality C.A.D. outputs.

This technique enables design contractors to take existing plant drawings and to quickly use them for plant revamp design, without the need to input the whole design piece by piece into the computer data base.

5.3 Computer Storage and Retrieval of Data

The ability to assign attributes to a line or item gives the designer and operator the ability to use a graphic presentation as the master document index on a project, and to call up at will all back-up data to that item. For example, all maintenance information, including instructions, location of parts and data on last maintenance could be called up by pointing out a specific item on the screen. Similarly, full details of the design history of the individual items can be called up using the same facility.

This clearly has applications for offshore maintenance operations, as well as in the control of the whole design process. This whole scenario is only limited by the computer storage capacity, and our ability to organise the data so that it may be retrieved easily.

We are already at a stage where the stored data is potentially so vast in scope that retrieval could become a significant problem unless the information has been stored in a carefully structured manner.

6 FUTURE TRENDS

In the short to medium term, we expect to see more powerful local workstations gradually reducing the dependency on the central computer processing power.

Gradually, as costs narrow, all workstations will be colour terminals. This will assist the engineers in the design, but will not immediately lead to the widespread generation of coloured engineering data, due primarily to the high differential cost of reproduction between monochrome and colour. A major technical breakthrough in colour copying, leading to major price reductions, may cause us to radically re-think our method of presentation, but until this occurs, the vast bulk of engineering drawings and data will continue to be in monochrome.

The faster processing powers and capacities of desk top machines are bound to lead to more and more sophisticated design programmes being available on PC's.

However, my own view is that on a major project, either a network of PC's or a series of mini "rings", linked together, will be the normal arrangement.

In the longer term, the development of parallel processing computers, (1) still very much in the embryonic development stage, must be expected to have a large impact on engineering design and analytical programmes, for which they seem to be tailor made. For example, our present clash detection routines would be greatly speeded up by this technique, to the point, possibly, where the machine would be constantly running such checks, and would immediately highlight when such a clash was attempted by the designer.

The enormous potential which parallel processing holds out may well lead future generations to regard our present computer hardware as no more than the fore-runner of "true" computers.

Against this vast potential, there are still areas of computer design which appear to have been static for a long time, without any clearly visible new development on the horizon.

(2) The man machine interface is perhaps the least clearly defined future development area, although some researchers are now beginning to look at this key area.

The QWERTY keyboard, still the main method of access into C.A.D. systems, has not really developed since the days of the first typewriters.

We are already finding that the man/machine interface (i.e. screen and keyboard or puck (mouse)) is often the limiting factor, not the machine power or size.

It is possible that recent developments in voice recognition techniques together with these substantially more powerful computer systems will take us straight into direct input via voice.

Perhaps then all designers will need is a crash course in elocution!

In the area of data storage and retrieval, the present magnetic storage of data, whilst generally very reliable, does have certain inherent problems, basically associated with its ease of alteration, and its vulnerability to stray magnetic fields.

I believe that the future developments of laser disc storage systems will have inherently less problems in this respect, and so will be far more suitable for long term storage and archiving of data once a design is essentially "fixed".At present, in my view, magnetic media is not really suitable for long term (10 years+) archiving of data, since it is necessary to regularly re-audit the data to confirm its continuing viability.

As the storage systems develop into new areas:- with such things as bubble memory or "juke box" laser disc storage systems, the potential storage capacity for data will vastly increase, to a point when it will be theoretically possible to store **all** data, including **all** documents, specifications and drawings on a single computer system. Of course, the volume of data stored may well completely swamp the record keeping systems, and files will be irretrievably lost if these systems are not managed effectively.

The latest trends in Artificial intelligence or Expert Systems may well lead us to use rule based techniques as a substitute for our designers and engineers in a wide range of disciplines. We are presently looking at the way the market and products are developing, and expect to be moving into this area of development in the next 18 months.

We already have certain programs which use expert systems architecture, for example in the field of fracture mechanics analysis.

As with all our computer developments, the Computer Systems Department personnel are much more in touch with the latest market status and products available, and to a large extent they will lead this area, knowing what our general requirements are within the Company.

As with all computer developments used in an active design production environment, any development will be on a step by step basis and cannot afford to change our whole design routines by a major or radical switch to a new system even if it has been fully proven elsewhere - it may not work for us. We must learn to walk **well** before we learn to run.

7 PITFALLS AND PROBLEMS

Lest you should all go home believing that C.A.D. techniques are the answer to all your problems, let me just touch on the problems and pitfalls.

First, it is expensive, not just in hardware, but in support services, and training of personnel. Training has to be an on ongoing, (continuous) process. We have been using C.A.D. techniques for about 10 years now, and are still learning.

Reliability of the hardware and software, whilst generally improving, is still far from 100%, and can lead to disastrous last minute panics. The equipment and/or software always fail when it is most heavily needed.

From painful experience we have found that there are no clear rules about system capacity when so many different programs are run on systems simultaneously.

The first indication that we are approaching system capacity is a deterioration in system response time. This does not always appear to be predictable, but it can to some extent be alleviated by avoiding day time computer intensive activities and removing from the system any old or unused data files.

The C.A.D. supervisors have to strike a balance between leaving old but occasionally used files on the system, and seeing the response times deteriorate.

8   SUMMARY OF CONCLUSION AND BENEFITS

C.A.D. can offer major benefits in design, but it brings with it new problems, not previously encountered. On some projects, a crisis of confidence occurs when the Project Personnel do not see drawings and data produced at the same rate as on a conventionally designed or "manual" Project. It takes some confidence in the system and the people to convince the project personnel that they will get their drawings on time, when you are approaching the design deliverable deadlines, and no actual drawings have been issued.

Our experience is represented Graphically In Appendix XIII.

This illustrates the critical nature of fast reliable printing equipment upon both the visibility of the design and the credibility of C.A.D. techniques in general.

You will note, however, that we actually get our designs out earlier, using C.A.D. techniques.

Whilst I recognise that C.A.D/C.A.M. techniques hold out the prospect of bypassing this intermediate paper drawing step, this is not yet the case on major projects, and it is difficult to see when such a position will become viable.

The man who invents the Robot machine that will dig the hole, pour the concrete, erect the steel, weld the pipe, lay cables and paint the plant will deserve every penny of the fortune he will surely make. Until that time arrives, the future developments in the use of C.A.D. for large projects will be in automating specific parts of the C.A.D./Fabrication interface, and substituting magnetic media or electronic instructions for drawings and specifications on a piece meal basis.

However, what we can now achieve using C.A.D. techniques for Large projects significantly improves the quality of the design documentation, reduces the design cycle, and provides real cost savings in the Engineering Design process.

In addition, the use of C.A.D. gives us the ability to carry out checks on our design and to provide data to the construction team that would have been prohibitively expensive by manual means. For example, Colour 3D Views, Sequence Build Drawings and regular up to date 3D views of the plant as the design proceeds. These features make the construction tasks easier and again lead to real cost reductions. Finally, the "Review" facilities now provide superior visual data for project model reviews to that previously provided by plastic models and has effectively made plastic models obsolete as a design tool.

REFERENCES

1. Advanced Computer Architectures by G.C. Fox and P.C. Messina
    Scientific American - Trends in Computing - Special Issue Volume 1
    1988.
2. Interfaces for Advanced Computing by J.D. Foley
    Scientific American - Trends in Computing - Special Issue Volume 1
    1988.

ACKNOWLEDGEMENTS AND GRATEFUL THANKS
ARE MADE TO THE FOLLOWING COMPANIES

-   PAFEC LTD. for references to DOGS.
-   CAD CENTRE Cambridge for reference to PDMS, SCHEMA and REVIEW.
-   DAVY COMPUTING LIMITED for references to COMDACE, COMPIPE.
-   AMOCO (UK) LTD for permission to publish 3D View of the ARBROATH
    PLATFORM (Appendix XII).
-   BP INTERNATIONAL for permission to publish 3D View of BP Bruce
    Platform (Appendix II).
-   CONOCO (UK) LTD. for permission to publish 3D View of Mablethorpe Gas
    Terminal (Appendix III).

9   APPENDICES

APPENDIX I	DIAGRAM OF JBE&C COMPUTER SYSTEM
APPENDIX II	3D PICTURE OF OFFSHORE PLATFORM
APPENDIX III	3D PICTURE OF ONSHORE GAS PROCESSING PLANT
APPENDIX IV	PORTION OF TYPICAL CABLE TRAY ROUTING DRAWING DEVELOPED FROM 3D C.A.D. MODEL
APPENDIX V	TYPICAL ISOMETRIC DRAWING
APPENDIX VI	TYPICAL OFFSHORE PROJECT PLAN
APPENDIX VII	DIAGRAMS TO SHOW HOW CORRIDORS ARE ALLOCATED DURING INITIAL DESIGN (2 Sheets)
APPENDIX VIII	TABULAR STATUS REPORT OF PDMS MODEL DEVELOPMENT
APPENDIX IX	TYPICAL MAINTENANCE ACCESS MODELLING (3 Sheets)
APPENDIX X	DIAGRAM OF SOFT CLASHES ILLUSTRATING DESIGNER JUDGEMENT DECISIONS
APPENDIX XI	TYPICAL HYDROTEST SYSTEM DRAWING
APPENDIX XII	SLIDE OF CLOSE IN SHOT USING "REVIEW" MODULE OF PDMS
APPENDIX XIII	GRAPH OF DRAWING PRODUCTION WITH TIME, FOR MANUAL AND CAD PROJECTS OF EQUAL SIZE.

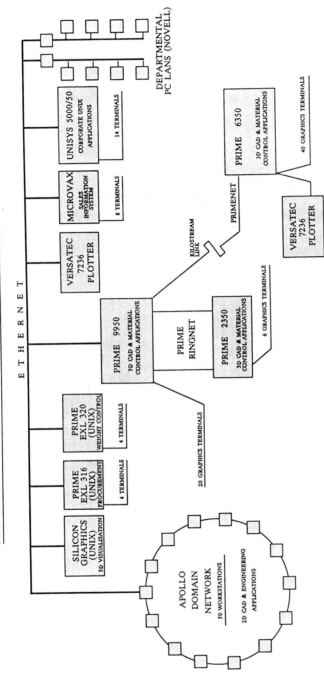

APPENDIX I – JOHN BROWN LONDON COMPUTER FACILITIES

JBNET – HARDWARE

880

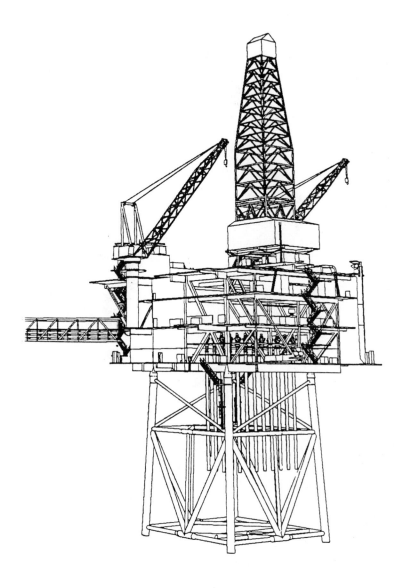

APPENDIX II

3D VIEW OF OFFSHORE PRODUCTION PLATFORM

APPENDIX III
3D VIEW OF ONSHORE GAS PROCESSING PLANT

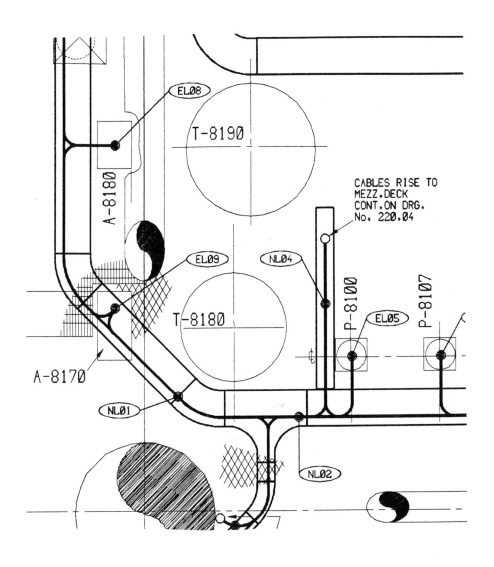

APPENDIX IV
PORTION OF TYPICAL CABLE TRAY ROUTING
DRAWING DEVELOPED FROM 3D C.A.D. MODEL

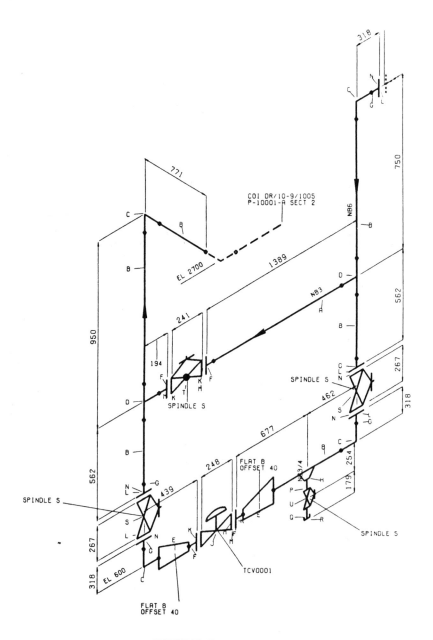

APPENDIX V
TYPICAL ISOMETRIC DRAWING

MAJOR EVENTS IN THE DESIGN OF AN OFFSHORE PLATFORM

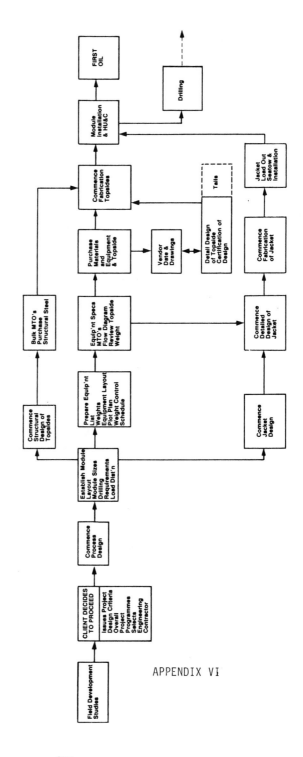

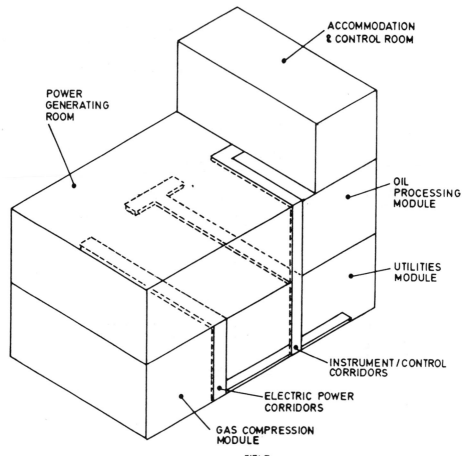

ACCOMMODATION & CONTROL ROOM

POWER GENERATING ROOM

OIL PROCESSING MODULE

UTILITIES MODULE

INSTRUMENT / CONTROL CORRIDORS

ELECTRIC POWER CORRIDORS

GAS COMPRESSION MODULE

TITLE

DIAGRAM TO ILLUSTRATE TYPICAL
POWER & CONTROL CABLE CORRIDORS
DESIGNATED DURING THE DESIGN PHASE.

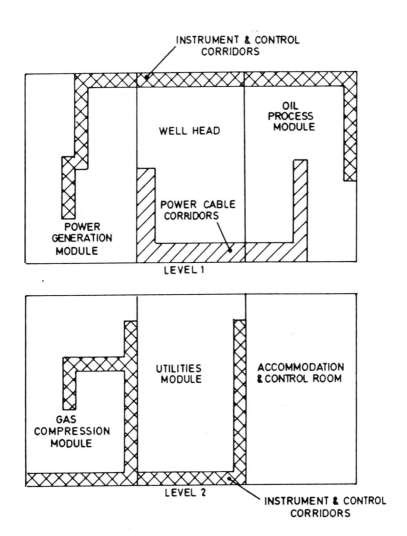

INSTRUMENT & CONTROL CORRIDORS

OIL PROCESS MODULE

WELL HEAD

POWER CABLE CORRIDORS

POWER GENERATION MODULE

LEVEL 1

UTILITIES MODULE

ACCOMMODATION & CONTROL ROOM

GAS COMPRESSION MODULE

LEVEL 2

INSTRUMENT & CONTROL CORRIDORS

TITLE
DIAGRAM TO ILLUSTRATE TYPICAL
POWER & CONTROL CABLE CORRIDORS
DESIGNATED DURING THE DESIGN PHASE.

APPENDIX VII

SHEET 2

887

COMPUTER MODELLING PROGRESS CHART

ACTIVITY	MANHOURS	No OFF ESTIMATE	No OFF MODELLED	INTERMEDIATE CLASH CHECK	FINAL CLASH CHECK	WEIGHT FACTOR	ACTIVITY PROGRESS	OVERALL PROGRESS
MODEL BUILD								
EQUIPMENT								
CIVILS								
STEELWORK								
CATALOGUE & SPECS								
CABLE TRAYS								
HVAC								
PIPES								
PIPE SUPPORTS								
ADMIN								
DRAWING PRODUCTION								
PLOT PLANS								
KEY PLANS								
EQUIPMENT LAYOUTS								
ESCAPE ROUTES								
PIPING GENERAL ARRANGEMENTS								
3D VIEWS								
ISOMETRICS								
TOTAL								

APPENDIX VIII

TABULAR STATUS REPORT OF PDMS
MODEL DEVELOPMENT

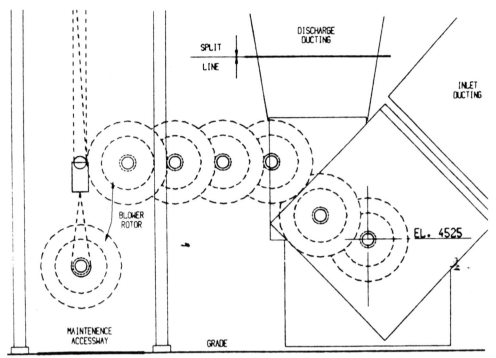

SPLIT
LINE

DISCHARGE
DUCTING

INLET
DUCTING

BLOWER
ROTOR

EL. 4525

MAINTENENCE
ACCESSWAY

GRADE

ELEVATION LOOKING EAST

APPENDIX IX

MAINTENANCE ACCESS ROUTE

FOR ROTOR REMOVAL DEVELOPED IN 3D MODEL

SHEET 1

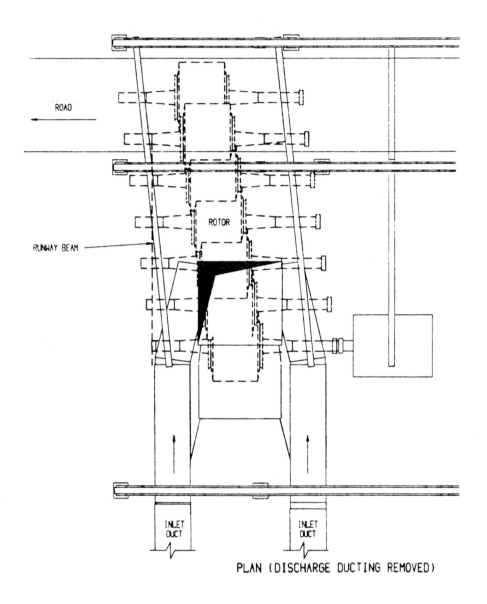

ROAD

ROTOR

RUNWAY BEAM

INLET
DUCT

INLET
DUCT

PLAN (DISCHARGE DUCTING REMOVED)

APPENDIX IX

MAINTENANCE ACCESS ROUTE

FOR ROTOR REMOVAL DEVELOPED IN 3D MODEL

SHEET 2

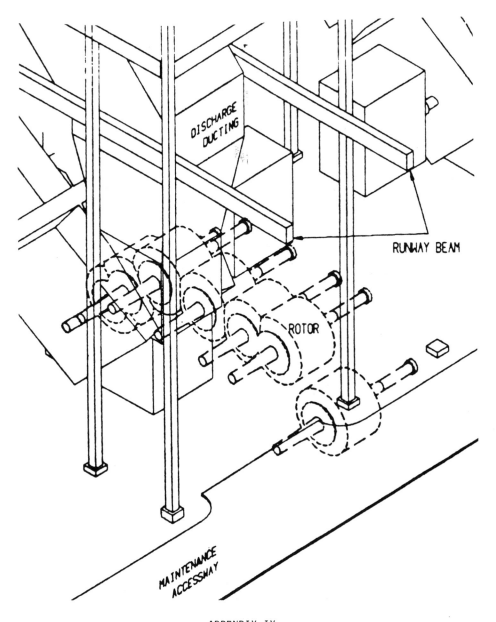

DISCHARGE DUCTING

RUNWAY BEAM

ROTOR

MAINTENANCE ACCESSWAY

APPENDIX IX

MAINTENANCE ACCESS ROUTE
FOR ROTOR REMOVAL DEVELOPED IN 3 D MODEL

SHEET 3

EXAMPLES OF TYPICAL
CLASHES ENCOUNTERED AND
DESIGN DECISIONS REACHED

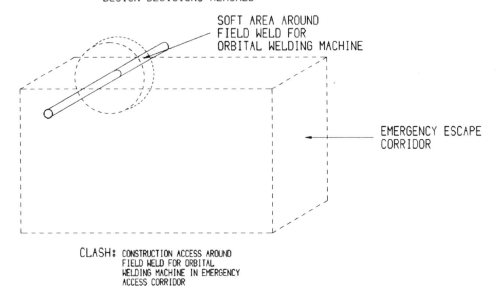

SOFT AREA AROUND
FIELD WELD FOR
ORBITAL WELDING MACHINE

EMERGENCY ESCAPE
CORRIDOR

CLASH: CONSTRUCTION ACCESS AROUND
FIELD WELD FOR ORBITAL
WELDING MACHINE IN EMERGENCY
ACCESS CORRIDOR

DECISION: ACCEPTED - NO ACTION

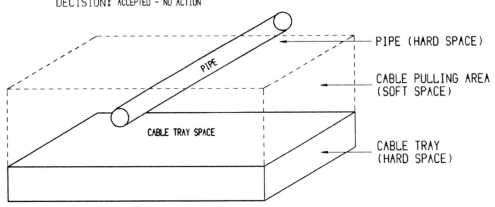

PIPE (HARD SPACE)

CABLE PULLING AREA
(SOFT SPACE)

CABLE TRAY SPACE

CABLE TRAY
(HARD SPACE)

CLASH: SINGLE SMALL PIPE INTRUDES
IN SOFT SPACE RESERVED
FOR CABLE PULLING WORK
ABOVE TRAY

DECISION: ACCEPTED - NO ACTION

APPENDIX X

SHEET I

EXAMPLES OF TYPICAL
CLASHES ENCOUNTERED AND
DESIGN DECISIONS REACHED

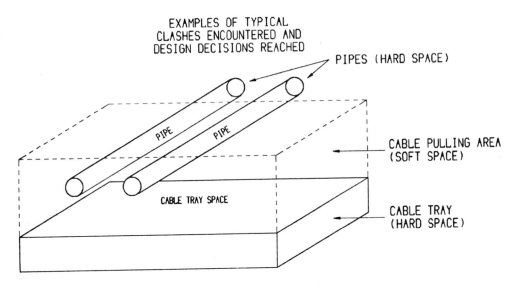

PIPES (HARD SPACE)

CABLE PULLING AREA
(SOFT SPACE)

CABLE TRAY
(HARD SPACE)

PIPE

PIPE

CABLE TRAY SPACE

CLASH: SEVERAL PIPES INTRUDE
IN SOFT SPACE RESERVED FOR
CABLE PULLING WORK ABOVE
TRAY

DECISION: ACCESS NOW UNACCEPTABLY
RESTRICTED - MOVE TRAY OR
ONE OF PIPES

APPENDIX X

SHEET 2

APPENDIX XI
TYPICAL HYDROTEST SYSTEM DRAWING

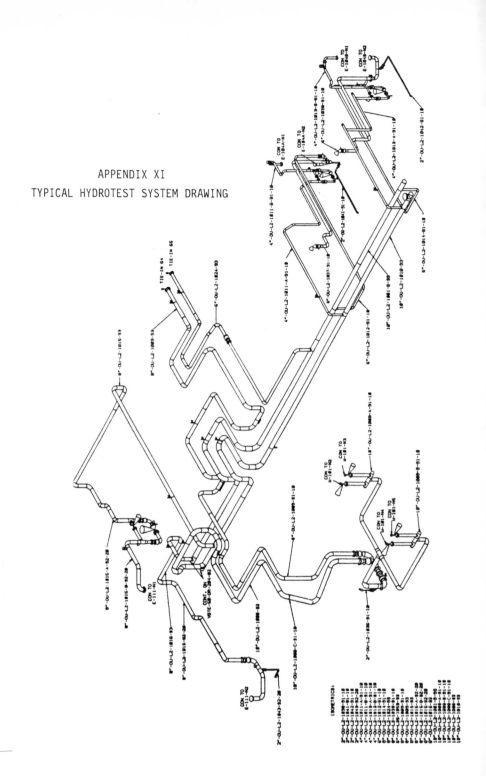

894

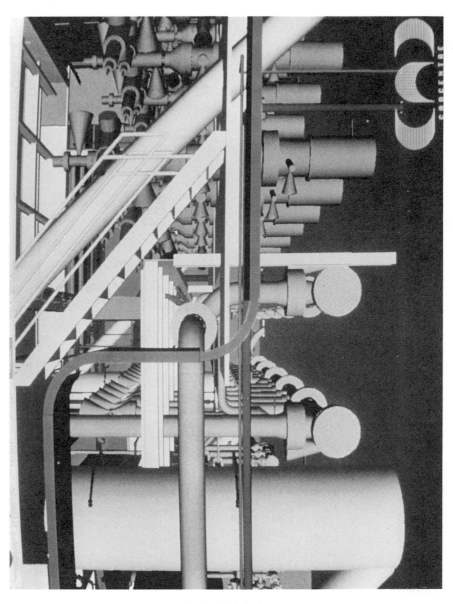

APPENDIX XII
3D VISUALISATION OF PLANT

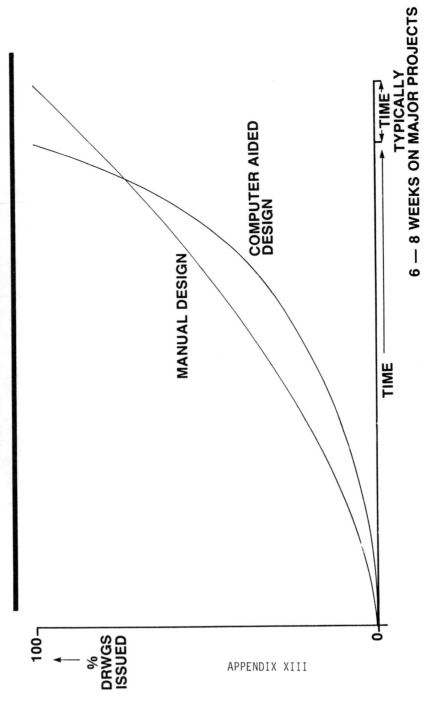

ISSUE OF DRAWINGS

MANUAL DESIGN

COMPUTER AIDED DESIGN

100

%
DRWGS
ISSUED

0

TIME

TIME

TYPICALLY
6 — 8 WEEKS ON MAJOR PROJECTS

APPENDIX XIII

C377/148

Integrated use of CAD/CAE methods in the development of new engine concepts

G SCHMIDT and **P REINDL**
BMW AG, Munich, West Germany

Introduction

In view of the increased requirements placed on the development and production of internal combustion engines, especially those of economy, low emissions, acoustics and low weight, coupled with the increased pressurs of costs and deadlines, the use of new methods in engine development in is urgently required.

For these reasons, the methods of computeraided design (CAD) and engineering (CAE) are an integral part of the development of new engine concepts.

The efficient use of CAD/CAE places particular importance on the following strategic objectives:

- increase in the development quality
- reduction of the development time
- rationalization of design activities
- relief from routine activities
- control of the design process

in conjunction with the necessary consistency of development data for production and quality assurance.

Methodical working basis

The information flow of current data and the resulting reliability of the "engineering documents" with reliable data storing and fast data access forms the basis for a high degree of target attainment in the development of engine projects.

User concept

In the context of the project-oriented CAD/CAE application, all participating employees, i.e. designers, computing engineers and the employees of the release departments are part of an integrated user concept.

This controls the internal and inter-departmental cooperation within engine development and defines the interfaces to "outside", i.e. to the departments responsible for chassis and body development, to parts scheduling, and to production centres and suppliers.

Fig. 1 clearly shows the project-oriented cooperation based in the exchange of CA data. In order to guarantee smooth exchange of data at the early stage of development, the function "CAD projekt coordination" was defined.

The CAD projekt coordinator manages all project related data up to test release. He is responsible for the continuous updating of the data base - sketches, drafts etc. - and is the central liaison person regarding the data transfer of all those participating in the project.

The project-related CAD data exchange in the vehicle development environment is represented in fig. 2. Current 3D models from body design - e.g. the engine bonnet - and from chassis design - e.g. the front axle - are used for the design of the radiator and its arrangement in the vehicle.

In addition, CAD data of the engine and vehicle electrical sysemts are used, for instance, for the wiring harness routing.

The concept is based on the release of the CAD model data, consistent with the drawing document and the master reference data for parts.

The release procedures of CAD data are shown schematically in fig. 3. In the past, CAD component data was intended only for private use. This meant that CAD model data was processed to produce a single CAD drawing. The CAD drawing was released as a representative information medium. The CAD data was not binding. The objective was to suppplement the information medium (drawing) by a binding CAD component model in the context of corporatewide communication.
Today, the CAD model data is released in conjunction with the CAD drawing. In addition to matching of the CAD geometric data with the master parts reference data and the drawing by the release departments, the data - which is now binding - is archived, distributed and made available to users form other systems.

Basic and application models

In addition to the reliability of the CAD component data, steps must also be taken to ensure its usefulness in the context of corporate-wide communication. This is guaranteed by a CAD design being based on basic and component-specific application models which are developed for complex engine components such as crankcases, cylinder head etc.

The designer thus works in pre-structured CAD models which are structured in such a way as to facilitate process chain continuity.

Fig. 4 shows a structure specification as exemplified by the application model "camshaft".

Further structure specifications are

- a standard CAD model name with is read by administrative databases when the models are archived and is used for automatically generating organisational data.

- standard **coordinate/reference** **systems** for easy and safe utilisation for components and arrangements

- **standardised parts geometry** of a component (e.g. undercuts) in order to limit design freedom where production standards have to be observed.

- **standardised drawing documentation,** with standard arrangement of view and cross sections and standardised dimensioning in order, for instance, to eliminate incorrect interpretations of blank and finished part drawings.

- **layer structuring,** to free the designer from the "CAD organisation" of his component.

Parametrics linking CAD/CAE

Parametrics makes it possible to link "insular" solutions, i.e. singular, non-coordinated methods, to a unified development system.

In the context of component or group-specific parametric programs, the originally pure geometry-orientated CAD applications are now integrated with other computer programs & methods e.g. engine mechanics and stress analysis.

In addition to standardising the design process the greater use of parametric programs reduces the number of versions, for example of camshafts, and results in standardised components.

In total requirement of component-specific and group-specific applications is shown in fig. 5.

Development process with CAD

In the wide application spectrum of engine development - from the hexagon bolt to the complex cast part - CAD methods have a profound influence on the development process.

The following applications are important:

Assembly studies

In engine/vehicle packaging, complex assembly studies and collision investigations are carried out on the basis of three-dimensional CAD component models. This results in decisions being taken with a greater degree of certainty in the early development stage, a reduction of effort in constructing mock-ups and a saving in prototype parts.

Fig. 6 shows the example of the pipe routing of an exhaust manifold in relation to adjoining body and chassis parts such as front bulkhead and engine support. Ease of maintenance and assembly of a layout is checked by simulating the access, for example, of mulitple-tightening devices, in the three-dimenisional representation of the engine compartment.

Design concepts

The development of engine design concepts and alternatives in the form highly realistic shaded images makes it easier to evaluate design alternatives and considerably reduces the number of models.

This design concept is illustrated with the example of the covering for a throttle housing assembly: starting with conventionally generated renderings, cf. fig. 7, highly detailed shaded images, derived from the 3D surface model, are generated, and then form the basis for any decision - cf. fig. 8. Selected alternatives are then modelled by NC processing of the available 3D surfaces. On the basis of this, the component reaches its final form, as shown in fig. 9.

Kinematic simulation

With the simulation of a complex kinematic process - for instance throttle operation - a high degree of design certainty can be attained.

Fig. 10 shows a throttle kinematic process, consisting of throttle, operating lever and guide as a three-dimesional wire model in the zero position.

Fig. 11 shows the corresponding motion pattern up to maximum load. The desired design - Bowden cable movement against throttle angle, cf. fig. 12, can be represented quickly for different kinematic variants.

Drawing documentation

Drawing documentation by means of CAD enables component-specific standardisation of the workshop drawing and rationalisation of the drawing activity with new and modified designs, which goes beyond conventional component description.

Data transfer

The process chain oriented transfer of the CAD component data to production resources planning, qualitiy assurance etc. makes high quality components possible though unambiguous component descriptions, as the exambles combustion chamber surface and volume show.

Design support through CAE methods

The product development process is based on the iterative data exchange between CAD and CAE applications. Fig. 13 shows the process in its basic form. Starting with a geometric concept and first sketches, a simplified functional model is created which in turn is then transferred to a three-dimensional rough draft. The target attainment of the functional parameters is checked in an detailed functional model which finally forms the basis for the manufactured component model.

After the corresponding prototyps have been released and made available, test results comparing target with actual performance values may show that checking and possibly modification of the functional model is required. Production-related factors could result in the component having to be redesigned.

The support of the design using CAE methods is illustrated here by the topics of gas exchange and combustion computation.

Gas Exchange Process

By simulating the dynamic flow processes of the intake and exhaust systems on the basis of purely computational models, a functional design of the gas flow components on the intake and exhaust side is possible even for the first prototyps. Target attainment for the initial design is thus higher and the number of iterations steps required is minimised.

In order to attain sufficient accuracy with the simulation computation - fig. 14 shows a comparison of the torque curve between calculation and measurement - the entire engine must be modelled.

Fig. 15 represents such a simulation model. The detailed analysis of the pressure waves in the intake and exhaust pipes - fig. 16, 17 - or the massflow in or out of the cylinders - fig. 18 - enables efficient optimisation of the entire system.

Combustion Computation

Only by understanding the interaction between the flow processes and the flame propagation in the cylinder is it possible to optimise the combustion chamber geometry, taking account of all targeted requirements such as consumption, performance and emissions. Using modern methods the spatial flame propagation can be simulated.

The entire combustion chamber surface must be reproduced in fine detail, with the enclosed volumes subdivided into individual cells. The example of the BMW 4- and 12-cylinder combustion chamber - fig. 19 - shows that the representation of real combustion chambers, as shown in fig. 20, is to a large extent feasible today.

Further, results of the three-dimensional flow and combustion computation affords greater insight into the internal engine processes.

Both the position of the flame front in the combustion chamber during heat release (fig. 21) and the resulting flow field (fig. 22) can be observed at any given instant.

Application examples of integrated processing for complex engine topics

Integrated design for valve timing

In the application "integrated design of cylinder head and valve timing", the following processing steps are coupled via defined interfaces:

- cylinder head concept with specification of combustion chamber, and intake and exhaust ports

- geometric arrangement of cylinder head

902

- gas exchange process

- camshaft design

- dynamic investigation of timing gear.

This coupling is explained in more using as an example, a building block of the above application: the combustion chamber design - cf. fig. 23.

Starting with an initial concept which includes design criteria such as valve angle, outer diameter of the combustion chamber etc., the requirements of valve lift and diameter, based on the gas exchange calculations, are then incorporated.

Using these basic parameters a rough 3D-parametric model is generated within the CAD program. After transferring the corresponding 3D surface data to the CAE system, a function-oriented specification of the CAD surface model. Fig. 24 shows such a three-dimensional combustion chamber model together with the intake and exhaust ports.

The results of the actual combustion computation lead to correction of the combustion chamber model. A production-related evaluation of the component data is carried out in the context of the eventual NC programming and production. Test-bed trials indicate the level of agreement between calculation and measuremant.

The optimisation process of different design versions is carried out with continuous matching of data from different systems.

The data flow to production and the transfer of corresponding measurement data back to engine development (e.g. spherical sensoring) is also integrated into the application.

Integrated component Stress Ananlysis

In the area of component stress analysis the CAD data (geometric model) is used for direct finite element mesh generation. Duplication in generating the FE-model is therefore reduced or avoided.

Fig. 25 shows the drawing of a camshaft. The respective three-dimensional CAD-model is represented in fig. 26. The model is generated in such a manner that the subvolumes needed for the FE preprocessor are already present after being transferred through the interface. Meshing of the model and the inclusion of the boundary conditions and loads then taken place in the pre-processor. Fig. 27 shows the corresponding stress levels in the flange area of the camshaft.

Fig. 28 shows the CAD rough draft of the crankcase in the form of a three-dimensional wire model. It provides the necessary geometric data for a dynamic model which enables a computational evaluation of the first draft. Rough models of cylinder head, sump and gearbox bell-housing are produced in the same manner and integrated with the crankcage model, see fig. 29. Fig. 30 shows the first natural mode of the drankcase; this is the first torsional.

Outlook

Seen as a whole, a wide spectrum exists for computer-aided methods in the area of engine development.

These have a profound influence on the development process and series improvements in the life cycle of an engine. It is therefore of considerable importance to link up the individual solutions that still exist today and to promote the use of unified CAD/CAE methods.

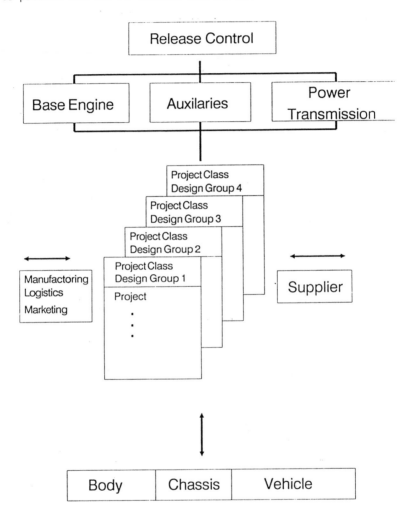

Fig 1   Project orientated CA-data exchange

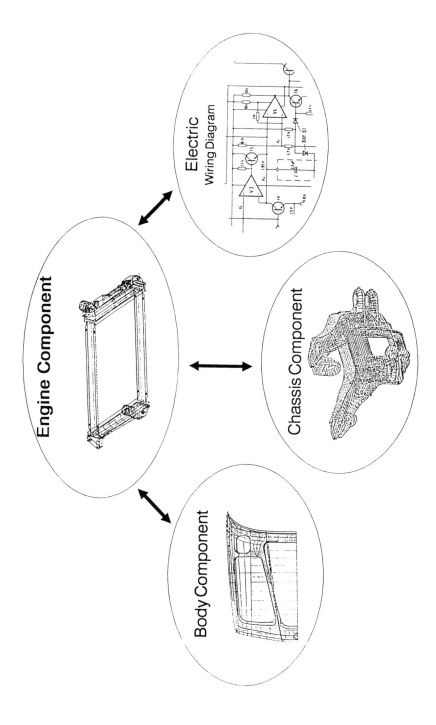

Fig 2   CAD-exchange in vehicle development

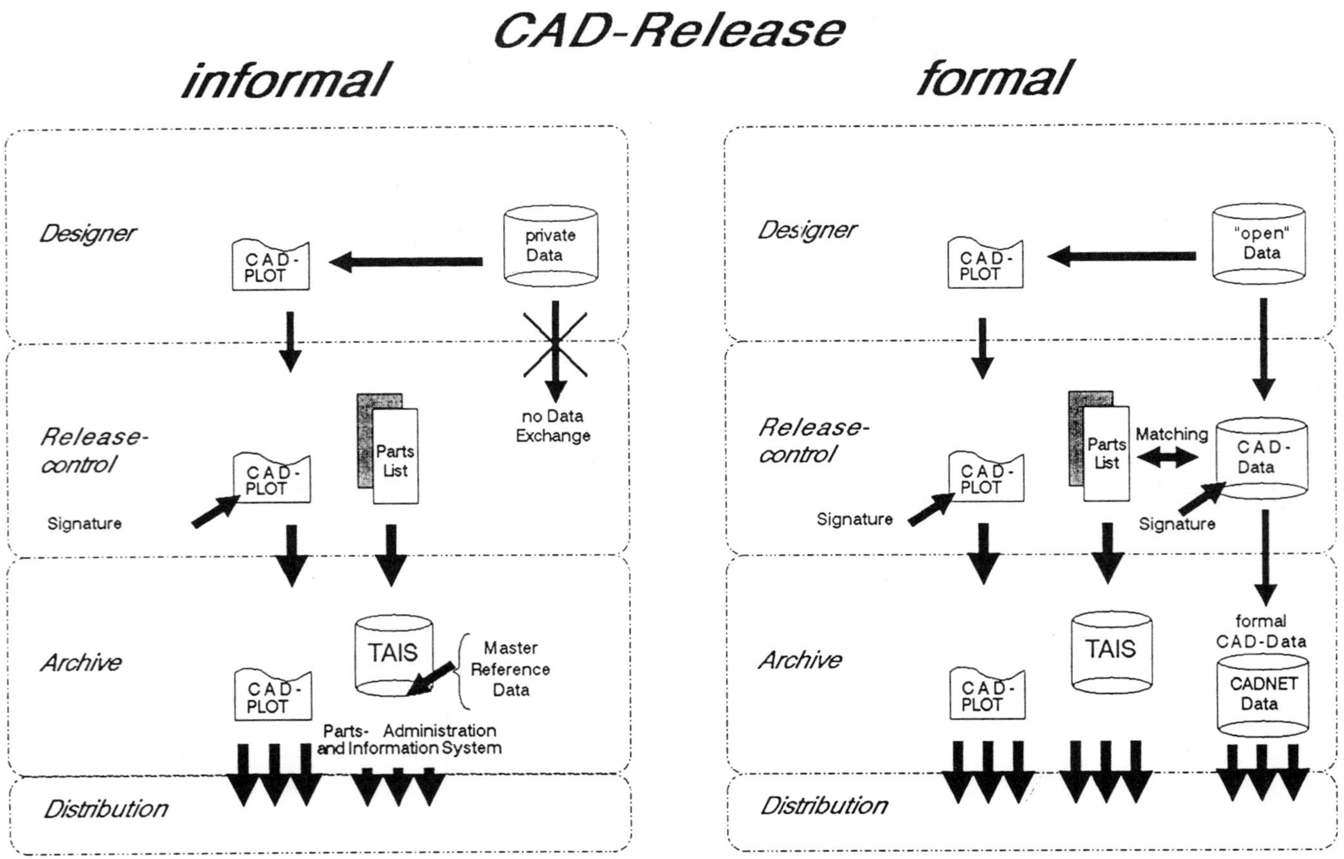

CAD-Release

informal

formal

Fig 3   Release of CAD-data

906

Standardisation

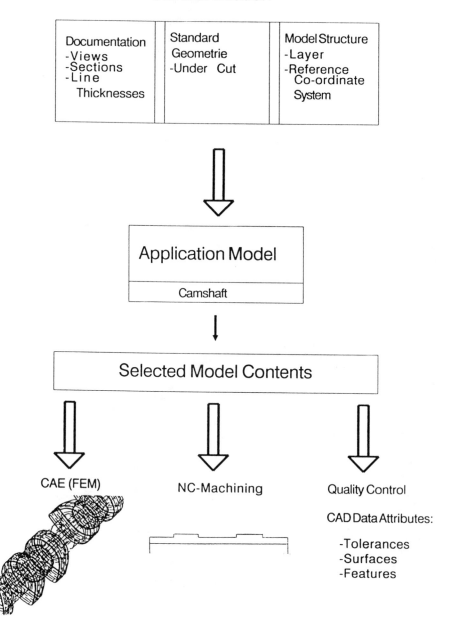

Documentation
-Views
-Sections
-Line
   Thicknesses
	Standard
	Geometrie
	-Under Cut
		Model Structure
		-Layer
		-Reference
		   Co-ordinate
		   System

Application Model

Camshaft

Selected Model Contents

CAE (FEM)

NC-Machining

Quality Control

CAD Data Attributes:

-Tolerances
-Surfaces
-Features

Fig 4   Model structure for camshaft example

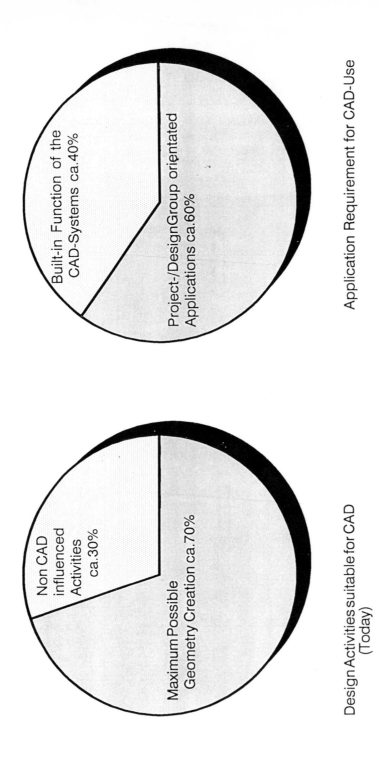

Non CAD
influenced
Activities
ca.30%

Maximum Possible
Geometry Creation ca.70%

Built-in Function of the
CAD-Systems ca.40%

Project-/DesignGroup orientated
Applications ca.60%

Design Activities suitable for CAD
(Today)

Application Requirement for CAD-Use

Fig 5   Requirement for project-/design group applications — target 1990

908

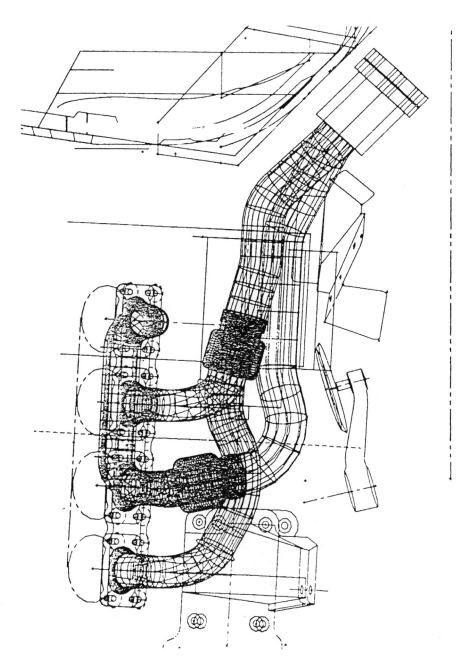

Fig 6　Layout investigation for the exhaust manifold

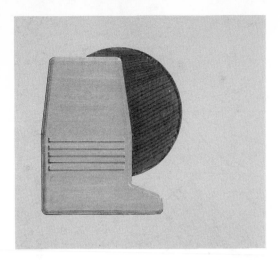

Fig 7   Cover for throttle housing – rendering

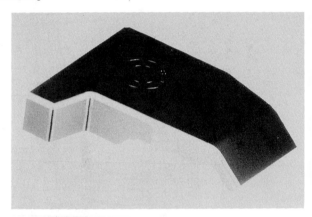

Fig 8   Cover for throttle housing – 3D-surface model

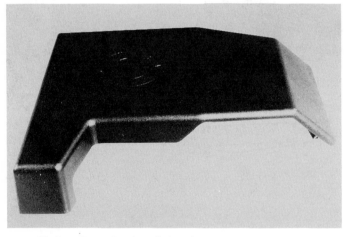

Fig 9   Cover for throttle housing – component

910

FINAL PARM	VARIATION	CUR PARM
		0.000

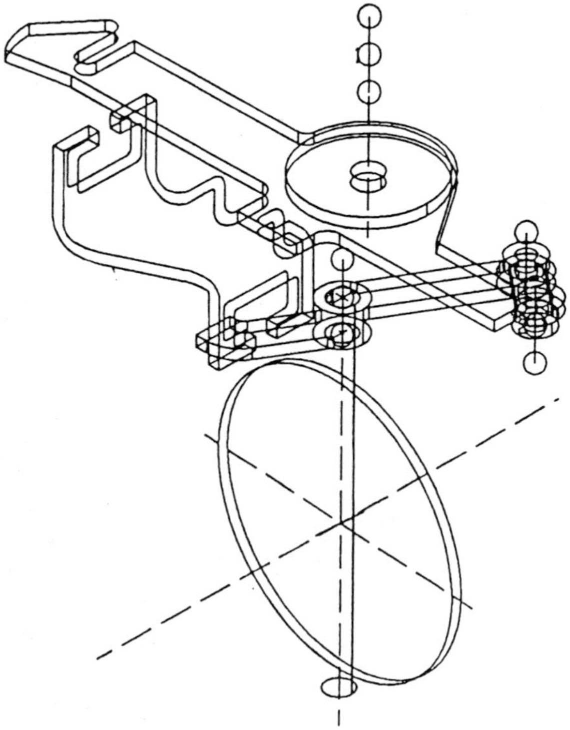

Fig 10   Throttle valve kinematic — simulation

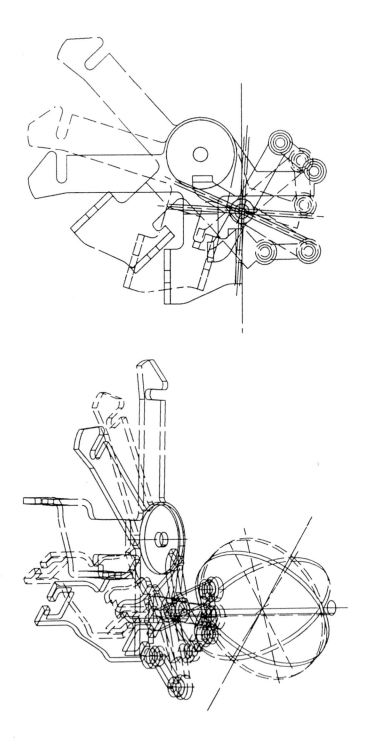

Fig 11   Throttle valve kinematic – movement

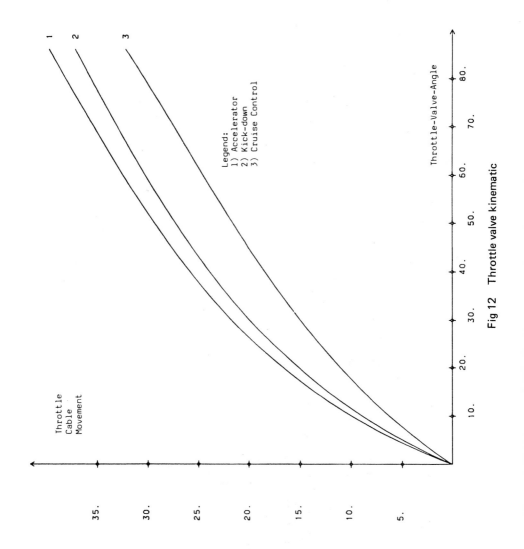

Fig 12   Throttle valve kinematic

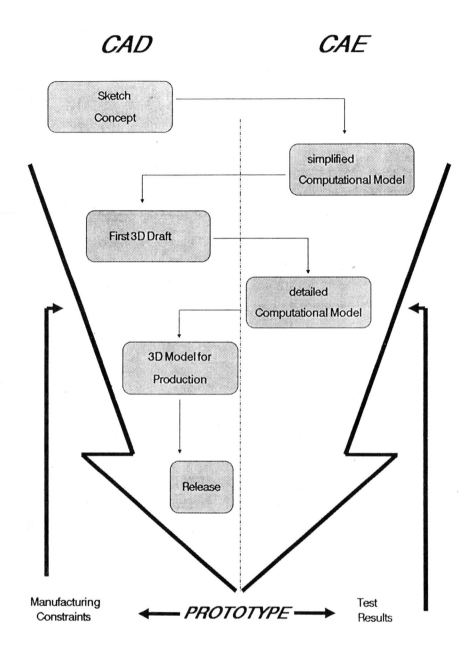

Fig 13   Iterative product development process with CAD/CAE

914

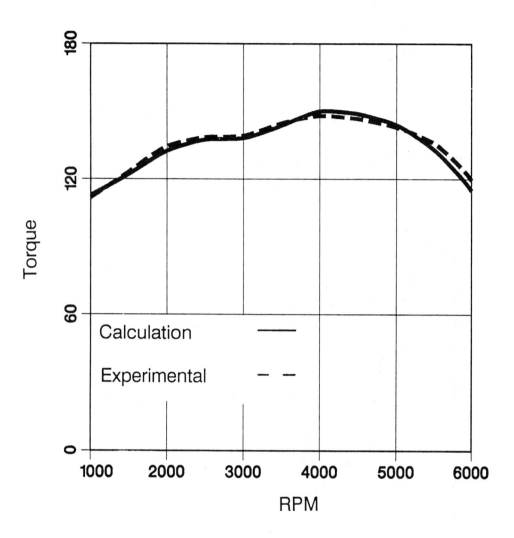

Fig 14 Comparison calculation — experimental

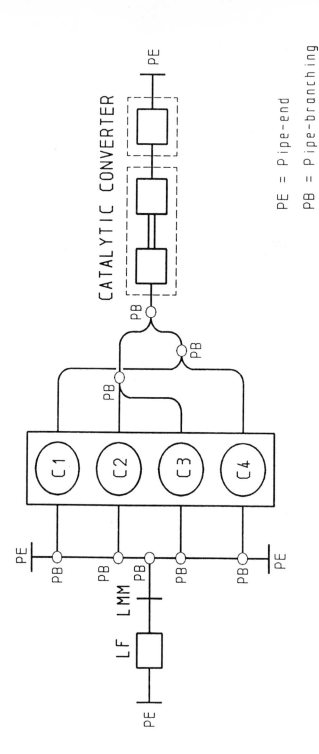

PE = Pipe-end

PB = Pipe-branching

Fig 15   Gas exchange schematic model

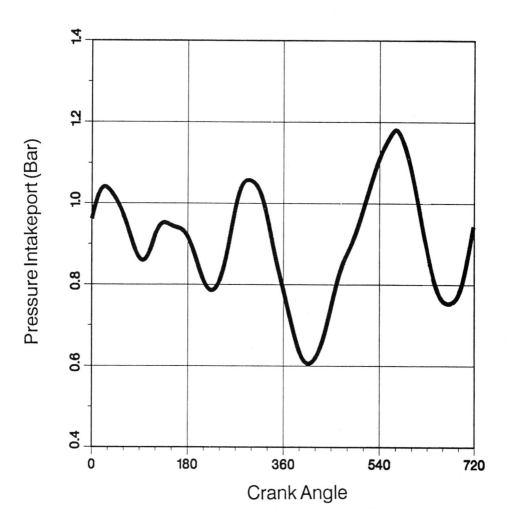

Fig 16   Pressure in intake port

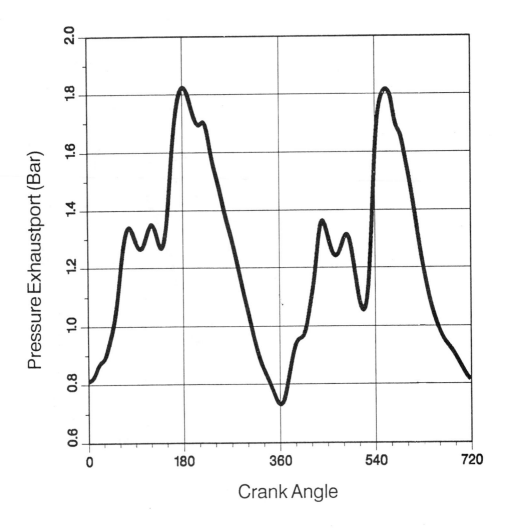

Fig 17   Pressure in exhaust port

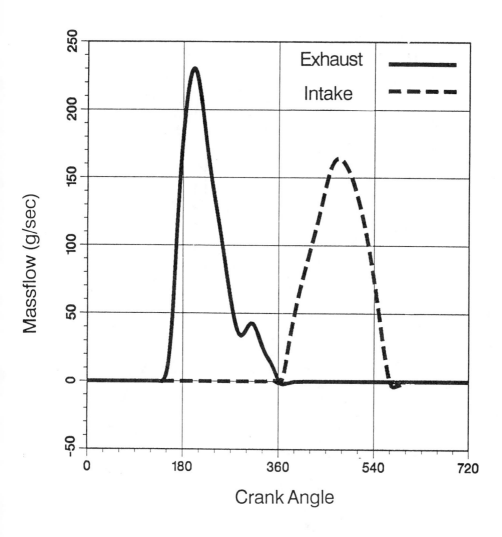

Fig 18   Massflow

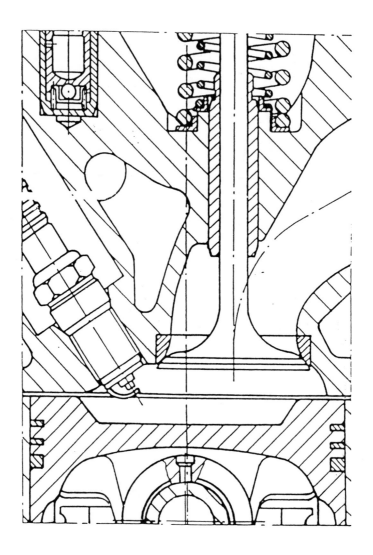

Fig 19   Combustion chamber 4- and 12-cylinder engine

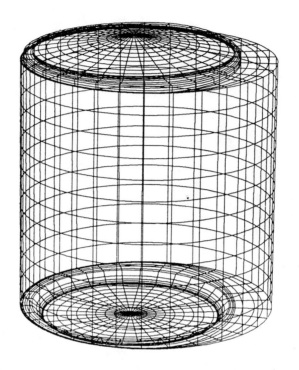

12608   Volume Cells

Fig 20   Surface mesh combustion chamber/cylinder

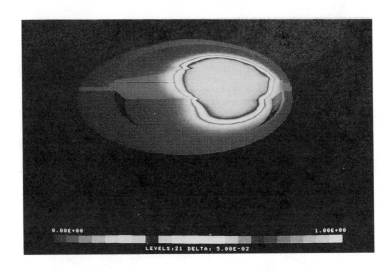

Fig 21   BMW combustion chamber — flame propagation

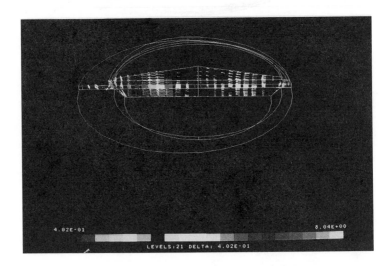

Fig 22   BMW combustion chamber — flow field

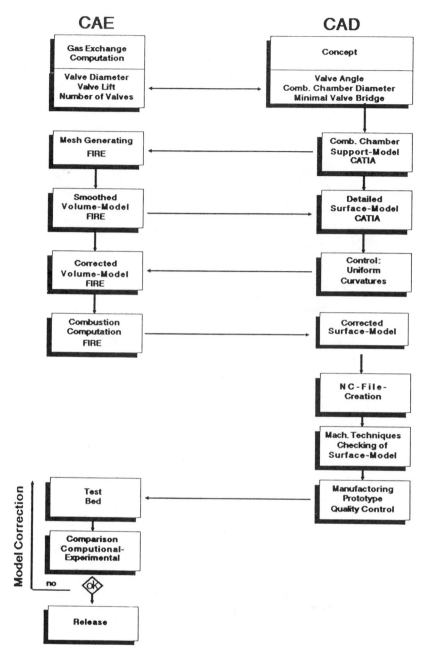

Fig 23   Combustion chamber: CAE-orientated development

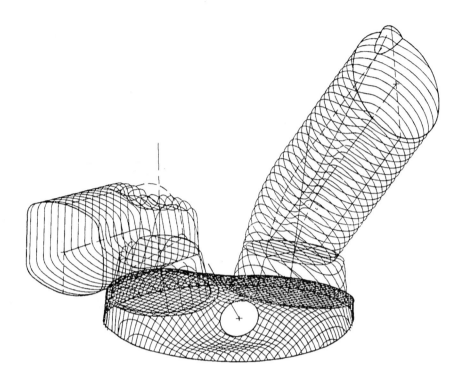

Fig 24   Combustion chamber with intake and exhaust ports

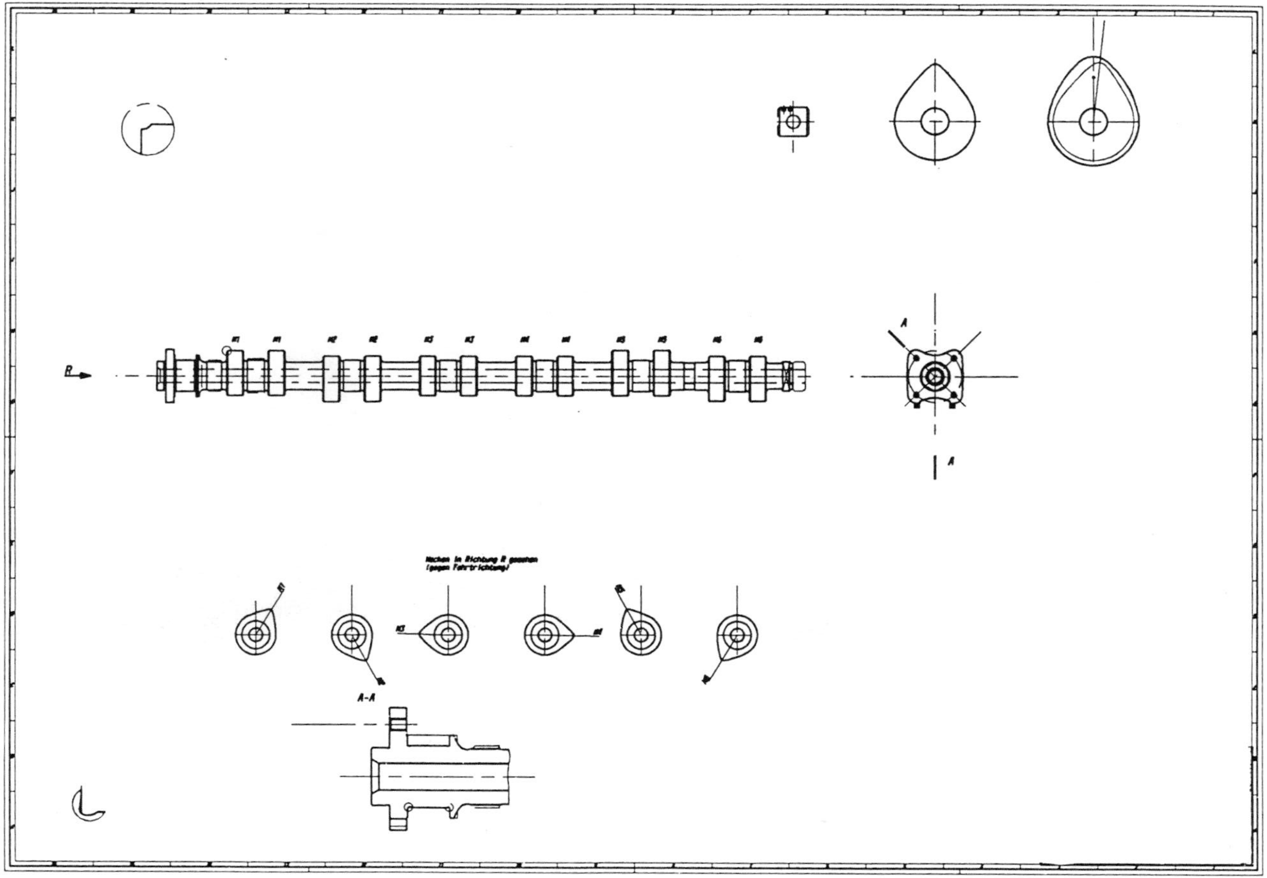

Fig 25   Camshaft CAD-drawing

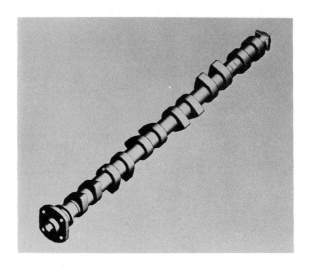

Fig 26   Camshaft 3D-surface model

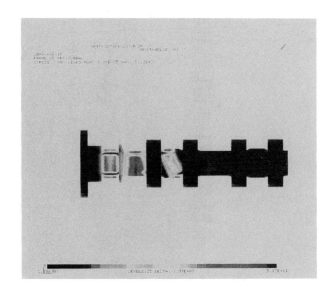

Fig 27   Camshaft – stress analysis in flange area

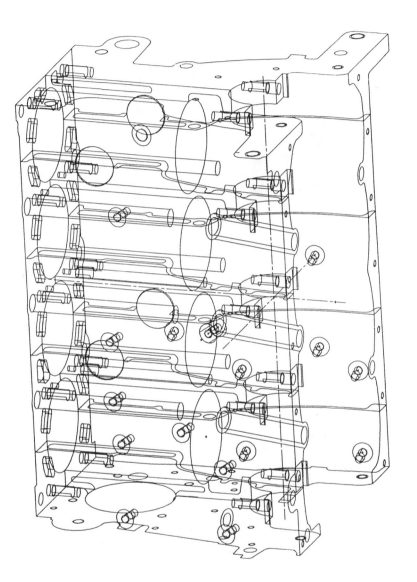

Fig 28   Crank case 3D-wire model

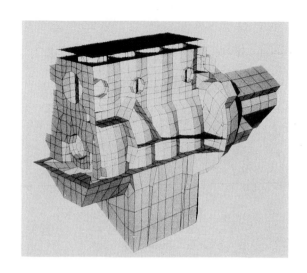

Fig 29   Crank case FE-model

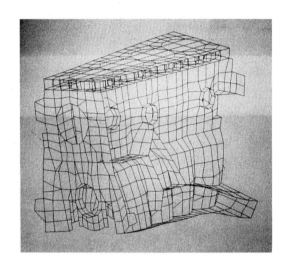

Fig 30   Crank case — 1. Torsional mode

C377/098

Concept embodiment: a methodological basis for effective computer aided design

I BLACK, BSc, MSc, CEng, MIMechE and **J L MURRAY**, BSc, MSc, CEng, FIMechE
Department of Mechanical Engineering, Heriot-Watt University, Edinburgh

This paper principally focuses on the embodiment of solution concepts, a 'formal' design activity in which it is argued that Computer-Aided Design (CAD) methods can be favourably deployed, hence increasing their application to conceptual work. An industrial case study is cited which illustrates the utilisation of concept embodiment within a particular design project undertaken by a small-batch, mechanical manufacturing company. The case study also attempts to show that the effective application of CAD, through concept embodiment, is significantly dependent on both multi-disciplinary team working and a technically-competent CAD applications esource.

1. BACKGROUND

It is generally recognised that most CAD systems give the design engineer little or no assistance in performing conceptual studies during the product design process, as there may be a significant requirement for original or exploratory thought (1), (2). However, though CAD systems may at present be unable to share the design engineer's understanding of the 'grey' and nebulous procedures which constitute 'pure' conceptual design, they are eminently suited for the creation, manipulation and storage of graphical data. It is this type of information upon which design engineers mainly rely in order to translate requirement specifications into a finished product, and which CAD systems have the unrivalled capability for storing and manipulating.

A number of studies have dealt with applications philiosophies for CAD and/or the major considerations which must be taken into account when applying CAD technology - refer, for example, to (2), (3), (4), (5), (6) and (7). Common to all of these is the establishment of a prescriptive model of conventional design practice as the foundation for CAD application, which appear to take the two approaches mooted here. For the purposes of this paper, these can be viewed as either 'sequentially-' or 'spectrally'-based. By sequential, it is meant that the information flow through the design process is considered the key parameter in a model, whereas spectral approaches show the type of design occurring across the

929

design process, dependent on the amount of creative (or routine) activity inherent in the design work. These two approaches can be best illustrated by reference to Figures 1 and 2.

Figure 1 shows a 3-D view of a general sequential model which combines and enhances several extant philosophies (after (8), (9), (10), (11) and (12)), and attempts to illustrate the cyclical feeback and feedforward nature of the information flow which occurs throughout a formally-based process structure. Figure 2 also presents a 3-D view of a model which attempts to portray a spectral approach (after (1), (8), (13) and (14)). By travelling around the perimeter of the envelope (in the direction of the arrowed line) from position A, say, the amount of inherently creative work in design steadily decreases to B, reaching a minimum at C. It then increases from C to D, reaching a maximum at A. The converse applies for the amount of routine activity present in design work.

Both of the above models are complementary to one another, in that they show design activity coupled with the effects of design creativity. Most manufacturing companies, with the possible exception of those solely undertaking research and development work, will be involved in designing, developing or changing known products for different needs or tasks. That is, they adopt an 'evolutionary' approach rather than a more 'innovatory' approach (15).

However, from inspection of Figure 2, the degree of creative activity (whether it is innovatory or evolutionary) present in any particular product design is dependent on the design category that best describes the design concept, i.e. whether the concept is associated with innovative design, adaptive design, variant design or order execution work (1). The authors propose that the spectral analysis of design is of extreme significance when considering the effective deployment of CAD. By attempting to rationalise the degree of conceptual practice inherent in each category of design, spectral models may indicate where and when he application of CAD will prove most effective for any particular product design.

2.   CONCEPT EMBODIMENT

Although a spectral analysis should indicate where and when CAD will prove most effective, it requires a sequential analysis to show how CAD techniques can be best applied. In this connection, it is the authors belief that due to their rigorous and comprehensive nature, 'formal' or 'systematic' design methods - see (8), (9), (16), (17), (18) and (19) for example - provide the most logical structure for the deployment of CAD techniques.

Systematic design practice is chracterised by four main phases, i.e.

(1)   Task Clarification; which involves the collection of information about requirements and constraints, followed by the generation and subsequent elaboration of detailed specifications.

(2) Conceptual Design; whereby solution concepts are reached satisfying the requirement specifications.

(3) Embodiment Design; in which selected concepts are successively developed and refined.

(4) Detail Design; where the arrangement, form, dimensions and surface properties of all parts are finally laid-down, the materials specifed, the product feasibility re-checked and all production documentation produced.

Though these phases can be thought of as constituting a well-defined, sequential process, in practice there is a great deal of 'overlap' and transition at the boundaries (refer to Figure 1), since at every step of the total process a decision has to be made as to whether previous steps have to be repeated in order to proceed to the next activity (so -called 'iterative cycles').

It is important to note that three central problems in systematic product design have been highlighted (10). These are

(a) the generation of good concepts,
(b) securing the 'best' embodiment of those concepts, and
(c) the evaluation of alternatives (feedback and feedforward between conceptualisation and concept embodiment).

Although (a) will always prove a difficult task no matter what assistance is available to the design engineer, CAD can beneficially assist the activities outlined in (b) and (c). It is this proposition which will form the nucleus of a methodology for the effective application of CAD techniques. However, it will first prove useful to look at the embodiment design phase in a little more detail.

2.1 Philosophy behind concept embodiment

During the embodiment phase of the formal design process, the design engineer, starting from the selected concept, determines the preliminary design configuration of the product (i.e. geometry, form and material) and develops a product definition in accordance with the considerations of

(i) Functionality;

(ii) Cost;

(iii) Manufacturability; and

(iv) Standardisation

Embodiment design therefore effectively forms a bridge between the conceptual and detail phases, through the translation of solution concepts into geometrically precise layouts.

Frequently, the evaluation of individual layout variants may lead to the selection of one that looks particularly promising but which may nevertheless benefit from, and be further improved by, incorporating ideas and solutions from the others. By appropriate combination and elimination of weak links, the best layouts can thus be obtained. The result of concept embodiment should therefore be the definitive layouts which provide a check of the factors stated above in regard to the proposed product design.

From inspection of Figure 3 (19), (20) it can be seen that the proportion of finance spent on a project during the conceptual phase is relatively minor with regard to total project spend. However, the amount of costs committed to product manufacture at those early stages is relatively large. Hence, although design is at the 'low-spend' end of a project, it has extensive implications for manufacturing costs. Investing in a more rigorous, 'right-first-time' approach within the conceptual phase, together with the incorporation of information about production methods and processes, will yield significant returns during manufacturing. It is those returns that may be assured through proper concept embodiment using CAD techniques.

2.2 Proposed methodology of concept embodiment

Figures 1 and 2 have shown sequential and spectral models of design practice respectively, and it is these models which form the general philosophical core of the CAD methodology outlined here. The central strategy behind thismethodology is the utilisation of formal design practice as a vehicle for the application of CAD technology, using the 'catalyst' of multi-disciplinary working coupled with technically-competent applications expertise; remembering that, as has been previously mentioned, the effectiveness of CAD application is also dependent on the conceptual status of a product design.

The actual methodology of concept embodiment principally exploits the precise description of geometry that a CAD system can store. Specifically, this requires that component geometry be defined full-scale and mid-size. Design engineers can therefore assess the geometric (and attribute) data held by the CAD system which precisely describes the developing product form. The essential point to note is that the CAD system assists with the unambiguous translation of solution concepts (mainly generated by the design engineer with conventional 'paper-and-pencil' methods) to geometrically correct layouts stored within the CAD database - it does not replace the design engineer's intuitive ability for arriving at or initiating changes to those layouts. Attention to detail and close inter-disciplinary liason (principally between the design and production functions) are also required throughout concept embodiment. This enables the detail design of layout componets to be accomplished at a stage where the configurations of various technical systems are being determined, thus reducing the incidence of expensive mistakes and/or changes occurring later on. At the end of the day, the results of the above methodology should be

(a) a design which matches the requirement specifications formulated during task clarification (thus ensuring quality is being 'designed-in');

(b) a design which incorporates the appropriate standard components and sub-systems;

(c) a design which can be capable of being manufactured in a cost-effective manner; and

(d) a design which can be modified quickly and precisely to any changes in configuration and/or form.

There are also significant implications for project lead-times (see Figure 4), meaning that

(i) the same number of iterative cycles, feedforward and feedback loops are possible in a shorter lead-time (leading to increased market competitiveness), or

(ii) the number of such cycles and loops are increased in the same lead-time (leading to increased technical excellence).

It is important to recognise that during concept embodiment any product design being tackled with CAD techniques will have a growing database of graphical information associated with it. This database will act as part of a central 'core' of product data which will, ultimately, completely describe and define the product form. It is essential therefore that this database contains coherent and precise product information which is directly re-usable by other functions. With reference to geometry, this condition is mainly dictated by the configuration of the layouts, since these contain the unique geometric information pertaining to a particular product.

3.   CASE STUDY

The philosophy and methodology outlined previously will be exemplified through an industrial case study involving a contemporary heavy/medium, mechanical engineering company. This company is principally a small-batch manufacturing concern, and much of its work is done on a contract basis. In addition, the company is currently attempting to evolve modular ranges of products involving as many standard components and/or systemsas possible; however, adaptive design work is performed when a significant degree of product customisation is required.

The company currently operates a centralised, minicomputer-based CAD/CAM systemcomprising of three monochromatic workstations and associated plotter, printer and peripheral alphanumerical terminals. This configuration principally supports 2-D definition and NC part-programming software, together with an extensive range of user-extension libraries which considerably enhance the usefulness of the basic CAD system.

933

3.1 Introduction

A substantial amount of investigative work on engineering design philosophy and CADapplication within the company had been undertaken as part of a SERC/DTI Teaching Company Programme between the company and Heriot- Watt University. The results of those findings (20), (21) and (22) provided the foundation for a methodology, involving concept embodiment, which would effectively utilise the company's CAD system. After much deliberation, the decision was taken to put a substantial part of this methodology to the test. A pilot application was identified which would act as a company benchmark.

The company has long recognised the importance of proper conceptual design, and 'computer-based' techniques may begin to get involved to some extent at this stage (but not the CAD system). Programmable calculators and microcomputers, using either specially written or general purpose software, are used to estimate functional or duty requirements for the various technical systems and sub-systems forming part of the solution concepts under consideration. Once this preliminary simulation work is well underway it is then possibe to start defining the basic shape of components on the CAD system; marking the beginning of concept embodiment. Product geometry is passed from nominal-size 'drawdowns' or general arrangement drawings (done conventionally' on the drawing board') to the CAD system in the form of 2-D geometric layouts.

As far as the company is concerned, during the initial stages of concept embodiment two major issues are considered:

(i)   The question "can standard items be used ?" is raised. Standard items may be bought-out parts of components already held in inventory for other products that the company manufacture. If the answer to this question is "yes", then these items will become part of the product design at this stage and will be fully incorporated into the layout(s) under consideration. in addition, parametric component definitions, such as bolts, capscrews, external threads, etc. that are held within the CAD system are utilised as far as possible.

(ii) Production engineers look at the proposed design configuration (i.e. the layouts) in some detail. Since production engineering is very important specialist field in its own right - and, at the end of the day the object is to manufacture a product - this is perceived within the company as a critical function. By getting production engineers involved at a relatively early stage in the design process, proper recognition can be taken of likely problems in machine-tool usage, manufacturing tolerances, production scheduling, materials usage, process planning, etc.

Further to the last point, practically all components made by the company fall into two basic categories, i.e. those where manufacturing will have to cope with a fully defined and unalterable design requirement, and those where the design is flexible enough to accommodate the requirements of manufacturing. Fortunately, the former category usually accounts for only 0 - 5% of the component count in any product.

3.2 Pilot application of CAD methodology

The pilot project carried out within the company that 'benchmarked' the advocated CAD methodology involved design work on a test rig tht was intended to evaluate the performance of seals for use in an offshore, multi-pass fluid swivel. This particualr project was considered for three basic reasons:

(1)   It was project that took an adaptive design to a manufactured item.

(2)   It used the company's CAD/CAM system as a successful design (and manufacturing) tool.

(3)   It involved components that were of a relatively simple shape (i.e. axisymmetric).

The fluid swivel itself was of modular construction, comprising of stackable swivel units (see Figure 5) so that complete swivel arrangements could be offered for each particular application (refer to Figure 6). The individual units considered ranged in size from 1.0m up to and including 2.4m (mean toroidal diameter). All components had exacting surface finish and tolerance requirements.

Figure 7 shows an assembly view of the test rig with components shown in basic geometric form. The rig itself consisted of a stack of axisymmetric plates comprising an outer stator stack and an inner rotor stack. The inner stack rested on a bearing and load cell which was precisely shimmed. Because the seals under test had to with stand pressures in order of 10000 psi, account had to be taken of the seal extrusion gaps which were critical to efficient fluid swivel operation. These gaps had to be set at 0.6mm and 1.2mm to vertify the manufacturer's claims against the project requirement specification. Tolerances on these gaps had to be kept very tight (in order of +/- 0.01mm). Providing stresses were low enough the overall sizes of the plates were not important; only their relative sizes and the tolerance build-up between them were significant.

From basic sketches and then conventional drawdowns of the test-rig arrangements (at nominal size) the design engineers concerned used the CAD system to generate initial layout components at their basic sizes, that is, full-scale and mid-size (see Figure 8). Initial layouts could be generated semi-automatically using a user-extension program specifically developed to precisely locate individual layout components with respect to an overall coordinate datum. Becasue of the need to evaluate the check extrusion gaps, tolerance information (in the form of maximum and minimum conditions) was incorporated into the layouts during subsequent refinement. Refined (modified) layouts could then be rapidly and accurately generated using the appropriate user-extension program. Due to the test rig components having sinmple shapes, and the fact tht only 1-D tolerance build-up problems were being investigated, one basic definition was created for each layout component (i.e. at mid-size) and its geometry was edited as required. This technique allowed the design engineers to see the effects on build-up of deliberately introducing spacer shims to change the extrusion gaps, and to see what was going to happen when the production engineers informed them of the company's machine-tool capabilities.

Having successfully refined and optimised the definitive layouts, these could then be re-used by the project draughtsmen in order to produce the necessary manufacturing drawings. Because the requisite design information was held in aCAD database and had been defined (and refined) in a disciplined manner, detail draughting proved a relatively simple matter. Potential errors were minimised by directly re-using the geometric data of a 2-D layout to produce the required manufacturing drawings (assembly and piece-part), whilst appropriate 'custom-built' user-extension programs assisted with the rapid generation of the required textual and ancillary information (e.g. company drawing frame, notes, etc.). Hence geometry bound for in-house manufacturing retained its full-scale and mid-size definition. However, the time to stop refining a layout and start producing an annotated drawing set was not as definite as implied, due to particular company needs on sub-contract manufacture and procuring bought-out items.

The finalised manufacturing drawings for all components were manually checked in the usual way, and passed on to the CNC part-programmers to produce paper control tapes for manufacturing. The CNC part-programmers were able to proceed with full confidence in the geometric data available and generally only minor changes resulted, which were mainly associated with specific machining technique.

It cannot be emphasised too much how useful the ability to modify toleranced sizes using the CAD system proved. At the outset production engineers informed the design activity about the tolerances they could satisfy or expect. As embodiment design progressed the design engineers were able to see how they could cope with these tolerances and, if they couldn't cope, inform the production engineers of what tolerances were required to meet the project specification. In some particular instances, this practice was further extended into the manufacturing phase. The actual sizes of components which had been manufactured first could be fed back, via inspection, into the geometric layout of the CAD system to see what sizes had to be achieved on the next few components so that tolerance build-up effects could be kept to a minimum. This could generally be accomplished within minutes of a manufactured component size being known.

3.3 Project organisation

A primary reason for the overall success of the pilot project was the effective organisation of the project design team.   Judicious use of design reviews, value engineering methods and the establishment of a multi-lateral communication structure within the project team ensured a significant degree of design project control.   Figure 9 gives a diagrammatic representation of design team structure and the interfaces with other company activities.

Design reviews enabled regular and sensible communication to take place between production engineers and design personnel, and they also allowed representatives from purchasing, estimating, quality, and the workshop to exchange ideas, suggestions and recommendations on the test-rig design.   Value engineering played a large part in the communication between the design and production functions at these review meetings.

Project team organisation for design of the test rig essentially involved a structure which was based on 'skills matching' as far as possible. The project team itself consisted of

(a) a project manager who guided and controlled the project and reported to the technical manager on progress.

(b) 'innovators' who generated or expounded original ideas for the overall design.

(c) 'developers' who refined and developed the information emanating from the innovators, and other external inputs such as production engineering, purchasing, etc.

(d) 'detailers' who produced the hard-copy manufacturing drawings, parts lists, etc.

(e) 'administrators' who generated test specifications, purchase requisitions, etc.

Each identified skill was not unique to a particular team member.   Some persons could possess an aptitude for more than one skill.   The emphasiswas on lateral communication between team members, with each team member utilising and complementing the skills of one another.

Throughout the duration of the pilot work, the Teaching Company kept in close contact with the project team to monitor and assess the impact of the proposed methodology in a live situation.   In effect, the Teaching Company acted as an in-house CAD applications consultancy, formulating and recommending suitable   action on any relevant problems that arose throughout the project.   It is worth noting that both the Teaching Company and computer systems people had been intimately involved in the specification and generation of the requisite user-extension software for the CAD system that assisted layout development.   Thus the necessary expertise was immediately at hand to the project team as and when required.

4. CONCLUDING REMARKS

There are two principal criteria to recognise when considering the application of CAD methods to the design of products. First, in order for that application to be effective, it is essential that a thoughtful and systematic perspective is taken of the conceptual status of the product. Secondly, the principle of concept embodiment also demonstrates that effective application of CAD is a careful blend of methodology, technology and organisational resources. This principle can be best summarised by reference to the diagram given in Figure 10.

For the company concerned in the case study, a major benefit of a methodological approach was that proper planning and control of the project design work could be effected within the required lead-time. This was a novel departure from the traditional approach of attempting to address changes in design configuration on an 'ad hoc' basis, with the associated ill-effects this has for product quality. Manufacturing costs were therefore controlled throughout the design process, and the completed test rig was available within the specified timescale. The CAD applications methodology was therefore vindicated through very successful use of an integrated design (and manufacturing) approach to a company project.

In addition, the project emphasised that facilitating concept embodiment relies just as much on people as it does making good use of the available CAD technology. It also highlighted the invaluable contribution provided by technically-competent, in-house applications expertise in making the methodology work within a live manufacturing environment.

REFERENCES

(1) BLACK, I. and MURRAY, J. L. 'Conceptualisation and computer-aided design: an applications perspective '. Proceedings of third international conference - effective CADCAM'87: towards integration, London, 1987, 49-53 (Mechanical Engineering Publications Limited).

(2) MEDLAND, A. J. The Computer-based design process, 1986 (Kogan Page).

(3) BLACK, I. 'Applications and influences of interactive computer graphics within mechanical product design: a critial appraisal and evaluation'. Unpublished MSc thesis, Department of Mechanical Engineering, Heriot-Watt University, 1986.

(4) ENCARNACAO, J. and SCHLECHTENDAHL, E.G.
Computer-aided design; fundamentals and system architectures.   1983
(Springer-Verlag).

(5) KRAUSE, F. L. CAD-induced changes in design work. International journal of robotics and computer-integrated manufacture 1984, 1, 2, 173-180.

(6) RZEVSKI, G. The role of computers in engineering design. Computer-aided engineering journal, November 1983, 5-8.

(7) STARK, J. What every engineer should know about practical CAD/CAM applications, 1986 (Marcel Dekker).

(8) PAHL, G. and BEITZ, W. Engineering design, 1984 (Springer-Verlag/Design Council).

(9) HUBKA, V. Principles of engineering design, 1982 (Butterworth).

(10) FRENCH, M. J. Conceptual design for engineers, 1985, (Springer-Verlag/Design Council).

(11) LOVE, S. F. Planning and creating successful engineered designs, 1980 (Van Nostrand Reinhold).

(12) OAKLEY, M. Managing product design, 1984 (Weidenfield and Nicolson).

(13) PUGH, S. 'CAD/CAM - hindrance or help to design ?'. Proceedings of conference on conception et fabrication assistees par ordinateur, Brussels, 1984.

(14) BLACK, I. and LINTON, H. C. A rationalised approach to the application of CAD within mechanical product design. Computer-aided engineering journal, 1987, 4, 5, 213-216.

(15) ASIMOW, M. Introduction to design, 1962 (Prentice-Hall).

(16) LEYER, A. Machine design, 1974 (Blackie and son).

(17) ROTH, K. H. Foundation of methodical procdures in design. Design studies, 1982, 2.

(18) RODENACKER, W. G. Methodisches konstruieren, 1970 (Springer).

(19) ANDREASEN, M. M. 'The use of systematic design in practice'. Proceedings of international syposium on design and synthesis, Tokyo, 1984, 133-138.

(20) WILSON, P. M. BLACK, I. 'Computer-aided production engineering of embodiment design'. Proceedings of international conference on computer-aided production engineering, Edinburgh, 1986, 217-222.

(21) BLACK, I. Computers add the production aspect. Engineering, 226, 10, 754-755.

(22) BLACK, I. Assuring confident re-use of production drawing information by the NC programming activity. Computer-aided engineering journal, 1986, 3, 4, 154-163.

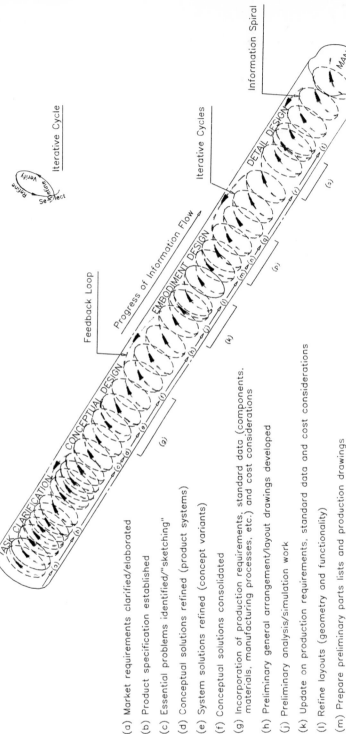

(a) Market requirements clarified/elaborated

(b) Product specification established

(c) Essential problems identified/"sketching"

(d) Conceptual solutions refined (product systems)

(e) System solutions refined (concept variants)

(f) Conceptual solutions consolidated

(g) Incorporation of production requirements, standard data (components, materials, manufacturing processes, etc.) and cost considerations

(h) Preliminary general arrangement/layout drawings developed

(j) Preliminary analysis/simulation work

(k) Update on production requirements, standard data and cost considerations

(l) Refine layouts (geometry and functionality)

(m) Prepare preliminary parts lists and production drawings

(n) Finalise on analysis/simulation

(p) Update on production, standards and cost

(q) Definitive layouts laid down

(r) Production drawings generated/parts lists prepared

(s) Finalise on production, standards and cost

(t) Issue bulk manufacturing documentation

Fig 1   General sequential model of design process

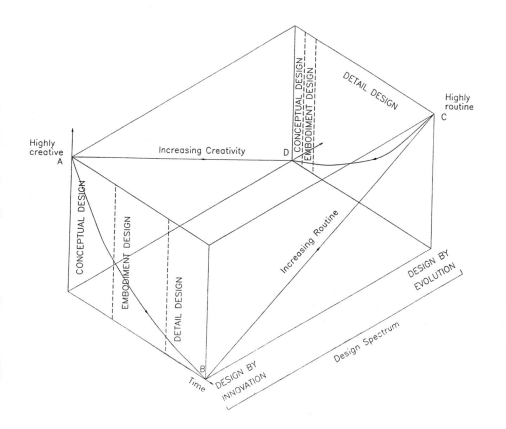

Fig 2   Spectral approach to design

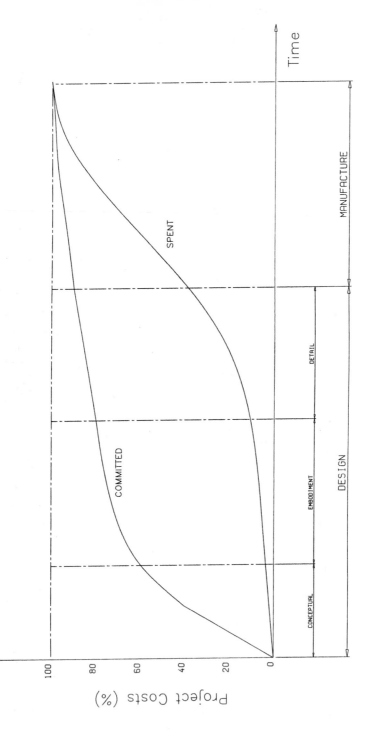

Fig 3   Proportion of finance allocated to a project

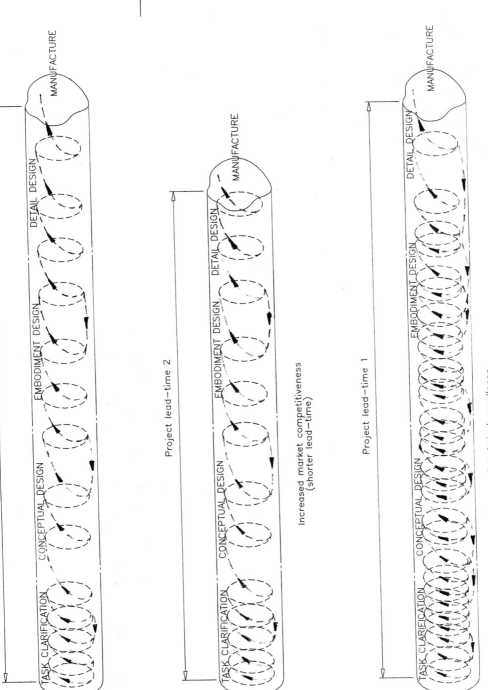

Fig 4   Effects of project lead-time

Application of concept embodiment using CAD methods

Project lead-time 2

Increased market competitiveness
(shorter lead-time)

Project lead-time 1

Increased technical excellence
(increased number of cycles)

TASK CLARIFICATION   CONCEPTUAL DESIGN   EMBODIMENT DESIGN   DETAIL DESIGN   MANUFACTURE

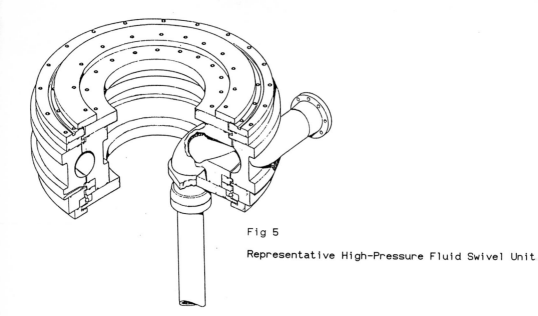

Fig 5

Representative High-Pressure Fluid Swivel Unit

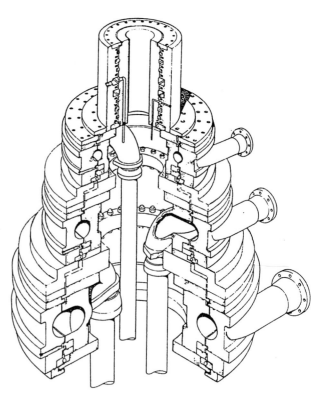

Fig 6   Stack of individual swivel units

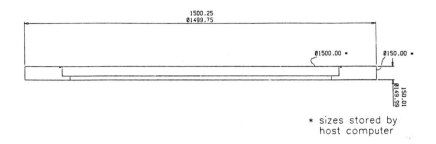

* sizes stored by
host computer

Fig 8   Precise component definition suitable for re-use

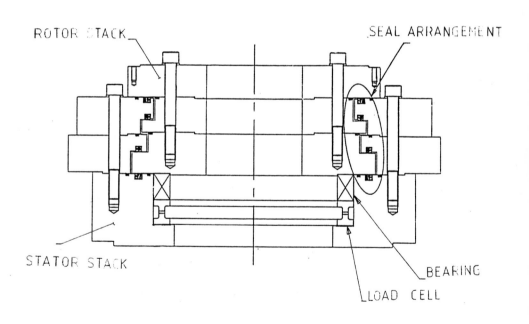

ROTOR STACK

SEAL ARRANGEMENT

STATOR STACK

LOAD CELL

BEARING

Fig 7   Schematic of test-rig assembly

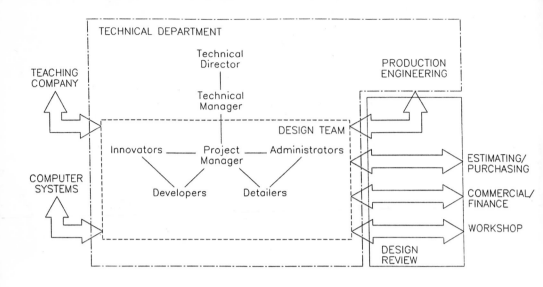

Fig 9   Design team organisation and interfaces

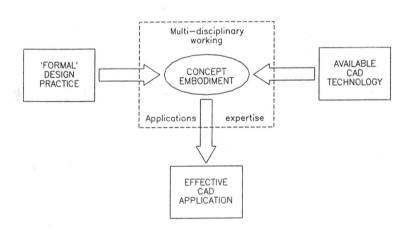

Fig 10   Concept embodiment : a methodological basis for effective CAD

C377/086

Development of the surface modeller swans

A A BALL, PhD, CEng, MIProdE
Department of Mechanical Engineering, University of Birmingham
J M HANDS, BSc
Pafec Limited, Nottingham

The surface modeller SWANS is based on the theory of
reparametrisation and has been developed by Pafec Ltd
in collaboration with the Geometric Modelling Group
(which has recently moved 'en bloc' from Loughborough
to Birmingham). The paper outlines the basic ideas
of parametric curves and surfaces and the role of
reparametrisation, and introduces the surface design
procedures of SWANS with a range of industrially
based examples.

1 INTRODUCTION

Surface modelling is of crucial importance to the application of CADCAM in
industries where the products or components are of a complex shape. It
provides the geometric data-base, not only for picturing an object but also
for the follow-on activities of analysis and manufacture. The numerical
modelling of 'free-form' or 'lofted' surfaces was pioneered in the shipbuilding, automotive and aerospace industries. But the technology is now
finding far wider application, particularly in the die and mould industry.

The key problem in surface modelling is the excessive time it takes
designers to define satisfactory numerical surfaces. Fundamental research in
the early Eighties related many of the difficulties experienced by designers,
in conception and realisation, to limitations in the underlying mathematics;
and led to the theory of reparametrisation [1, 2]. The work was adopted by
Pafec and developed via a collaborative research programme to form the basis
of the surface modeller SWANS [3].

This paper describes the collaboration with Pafec and is essentially in
two parts. The first part introduces the basic ideas of reparametrisation
with some examples, to explain why good designers can experience so much
difficulty with numerical surface definition. The second part introduces the
basic procedures of SWANS with a range of industrially based examples to
demonstrate the practical advantages of reparametrisation.

2 PARAMETRIC CURVES AND SURFACES

Parametric methods for the numerical description of curves and surfaces are
established and have been used for more than twenty years in computer-aided

design [4]. In this approach a curve or surface is represented as the locus of a point [x, y, z] moving in three-dimensional space. For a curve, the generic point is defined by a vector function of a single variable:

$$P(u) \equiv \left[x(u), y(u), z(u) \right]$$

For a surface, the generic point is defined by a vector function of two independent variables:

$$P(u,v) \equiv \left[x(u,v), y(u,v), z(u,v) \right]$$

In free-form design, one cannot generally find a single analytic expression for a whole curve or surface and it is necessary to build up a piecewise representation of bounded analytic functions. The parametric method accommodates bounding simply by bounding the range of parametric variables. Typically, $P(u)$ $0 \leqslant u \leqslant 1$ defines a curve segment and $P(u,v)$ $0 \leqslant u,v \leqslant 1$ defines a surface patch or tile.

3 PARAMETRIC CUBIC SEGMENT

The parametric cubic segment has been, in its various guises, the most widely used curve segment in computer-aided design. For convenience we use the equation form:

$$P(u) = (1 - u)^2 P_1 + 2(1-u)^2 uP_2 + 2(1 - u)u^2 P_3 + u^2 P_4 \qquad 0 \leqslant u \leqslant 1$$

It is easily checked that P_1 and P_4 are the end points of the segment, and P_2 and P_3 lie on the end tangents as illustrated in Figure 1. The precise location of P_2 and P_3 along the end tangents is significant and affects the internal shape of the curve: moving them closer to the end points P_1 and P_4 makes the curve flatter and moving them further away gives the curve more body.

Note that if $P_2 \equiv P_3$ then

$$P(u) = (1 - u)^2 P_1 + 2(1 - u)uP_2 + u^2 P_4 \qquad 0 \leqslant u \leqslant 1$$

and the curve is a parabolic segment with P_2 as the point of intersection of the two end tangents.

If in addition P_2 is the midpoint of P_1 and P_4 then

$$P(u) = (1 - u)P_1 + uP_4 \qquad\qquad 0 \leqslant u \leqslant 1$$

and the curve is a straight line segment.

Conversely, any straight line or parabolic segment can be defined by a parametric cubic segment.

4 BICUBIC PATCH

The equation of the parametric cubic segment can be adapted to a lofting definition of the bicubic patch by introducing a second parameter v:

$$P(u,v) = (1 - u)^2 P_1(v) + 2(1 - u)^2 u P_2(v) + 2(1 - u)u^2 P_3 + u^2 P_4(v)$$

$$0 \leqslant u,v \leqslant 1$$

where $P_1(v)$, $P_2(v)$, $P_3(v)$ and $P_4(v)$ vary as parametric cubic segments.

Let $Q(v)$ be the parametric cubic segment defined by $P_1(v)$, $P_2(v)$, $P_3(v)$ and $P_4(v)$ $0 \leqslant v \leqslant 1$. Then the patch is a blend from $Q(0)$ to $Q(1)$ defined by the variation of $Q(v)$ $0 \leqslant v \leqslant 1$ as illustrated in Figure 2.

Note that if

$$P_2(v) \equiv P_3(v) \qquad\qquad 0 \leqslant v \leqslant 1$$

then the patch is the locus of a varying parabolic segment, and if

$$P_2(v) \equiv P_3(v) \equiv \tfrac{1}{2}(P_1(v) + P_4(v)) \qquad\qquad 0 \leqslant v \leqslant 1$$

then the patch is the locus of a varying straight line segment.

However it is not true that the locus of a varying parabolic or straight line segment can always be defined by a bicubic patch, even when the variation is defined in terms of parametric cubic segments. In the next section we give an example.

5 LOUGHBOROUGH BENCHMARK

Consider the surface drawn in Figure 3 where all the curves are parametric cubic segments. The end sections lie in parallel planes and are composed of two straights and a parabolic blend; and as one would expect each inter-mediate section lies in a parallel plane and is composed of two straights and a parabolic blend.

The surface is defined totally in terms of parametric cubic segments and yet it cannot be matched by three (or more) bicubic patches [2]. This 'paradox' is fundamental to the difficulties generally experienced by designers when using surface modellers and the explanation is fundamental to our proposal to adopt reparametrisation.

We have noted that $P_2(v)$ and $P_3(v)$ can be assigned so that $Q(v)$ is a varying straight joining $P_1(v)$ and $P_4(v)$ $0 \leqslant v \leqslant 1$, but it is very unlikely that the parametric points $P_1(v)$ and $P_4(v)$ will 'line up' in a parallel plane as required. Figure 4 shows an example of two parametric cubic segments $P_1(v)$ and $P_4(v)$ which line up at $v = 0$ and $v = 1$ but do not line up anywhere in between. If $P_1(v)$ and $P_4(v)$ $0 \leqslant v \leqslant 1$ are not coplanar then the ruled surface will be physically different from the shape required.

It is possible to redefine the parametric cubic segments $P_1(v)$ and $P_4(v)$ $0 \leqslant v \leqslant 1$ so that the parametric points line up, but that involves changing the curve shapes which may not be acceptable to the designer. The real solution is to give the designer the curve shapes he wants and to change the parametric representation so that the parametric points line up. The process of changing parametric representation without changing shape is called reparametrisation.

The benchmark surface highlights another practical problem. We have noted that if $P_2(v) \equiv P_3(v)$ $0 \leqslant v \leqslant 1$ then $Q(v)$ is a varying parabolic segment but it must also blend the two adjacent straights. The mathematical conditions for the smooth assembly of bicubic patches are very severe and to satisfy these conditions the final section would need to be the shape shown in Figure 5! The fundamental problem is that P_2 and P_3 play two roles in the definition of a parametric cubic segment: to define the respective end tangent directions and to control the internal shape of the curve. Reparametrisation enables the two roles to be separated.

6 SWANS

The surface modeller SWANS is based on the theory of reparametrisation, but a designer using the system would be totally unaware of the fact (except that he might notice that the parameter lines were very regularly spaced). In essence the designer specifies a mesh of ' sections' and 'stringers' as shown in Figure 6 and the system then automatically regularises the respective curve parametrisations and interpolates a surface.

The curves in SWANS are built up of rational cubic segments [5], which are very versatile and encompass all conic segments ie arcs of circles, ellipses, parabolae and hyperbolae in addition to all parametric cubic segments. Each segment can be defined independently and the interpolating surface is guaranteed to be smooth. This is an important feature of SWANS and enables truly local changes. For example the effects of a change to one sectional segment can be confined to only two surface patches, which compares very favourably to the 'local' changes of cubic B-spline surfaces which spread over a 4x4 array of patches [6].

There is an option in SWANS for the user to control the variation of slope across a stringer by specifying a 'tangent stringer'. Figure 7 shows the configuration of section, stringer and tangent stringer segments for a single patch. In practice, however, most users allow SWANS to define the tangent stringers automatically.

The final figures show a selection of surfaces defined using SWANS.

REFERENCES

[1] Ball, A.A. How to make the bicubic patch work using reparametrisation. Proc. CAD82 Brighton, 1982, 315-322

[2] Ball, A.A. Reparametrisation and its application in CAGD. I J Num Meth Eng, 1984, 20, 197-216

[3] Hands, J.M. Reparametrisation of rational surfaces. The mathematics of surfaces II Ed R.R. Martin Oxford, 1987, 87-100

[4] Ferguson, J.C. Multivariate curve interpolation. JACM, 1964, 11

[5] Boehm, W. On cubics: A Survey. Comput Graph Imag Proc,1982,19,201-226

[6]   Tiller, W.   Rational B-splines for curve and surface representation.
    IEEE Comput Graph Applic, 1983, 17, 61-69

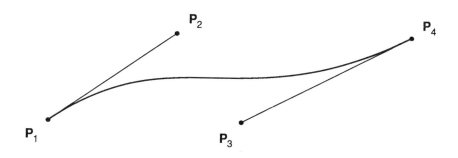

Fig 1

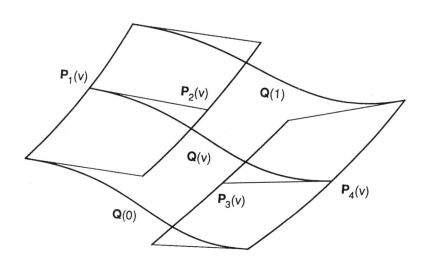

Fig 2

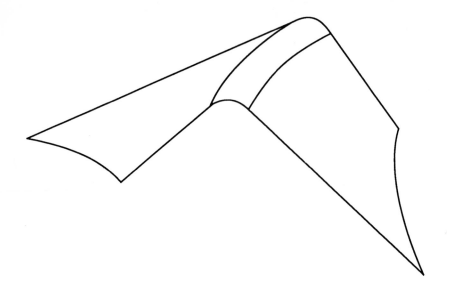

Fig 3

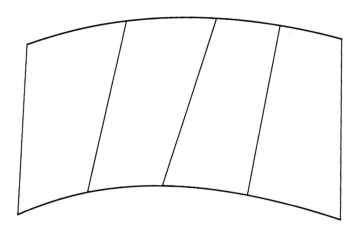

Fig 4

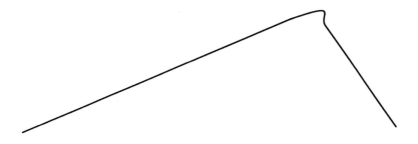

Fig 5

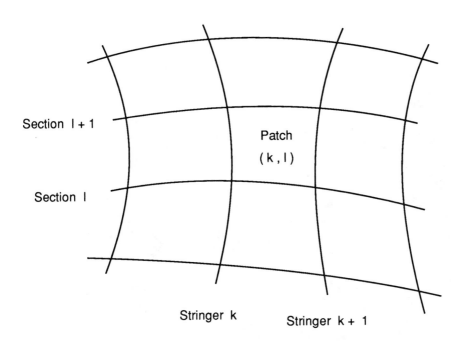

Section l + 1

Section l

Patch

(k , l)

Stringer k

Stringer k + 1

Fig 6

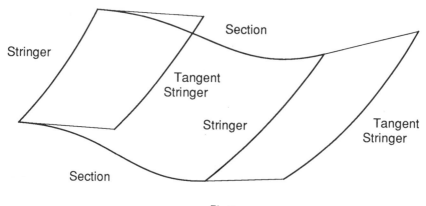

Fig 7

Fig 8

Fig 9

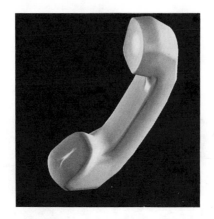

Fig 10

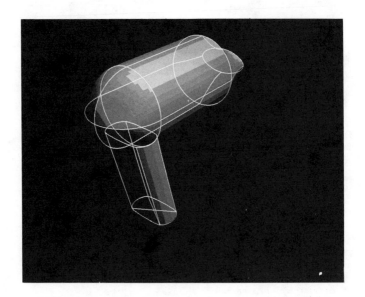

Fig 11

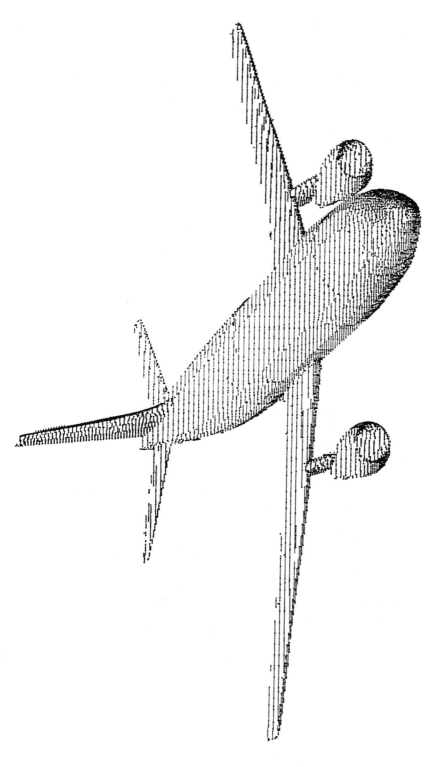

Fig 12

C377/088

An expert system to assist the design for manufacture of die cast components

J A J WOODWARD, BSc, MSc and **J CORBETT**, MSc, CEng, MIProdE, FIED
Cranfield Institute of Technology, Bedfordshire

SYNOPSIS This paper describes how knowledge based systems techniques can be
applied to a features oriented representation of a die cast component, in
order to provide a designer with the design for manufacture information that
is needed when designing an aluminium alloy die casting. The features
oriented approach has been previously applied to 'simple' machined geometries.
However, this technique is shown to be suitable for the design for manufacture
of complex components which incorporate the sort of shapes that can be
achieved with the die casting process.

1 INTRODUCTION

In virtually every field of engineering, the engineer's most powerful tool is
knowledge. An experienced engineer is generally seen as being more valuable
than a novice, by virtue of the depth of knowledge and experience he or she
possesses. The importance of this knowledge is clear from the existence of
a need to differentiate between various levels of competence, and the fact
that this differentiation is based on knowledge possessed.

Considering the design task and design for manufacture in particular, a
design engineer will tend to specialise in a specific aspect of problem
solving, where the experience gained from previous work can be carried
forward and applied to the current problems. Such a designer is more likely
to provide an acceptable or even optimal design much faster and with fewer
subsequent changes than a colleague who does not have the relevant
experience.

Unfortunately, real world design problems do not fall into categories
that correspond exactly with any one particular person's experience. This is
why a design team is generally more effective than a single designer working
alone. One of the most common problems affecting design for manufacture is
that design and production engineers cannot exchange information or
collaborate adequately across traditional departmental boundaries. There is
therefore a need for a more effective way of providing a designer with
production engineering expertise as he is working on an initial design, and
before any commitments are made which would be expensive if they were put
into production.

2 KNOWLEDGE BASED SYSTEMS

Knowledge based systems are a part of the area of computing called artificial intelligence (1). This is essentially concerned with using computers to deal with problems that traditionally computers have not been very good at, but which for most people are relatively straightforward.

They are computer programs which are able to deal with abstract concepts and uncertainty, by simulating the thought processes and methods of reasoning. This is achieved by building a symbolic representation of knowledge about a certain topic or domain, which can be supplied to a situation involving that topic, in order to reason about things which are not explicitly stated, and answer questions or give advice. In their simplest form they can be viewed as a collection of IF-THEN type rules, but there are also more sophisticated representation methods available which can be used to build a more accurate model of some human reasoning processes (2).

These are often called expert systems, but this is not really true until the answers that they give, and the apparent understanding they show are comparable with the answers and understanding of a person who is considered to be an expert in that particular field.

3 DESIGN FOR MANUFACTURE OF DIE CASTINGS

When designing a die casting there are a number of different and often conflicting aspects that the designer must consider. These include the part's primary function, its manufacturability, material properties and characteristics, how it is to be handled and assembled, and any subsequent operations such as machining or painting. It is the designer's job to find the compromise which achieves the necessary functional requirements at the minimum cost.

Manufacturability is concerned with the problem of whether a part can actually be made by a particular process, and is a fairly fundamental consideration. For die casting it means asking questions such as:
- Will the mould fill properly?
- Will there be any porosity?
- Can the part be removed from the die, after the metal
   has solidified?

Typical problems affecting the die casting manufacturability arise from the designer allowing insufficient draft, which usually results in the part sticking in the die, inclusion of unnecessary undercuts that then need costly moving cores to allow the part to be removed, over tolerancing and excessive variation in section thickness. These are all relatively easy problems to correct if they are detected in time, but can add greatly to manufacturing costs if they find their way through into production.

Procedures employed for checking for these faults vary from company to company, but the final check is usually done by the die caster, who draws on past experience to assess a proposed design. One of the drawbacks with this approach is the length of time taken for change requirements to come back to the designer, which can add considerably to the total product development time.

A knowledge based system which could provide a designer with design for die casting information at the initial design stage would provide the following advantages:
- To allow 'simultaneous' engineering by integrating the design and manufacturing functions.
- Designs which were closer to being right first time, with fewer changes needed to allow them to be satisfactorily produced.
- A means of finding the limitations of different design approaches without actually needing to make a test part.
- Less delay between realising that changes are needed and implementing those changes.
- Shorter total development times, but more time to spend on initial conceptual design.
- A means of training designers in good design for die casting techniques.

4 AIM

The aim of this work was to develop a means of helping a designer to produce an initial design for a die casting, which was right first time. This was to be achieved by using a knowledge based system to provide a designer working on a die casting, with the relevant experience and knowledge of an expert die caster as required.

5 KNOWLEDGE ACQUISITION

The first stage of the work was concerned with knowledge acquisition, that is, with gathering and organising the information which is needed to design a good die casting. In order to build an effective knowledge based system, it is necessary for the builder to have a reasonable understanding of the particular problem domain. Fortunately, die casting is a reasonably well documented process and so a lot of the initial domain knowledge was obtained from books, house design rules and other literature (3,4,5). This was then supplemented by interviewing experienced die casting designers and die casters and discussing specific design problems that they had encountered. All of this information was collated into a single, structured document, which then formed the basis of the computer design for die casting knowledge base.

6 KNOWLEDGE REPRESENTATION

The second stage of the work looked at how this knowledge should be represented in the computer. Analysis of the knowledge document indicated that most of the information could be written in the form of rules. This is a well established technique for representing domain knowledge in expert systems.

As an example:
IF: an undercut is detected
THEN: a moving core will be needed

IF: a moving core is needed
THEN: the die will be more expensive to make

A special control mechanism, known as an inference engine, allows the system to chain through a set of rules, and to deduce a particular conclusion from a given set of conditions. In the example given, if the system could show or be shown that there was an undercut present in a proposed design for a part, it could deduce that a moving core would be needed and that as a result, the die would be more expensive to make.

Additionally, it was found that most of the rules referred to geometric features on the parts, such as bosses, ribs and holes, etc. It was therefore decided that the analysis system should try and use a features oriented approach, with a CAD system used for direct input of the feature information.

7 FEATURE ORIENTED REPRESENTATION

When a designer is designing a part, he works with a higher and more abstract level of information than just a representation of its geometry, defined in terms of points, lines and surfaces. He is influenced for example by, certain implied assumptions about the function of various features, as well as constraints on how they can be combined, or interact with other features and parts. To a large extent this information is implicit and remains unspoken, but it is communicated between different people as a result of their common experience. Assumptions about function and design constraints are passed from the designer to the die caster by virtue of certain aspects of the design, without needing to be stated explicitly. For the analysis system to be able to make decisions about the manufacturability of a part therefore, the representation method used must incorporate this more abstract level of information.

Although solid modellers are currently the most complete method of representing component geometry in a CAD system, they can still only represent information about the space occupied by a part. They have no mechanism for representing the concept of possible functional differences between geometrically similar objects. The image shown on the screen is only significant to the user because of his interpretation of it. Some means is needed for extracting the higher level, implicit information from the designer's thought processes and representing it within the computer, in addition to the lower level geometric information. The mechanism for gathering this information has to be simple and natural for the designer to use, without slowing him down by asking directly for too much detail. At the same time it must collect all the relevant information as it is made available.

Current research related to this problem is divided between two main approaches for obtaining the relevant information. These are feature recognition (6) and designing with features (7). Both of these ideas appreciate the need for attaching implied information to the geometric representation of a component, and both see the part as a number of process specific features that affect how the part functions and how it is made. However, the methods that they use for actually obtaining it are quite different.

7.1. Feature Recognition

In the feature recognition approach, the designer first uses a conventional solid modeller to create a model of a part. This model is then submitted to a post processing operation, which tries to fit a library of predefined features that the computer has been taught, and effectively 'understands', against the given geometry, (Figure 1). From this, a representation of the part can be constructed that has a higher level or quality of information than the purely geometric representation used in the modeller. This representation is known as a feature base because it describes the component in terms of the specific features which it possesses.

The advantage of this approach is its generality: one model could be submitted to several different post processors, each for a different manufacturing process. The designer does not need to commit himself to any one process, and in theory does not even need to know anything about designing for a particular process while he is building the part.

Against this is the problem that if the post processor cannot recognise a feature, the information which is passed on to the expert system analysis stage is incomplete, and any decisions or recommendations made will probably not be valid. Given the freedom that current solid modellers allow the user in methods for constructing a model, it was considered highly likely that the designer would be able to include at least one feature which would not be recognised or correctly identified.

7.2. Designing with Features

The alternative is to use a designing with features approach, where the model is built up using a set of process specific parametric features, selected from a choice available to the designer on the screen (Figure 2). As the model is being built, the user can be prompted for any additional information about the feature, which the system knows will be needed for the expert system's analysis that follows. The feature base is built up in parallel with the modeller's own database, and so the modeller then becomes just a means for displaying the geometry on the CAD workstation screen.

The major advantage of this method is that as the model is built only from features that the system understands, using legal combination operations, it should not be possible for the designer to construct a model whose feature base the computer cannot subsequently evaluate.

The limitations are that the designer must know in advance what process is to be used to manufacture the part, and also a separate model needs to be built for each different process under consideration, which would be time consuming.

The main difference in actual implementation between the feature recognition and designing with features approaches is in the amount of knowledge required by the system. For designing with features, there is a finite, relatively small number of well defined features and combining operations available for use, which means that the size of the system can be kept within manageable proportions. For feature recognition, however, there are an almost infinite number of ways that the user can construct a model, which the system needs to be able to deal with. It was therefore decided to adopt the designing with features approach.

Apart from the representation of the part, there are also a number of problems as far as the designer is concerned in actually using the solid modeller.

Firstly, because solid modellers are intended to be used for describing a wide variety of real world objects, they are often complex to operate, which can get in the way of the process of actually designing anything. This is especially true when modelling die castings, where the curved and rounded shapes which are most suitable for this process are not easily modelled using the standard primitive shapes of blocks, cones and spheres. Although they can be modelled accurately, it takes a lot of effort from the designer to do this.

A second drawback is that a great deal of forward planning is needed by the designer to translate a mental image of a part into system primitives, and then to select a suitable way of combining these to form a model. It is quite common, especially for new users, to find that the initial approach used for building up a part leads to a state where some later features cannot be added, and it is then necessary to start again with a different method of combining the primitives.

It would be much easier for the designer if the general interface of a system could be adapted to suit a particular task, especially if it could be made to fit the natural way of thinking about and performing that task. For example, a die cast component can be viewed as a set of features required to perform certain physical functions, which are then positioned relative to one another at various locations in space and then joined up with metal.

It was decided that, by using the designing with features approach, it would be relatively straightforward to provide the user with a set of die casting primitives as an alternative to the more general primitives of most solid modellers. These features would correspond exactly to those functional features that a designer naturally thinks in terms of when considering a part produced as a die casting. By defining these features parametrically, it would also be possible to construct their geometry automatically, without the user needing to translate the shapes into general system primitives.

The next stage therefore involved defining a suitable set of die casting primitives, and specifying the parameters needed to describe them. The feature definitions proposed in (8) were considered, but these were found to be oriented more towards machining and fabrication operations and were not really suitable for use with die castings. The feature set developed was built up by analysing existing die castings, and trying to define them in terms of the features used in the knowledge representation rules. Figure 3 shows the working set of features that was developed, and Figure 4 shows an example of some of the parameters used for describing a boss. All of the features are defined in some way in terms of section profiles, which are then oriented and skinned to create a solid object in the modeller.

Although the number of features appears to be rather small, this set was found to be sufficient for most castings. If it were found that additional features were needed, the representation used is flexible enough to allow them to be added fairly simply. It was not considered appropriate to allow the designer to define his own personal feature set, because the subsequent analysis system would have no understanding of what it meant. This is the exact drawback associated with a feature recognition based approach that the designing with features method was intended to avoid!

Finally a simple test piece was defined, which included a typical selection of the different feature types available (Figure 5). This was then used in the initial development and testing of the knowledge based analysis system.

9 DESCRIPTION OF SYSTEM

The solid modelling software selected for development of the proposed system was CAEDS from SDRC. This was chosen, firstly, for its powerful in-built programming language called IDEAL, which was used to provide the modified user interface, and also for its integral relational database management system called PEARL, which was used to create and maintain the feature base under the control of an IDEAL program. Additionally, CAEDS is a full mechanical computer aided engineering (MCAE) design environment. This means that a solid model of a die casting can be passed to other modules within the system, which can be used to perform finite element analyses to determine its stress, thermal and dynamic behaviour, investigate its behaviour in a mechanism or even generate a CNC cutter path for producing the die cavity.

The system is used for detail designing a component, which it has already been decided will probably be produced as an aluminium pressure die casting. Factors such as size, production quantities and allowable costs will already give a good indication of this.

The designer logs on to a CAEDS session as normal and then runs an IDEAL program file. This modifies the user interface to allow the designer to work with a set of parametric die casting features instead of the general primitive shapes provided in the standard system. It also constructs and maintains the feature base as the model is being built, which is held as a PEARL database. A schematic diagram of the main components in the system is shown in Figure 6.

The user selects a die casting feature from a menu on the screen, and is asked to enter suitable values for the parametric attributes needed to fully define it. The underlying IDEAL program uses these values to automatically construct the geometry of the feature in the modeller and display it as a solid object on the screen. This feature can then be oriented in the desired position and moved to its relative location in the part. When the user decides that the feature is exactly as required, all the relevant information about it is written to a PEARL database.

Further features can be constructed and combined together using CUT and ADD operations in order to build up the final representation of the part, with the details of each new feature being added to the feature base automatically.

There are two additional points that should be made about the modified interface to the system.

Firstly, it aims to copy the general style and mechanism of the standard CAEDS interface. The user can still use any of the orientation and viewing

facilities that are available in the standard CAEDS environment. This is for
the benefit of users who are already familiar with the CAEDS system. The
main difference seen by the user is that instead of using blocks and cylinders
to create a model, actual process features can be selected.

Secondly, it has been possible to incorporate some design expertise
directly within the interface, rather than placing it all in the post
processor. This means that, depending on the operation being performed,
inappropriate values will not be accepted, e.g. when creating a boss, the
user will be informed that, in this particular instance, a draft angle of 2
degrees is not appropriate, and that a minimum of 4 degrees is required.
This helps to provide an even quicker feedback of design faults, since the
user doesn't need to invoke the post processor stage to be alerted to them.

When the user indicates that the model is finished the feature
information in the database is output to an ASCII text file, which is used to
create a frame based representation of the part ready for input to the
analysis program.

The knowledge bases accessed by the analysis system contain design for
die casting knowledge in the form of rules. These rules specify the
combinations of conditions and attribute values that will result in a
particular design fault, e.g.

IF: feature_has_insufficient_draft
THEN: part_may_stick_in_the_die

A control mechanism in the analysis system guides the application of
these rules by, firstly, assuming that a particular design fault exists, and
then trying to find evidence to support this assumption. It does this by
looking in the part feature base for attribute values which match the IF side
of the rule for that fault. If no such evidence is found, then it is
unlikely that the design fault would actually occur in the manufactured part.

However, if supportive evidence is found, a message is displayed to the
user. This states that a particular design fault is likely to occur, along
with the feature which is the cause of the problem. It also gives
information on the implications for manufacturing if the part is left
unaltered and, finally gives some advice on how the design needs to be
changed to overcome or avoid these problems. It is then up to the designer
to act on this information and return to the modeller and make any recommended
changes himself. The system has no capability for modifying the
representation of the part itself, since it is considered that responsibility
for the final design should rest with the designer. The build-analyse-modify
cycle can be repeated as many times as is necessary to achieve a satisfactory
design.

Each particular aspect of the casting design will eventually have its
own knowledge base e.g. manufacturability, handling, machining, etc. and in
turn each of these will be made up of modules which contain the rules for
detecting certain design faults. For instance, in the manufacturability
knowledge base there are individual rule modules for assessing the mechanical
complexity of a part, checking for complete/homogeneous mould filling and
determining whether the part can be removed from the die once the metal has
solidified. This structured modular approach means that development and
maintenance of the knowledge bases is made much easier than if all the rules
were lumped together in one large knowledge base.

Currently the analysis system is implemented in PROLOG on an IBM PC-AT, with CAEDS running on an IBM mainframe. The text file created in CAEDS which contains the feature base information is transferred between the systems via a data link. While this configuration is sufficient for demonstrating the concept of the features oriented approach when assessing the die casting manufacturability, it would almost certainly be inadequate for a system which considered all of the aspects associated with die casting design simultaneously. Future work will therefore involve transferring the analysis software to the CAEDS system mainframe, which will also provide an opportunity for achieving closer integration between the modelling and analysis systems.

10 CONCLUSIONS

Knowledge based systems have the potential to provide the designer with critical information immediately at the point of design. This will become increasingly important with product life cycles continuing to reduce, and the need to react swiftly to changing market requirements.

Although the current system is still only at a basic prototype stage, this work has shown that the proposed features oriented approach can be made to provide a high enough level of design information for a knowledge based system to be able to make satisfactory decisions about the manufacturability of a die cast component. It has also highlighted many of the support functions needed for knowledge based systems to interact effectively with CAD systems. Future work will be concerned with developing an improved feature modelling facility, extending the analysis capabilities of the knowledge system, knowledge based generation of change requirements and closer integration between the modelling and analysis stages. Additionally, the features approach is applicable to designing for other manufacturing processes as well as die casting, and could also be used to allow automated process planning of a finished design.

11 ACKNOWLEDGEMENTS

This work has been carried out as a PhD research project under the SERC's case award scheme, in collaboration with IBM (UK) Ltd's manufacturing plant at Havant.

REFERENCES

(1) HARMON, P, KING, D. AI in business: expert systems. Wiley Press, 1985.
(2) RAUCH-HINDIN, W.B. A guide to commercial AI. Prentice Hall, 1988.
(3) STREET, A.C. The diecasting book. Portcullis Press, 1977.
(4) BRALLA, J.G. Handbook of product design for manufacturing. McGraw-Hill Inc, 1986.
(5) BARTON, H.K. Product design for diecasting. Society of Die Casting Engineers, 1974.
(6) HENDERSON, M.R. ANDERSON, D.C. Computer recognition and extraction of form features: a CAD/CAM 1. Computers in Industry 5, 1984, 329-339.

(7) LUBY, S.C. DIXON, J.R. SIMMONS, M.K. Creating and using a features
     data base. Computers in Mechanical Engineering, November 86, 25-33.
(8) PRATT, M.J. Requirements for the support of form features in a solid
     modelling system. Final Report R-85-ASPP-01,CAM-i, Inc.

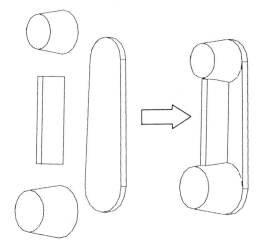

Fig 2   Designing with features based approach

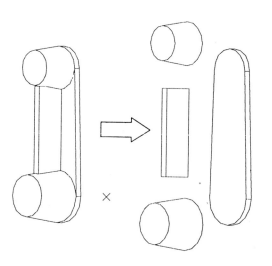

Fig 1   Feature recognition based approach

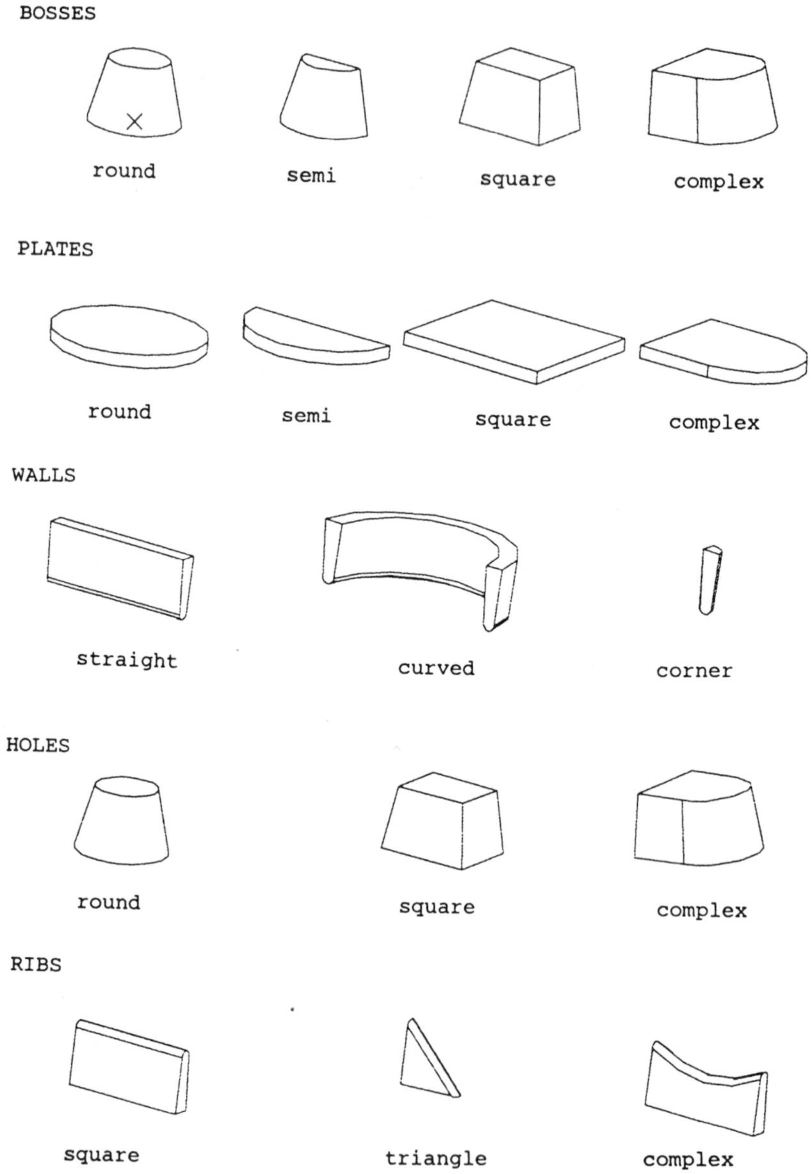

BOSSES

round   semi   square   complex

PLATES

round   semi   square   complex

WALLS

straight   curved   corner

HOLES

round   square   complex

RIBS

square   triangle   complex

Fig 3   Working set of die casting features

BOSSES

% Default orientation assumes correct production in top die half.
    NAME: bossN
    SHAPE: (round;semi;square;complex)   % ; indicates or
    TYPE: (solid;hollow{ list of holes })
% holes in hollow bosses should be created at the same time as the
% boss.
    KEY POINT: 0,0 of bottom profile
% 1. used for positioning feature and default rotation point.
% 2. always located on bottom profile.

MAIN EXTERNAL PARAMETERS: tol

    HEIGHT: (height)
% profiles are defined using a list of profile points,
  including 'fit' and 'close' indicators.
% Depending on the shape of the feature, profile generation
  is automatic.

    TOP PROFILE: (Pa,Pb,....Pn,c)
    BOTTOM PROFILE: (Px,Py,...,Pz,c)
    DRAFT ANGLE:
% edge radii are not modelled explicitly, but are required for
  use by the analysis system
    EDGE RADII: Outside Top
                    Outside Bottom
% absolute coordinates of Key Point
    POSITION: X
                 Y
                 Z
% absolute rotation about Key Point
    ORIENTATION: x
                   y
                   z
    DIE HALF: top - default unless changed by orientation

Fig 4   Parametric attributes for defining the feature oriented representation
of a boss

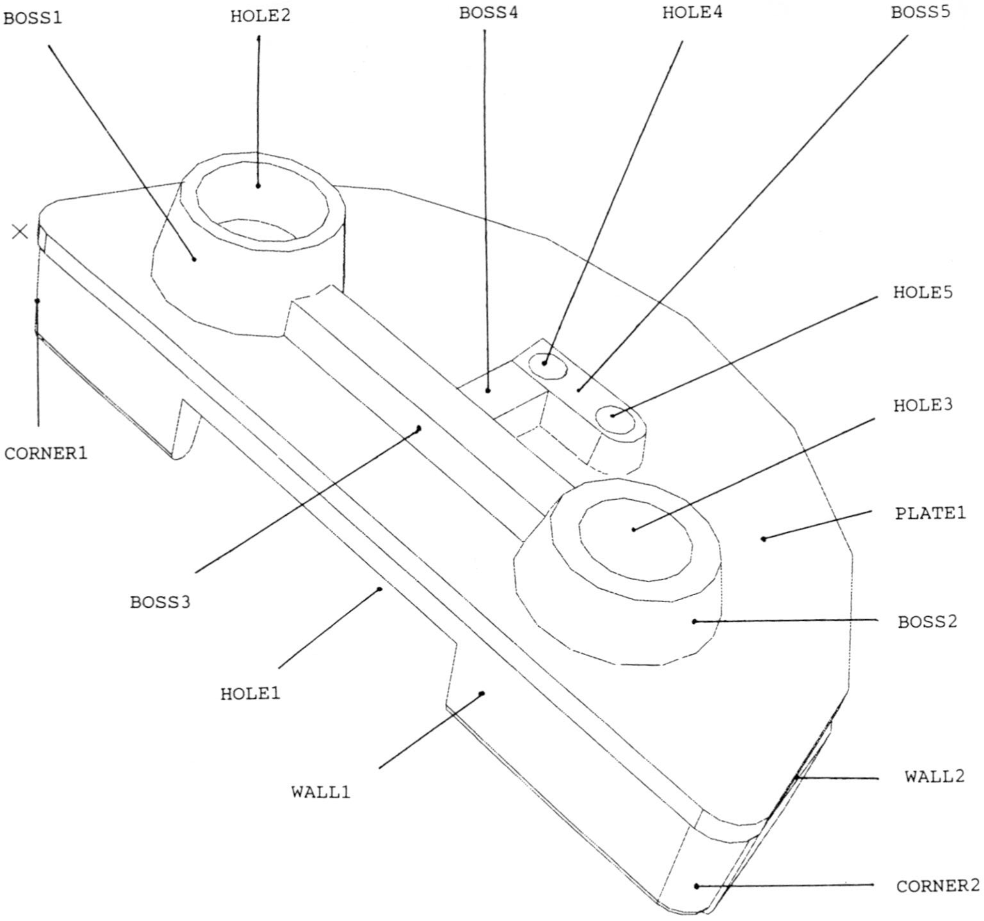

BOSS5

HOLE5

HOLE3

PLATE1

BOSS2

WALL2

CORNER2

HOLE4

BOSS4

HOLE2

BOSS1

CORNER1

BOSS3

HOLE1

WALL1

X

Fig 5   Test piece constructed using the designing with features approach

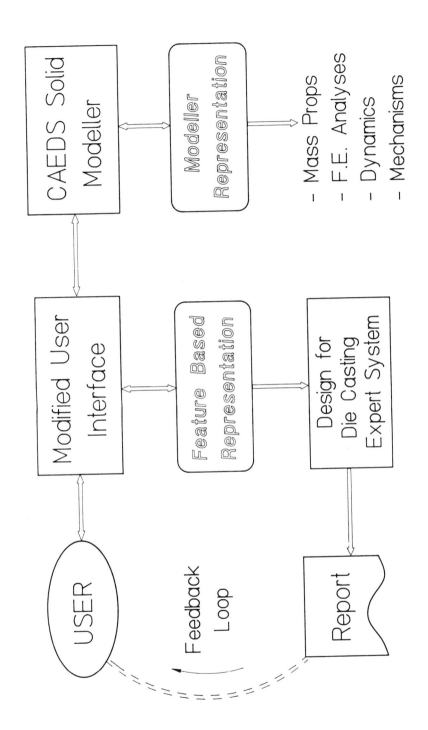

Fig 6   Schematic diagram showing main system components

973

C377/208

Development of an expert system approach to an engineering design procedure

S DAIZHONG, MSc and **R J FORGIE**, BSc, CEng, MIMechE
Design Division, University of Strathclyde, Glasgow

SYNOPSIS   A general approach is   described
which   can   be applied to a wide   range   of
design   problems   where   the   concepts   and
design methods are well defined but   making
decisions   to   reach the   best   design   is
still   complex.   The approach   covers   both
conceptual   and detail design   phases,   and
incorporates   symbolic reasoning   with   2–D
and   3–D   graphics   and   numerical
calculations.   A methodology is proposed by
which   basic   units   are   to   be   used   as
building blocks to form concepts which   are
then evaluated.

1   INTRODUCTION

The core for engineering design is shown in Fig.1 [1]. Normally,
the   design   procedure   is   as   follows: the   Product   Design
Specification (PDS) is generated first,   from which the   concept
evaluation   criteria   are extracted.   After evolving a number   of
concepts,   the next stage is concept evaluation during which the
dynamic   evaluation   matrix   [1]   is   formed.   Then   the   best
concept(s)   is   (are)   selected   ---   concept   modification   is
included   as   part   of the dynamic process   to   obtain   the   best
concept.   The   subsequent   phases   are   detail   design   and
manufacture.

The   expert system approach   is focussed on design   problems
in which the concepts are well defined but their range is   wide,
and in which the design methods exist but are still complex.   In
other words,   the decision-making to choose the best concept and
to   reach   the   best design is still complex   and   expertise   is
required. This class of design problems is common in engineering
practice.   It   is also called routine design [3] or   categorised
into   "B   boundary design" or near to "B boundary   design"   [2].
According   to   [2],   it   is   possible   to   apply   expert   system
techniques to it.

To develop the expert system approach for the design procedure, modules 1 -- 9 are given in Fig.2 in which the communications between the modules are indicated. The modules will be described in subsequent sections and, based on the modules, the approach is described in section 4.

It should be noted that this approach is a general one which can be applied to a wide range of design tasks. For illustration purpose, this study takes power transmission systems as the examplar vehicle.

2   PDS, CRITERIA AND BUILDING BLOCK STRUCTURE

2.1   PDS and Criteria Generation

For routine design (or `B boundary design') all of the PDS elements can be well defined, and so can the criteria related to the PDS. To illustrate this, Appendix A, PDS and Criteria for Power Transmission System, is a good example. Therefore, they can be put into the knowledge base directly, and it is not necessary to use the computer to generate the criteria from the PDS.

It must be mentioned that the listed PDS is for illustration purposes and not for a particular design, so it may be incomplete and lacking in detail. In the PDS for particular design problems, some values (e.g. transferred power = 4 KW), numerical ranges (e.g. for temperature, 20°C -- 50°C), and options (e.g. for quantity, large, medium and small) should be identified or given by the designer.

2.2 Building-Block Structure

After investigating the composition of some products' conceptual designs, it is found that the concepts are composed of basic units which are like building blocks and are well defined. The different arrangements of the basic units, or the different basic units, form different concepts.

For example, in Fig.3 some basic units for power transmission system are shown. Some power transmission system concepts constructed by them are shown in Fig. 4.

The function of a concept is determined by the following three factors:

(a)   Basic unit function --- different basic units form different concepts;

(b)   Basic unit position in the concept --- for the same basic unit, if its position in the concept changes, then the concept's function may change too; and

976

(c)   Concept arrangement pattern --- for the same set of basic
      units, their different arrangements form different concepts.

     The fact that the concept is formed by basic units is true
not only for power transmission system design but also for other
areas of designs. For example, the system design of buildings
used in architectural engineering, the design of chemical plant,
the ship-yard design, and so on . Therefore, the method using
basic units as building blocks to construct concepts can be
considered to be a general principle for some categories of
design.

3   THE EXPERT SYSTEM STRUCTURE

The   system   structure required by this approach is shown in
Fig.5, which includes five basic elements: inference engine and
black board, knowledge base, numerical calculation facilities,
engineering drawings, and data base. The inference engine
controls the other four elements and they exchange information
in the black board.

     In mechanical engineering design, the decision-making is
more involved with structures and numerical calculations, e.g.
how to determine the structures of the components, how to
arrange them, how to analyse the strength and other properties
of the components, and so on. To deal with them, the engineering
drawings and graphics, numerical calculation facilities and the
interfaces to them are required.

     In the data base, there are three kinds of data:

(a) Basic units and arrangement patterns --- which will be used
    to form concepts in the conceptual design phase, are kinds
    of 2-D graphic data stored in building block sub-base;

(b) The graphics sub-base --- which will   store   2-D and 3-D
    drawings of all candidate components and the initial and
    final layout for the detail design phase;

(c) Symbolic and numerical data --- such as PDS elements,
    criteria, calculation results, rating marks, the values of
    design parameters, and so on.

     In   the   numerical calculation   facilities,   built-in
programmes and software packages will be included. However,
since many software packages already exist, the expert system
should be designed to use them as much as possible by advising
the user when their use is relevant and using the results for
subsequent decisions. For some simple calculations and those
for which no software package is available, the system must have
its own built-in program.

     For engineering drawings, this approach will deal with them
in two ways:

At the conceptual design stage, the drawings are simple ones, so they can be produced using the graphic facilities built in the expert system.

At the detail design stage, the drawings are so complex that the built-in facilities of the expert system are inadequate to deal with them. So this approach will utilise dynamic drawing files created by the expert system and transferred to an external CAD package to perform the task.

4. THE DESIGN PROCEDURE REQUIRED BY THE APPROACH

4.1 For the Conceptual Design Phase

Based on the modules 1 -- 5 shown in Fig.2, the procedure involved in the approach for the conceptual design phase is shown in Fig.6. The following sections indicate the steps which will take place when a designer makes use of the system which is being developed, so the present tense has been used.

4.1.1 Identify PDS

When the design begins, the system lists all of the PDS elements, and the designer identifies the PDS for his particular design task from the list and put in the values for some PDS. When the PDS is identified, the criteria are automatically determined, since in the data base each PDS element associates with one or more criteria.

4.1.2 Construct Concepts

Based on the identified PDS, the expert system chooses basic units from the building block sub-base and various arrangement patterns to construct concepts, and shows them on the screen.

At this stage, after viewing the presented concepts, the designer may wish to generate some new concepts . If so, he can either put them into computer or, for the prototype expert system, keep them on paper.

4.1.3 Generate the Criteria Rating Mark of the Constructed Concept

As mentioned in section 2.2, the function of a concept is determined by three factors: basic unit function, unit position and arrangement pattern. A concept rating mark is composed of three sub-rating marks: unit mark, position coefficient and arrangement mark which reflect the above three factors respectively and are stored in the data base. In the data base, each basic unit and arrangement pattern is associated with its sub-rating marks for each criterion. When a concept is constructed by basic units in a certain pattern, its rating marks are also formed based on the sub-rating marks. The general principles how to synthesise the concept rating marks from the units' sub-rating marks have been developed [4].

For the concepts generated by the designer, he must provide their rating marks and input them to the data base. To help him to do so, the expert system will provide step by step instructions.

4.1.4 Evaluate the Concepts

At this point, the criteria have been determined, the concepts have been formed, and the rating marks of the criteria have been evaluated so the system can present the results of the evaluation.

At this stage, the best concept(s) is (are) presented with the criteria rating marks. The designer may make modifications or generate other new concepts again. If so, he can save them into the data base, and re-evaluate them with the existing concepts again. During the re-evaluation, the system should allow the designer to add or change the PDS and criteria if he wishes to do so.

As a guide to the designer, if he wishes to modify the best concept, those concepts are also shown for which the best concept's weakest criteria are their strong ones. If the better aspects can be incorporated then the concept is improved.

4.2 For Detail Design Phase

Based on the modules 6 -- 9 shown in Fig. 2, the design procedure of the approach for detail design phase is shown in Fig.7, which can be divided into five steps: determining design parameters, initial layout, component design, coordination and final layout.

In this phase of the approach, in comparison to conceptual design, many more drawings and numerical calculations are involved. On the other hand, compared to traditional design programmes, reasoning is involved.

4.2.1 Determining Design Parameters

In this stage, according to the information obtained from the conceptual design phase, the design parameters (e.g., for gear box design --- the gear diameters and width, the bearing diameters and width, shaft diameters and lengths, and so on) will be derived.

To obtain the parameters, the built-in programmes will be invoked to perform some simple calculations.

The parameters will be kept in a dynamic state, since they may be changed later as more knowledge is accumulated about the design.

4.2.2 Initial Layout

Although the arrangement of basic units has been generated in the conceptual design phase, their quantitative relationships have not been determined. In this stage, the initial layout, which gives the initial dimensions of the basic units and their positions, will be made based on both the arrangement and the design parameters. Graphic sub-base and built-in programmes will be invoked.

4.2.3 Components Detail Design

In this stage, the expert system will design each component in detail and assemble them into the positions determined by the above initial layout. More complex graphics and numerical calculations will be involved.

Because many software packages are available for component design, the expert system should be designed to use them as much as possible for obvious reasons. There are two ways to do so:

(a)   To build up the system's interface with the packages --- which would be a time-consuming development task; or

(b)   At certain stages, the system advises the designer which calculation should be carried out, what values should be used for the calculations and the values required on return to the expert system. The designer then follows the advice to do the calculation by running the relevant analysis software and then return the results to the expert system --- this is an easier system to implement. It is analogous to a medical expert system requesting that the doctor perform a test and feed back the results to the expert system.

The position of each basic unit has been determined in the initial layout stage. Each basic unit is composed of one or more components. The subsequent stage is to determine the shape and demensions of the component . To do so, the expert system will search the graphics sub-base to determine a suitable one.

In the graphics sub-base, there are 2-D and 3-D drawings of the candidates for components. When designing a component, the system selects the best structure from all candidates and shows it and, if required, all others on the screen. If the designer is not satisfied with the selected one, he is allowed to modify it on the screen (this may cause some difficulties, so the interaction between the screen and the graphics base must be very effective). Any problems with the designer's choices should be reported. This interactive aspect will be incorporated at an advanced stage of the development of the Expert System.

4.2.4   Interaction of System Components

It is highly probable that when the detail-designed components are assembled together, some conflicts may emerge. Therefore, the interaction between them must be checked. At this stage, the expert system can detect and report any conflicts which emerge during assembly of the selected components, and then make a decision (e.g. to modify the concept or design parameters, or re-design components) to resolve the conflicts.

4.2.5   Final Layout

After any conflicts have been solved, the final layout is formed. Then it can be saved into the graphic sub-base and drawn out.

5   A PROTOTYPE EXPERT SYSTEM

AS an application of the approach, an expert system for power transmission system design is being developed. For its conceptual design phase, many kinds of gear pairs (such as spur gear, bevel gear, helical gear and so on) will be given as basic units to form concepts. For its detail design phase, the sub-systems of gear design, shaft design, lubrication design and housing design will be included.

6   CONCLUSIONS

(a) The expert system approach proposed covers both conceptual and detail design phases. It can be applied to a wide range of design areas in which concepts are reasonably well defined and design methods are well understood, but the decision-making to reach the best design is still complex.

(b) The methodologies to form concepts from basic units (building blocks) in various arrangement patterns and to evaluate them are proposed.

(c) The approach incorporates symbolic reasoning with both engineering drawings and numerical calculations. For the engineering drawings, not only 2-D but also more complex 3-D drawings are considered; and for numerical calculations, a method is suggested whereby existing packages are used as much as possible for a particular design.

(d) Considering the dynamic nature of engineering design, the approach proposed includes a flexible method which allows the user to interact when running the expert system. In this way new basic units, arrangement patterns, concepts and ratings can be incorporated.

7   ACKNOWLEDGEMENT

Mr. Su is sponsored by an ORS Award, David Livingstone Bursary, Henry Lester Trust and University of Strathclyde R & D Budget Award. He gives many thanks to the organizations offering the awards. Many thanks are also given to Mr. R. J. Forgie and Professor K. J. MacCallum for their efforts to get the awards and to professor S. Pugh for his advice and guidance.

REFERENCES

[1]   Pugh, S.   **Introduction to Total Design**, Design Division, University of Stathclyde, 1986.

[2]   Pugh, S.   **Knowledge Based System in the Design Activity**, Proc. International Seminar, Norway, June 1988.

[3]   Brown, D.C. and Chandrasekaran, B. **An Approach to Expert system for Mechanical Design**, Proc.Trends and Applications, IEE Computer Society, 1983.

[4]   SU Daizhong   **Generation of Concept Rating Marks**, D.D./SU/ Research Report Number 03, Design Division, University of Strathclyde, 8th November, 1988.

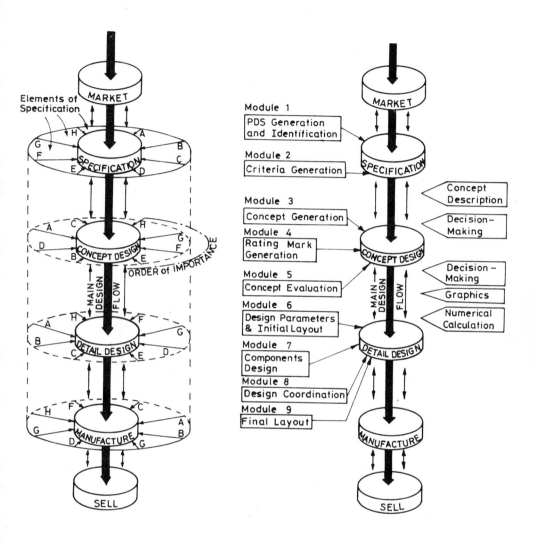

Fig 1

Fig 2

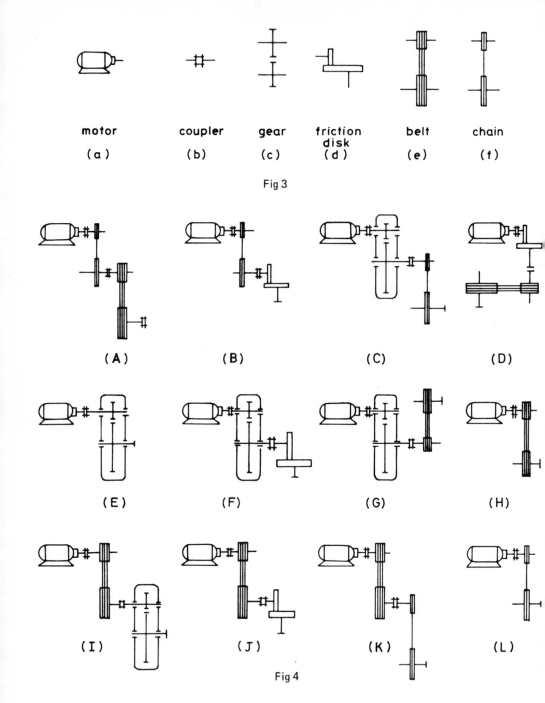

motor   coupler   gear   friction   belt   chain
disk

(a)   (b)   (c)   (d)   (e)   (f)

Fig 3

(A)   (B)   (C)   (D)

(E)   (F)   (G)   (H)

(I)   (J)   (K)   (L)

Fig 4

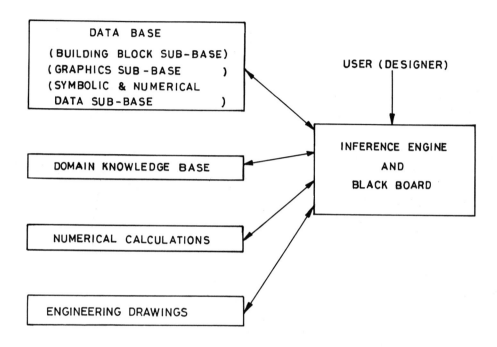

Fig 5

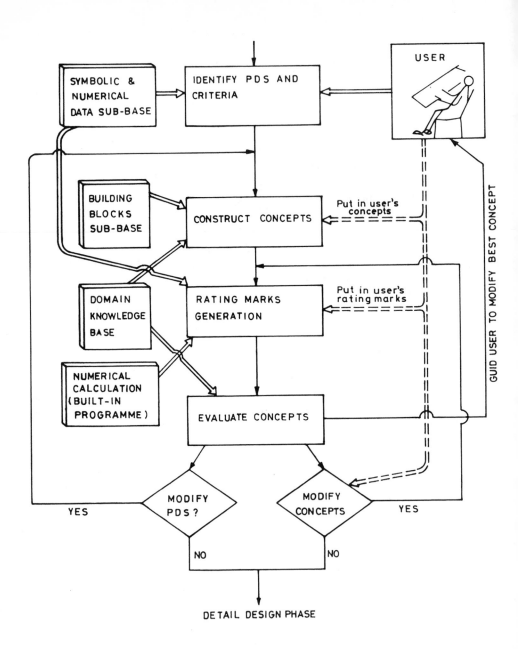

Fig 6

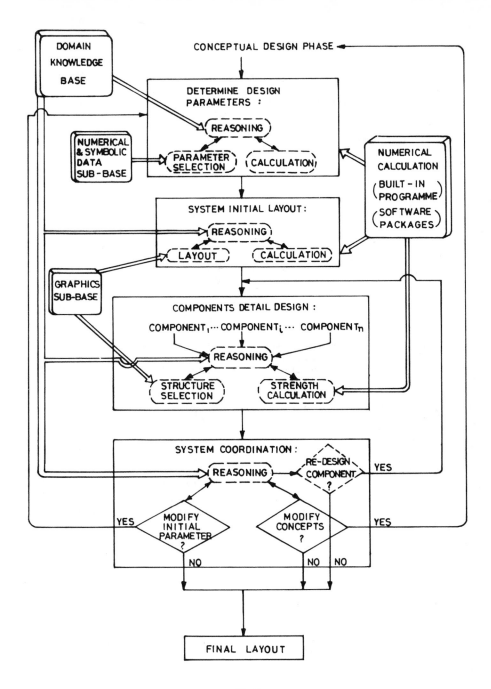

Fig 7

APPENDIX A

PDS AND EVALUATION CRITERIA OF POWER TRANSMISSION SYSTEM

(Note: The blank spaces in the left column will be completed by
the expert system user for a particular application.)

PDS Elements	Evaluation Criteria
1 Performance	
1.1 Centre distance	Tolerance of centre distance
....	....

1.2 Power transferred (input, output, efficiency)	a. Efficiency
	b. Surplus power
	        rated p.- required p.
	    $S.P. = \dfrac{\text{rated p.- required p.}}{\text{required power}}$

....	....
1.3 Speed ratio	Accuracy of speed ratio
....	....
1.4 Fluctuation of instantaneous speed ratio	Fluctuation of instantaneous speed ratio
....	
1.5 Smoothness of transmission	Smoothness of transmission
....	....
1.6 Relationship between input and output	Arrangement of input and output shafts
....	....
1.7 Flexibility to change centre distance	Flexibility to change centre distance
....	....
1.8 Box stiffness	Tolerance of the stiffness
2. Environment	
2.1 Temperature	Suitability to temperature
....	....
2.2 Humidity	Resistance to humidity
....	....
2.3 Dusty atmersphere	Resistance to dust
....	....
2.4 Linkage to input and output machines	a. Linkage reliability b. Linkage convenience c. Linkage safety
3. Life in Service (Some components need to be changed during the system's service period)	a. Number of components required to be changed b. Change frequency required
4. Maintenance	Convenience for maintenance
5. Target Product Cost	Cost

6. Competition	
7. Packing	Convenience for packing
8. Shipping	a. Convenience for shipping b. Safety for shipping
9. Quantity	Quantity
10. Manufacturing Facility	a. Special facility required b. Number of fac. required c. Facility cost
11. Size	Size
12. Weight	Weight
13. Aesthetics/Appearance/Finish	Appearance
14. Materials	Availability of materials
15. Product life span	Product life span
16. Standards and Specifications	
17. Ergonomics	Ergonomics
18. Customer	
19. Quality/Reliability	Quality/Reliability
20. Shelf Life (Storage)	
21. Processes	Cost of process
22. Time Scale	
23. Testing	Testing efficiency
24. Safety	Safety
25. Company Constraints	
26. Market Constraints	

C377/252

Expert system for process plant design

A O RIITAHUHTA, Dr Tech and **K J AHO**, Dr Tech
Tampella Limited, Tampere, Finland

Utilizing the expert system technology, methodology
for the boiler plant design has been developed on
the basis of VDI 2221. The use of the expert system
within the systematic design method will particularly clarify design and considerably reduce the
delivery time of a power plant. - The expert
system has also proved to be an excellent tool for
training engineers for diversified operations of
plant design.

1 INTRODUCTION

One vision of the future of the heavy engineering industry is that in five
years only those engineering workshops which invest around five per cent
of their turnover in research and development will operate as independent
units. Five per cent, compared with the current average of half a per cent
in the engineering industry, is rather a high percentage. When this money
is granted by business administration, a question arises how research and
development can utilize it profitably.

This paper deals with the methodology which has been developed for
boiler plant design on the basis of VDI 2221. This methodology development
supports, and also promotes, research and development of the products,
i.e. plant processes and structural modules. The methodology has been
developed utilizing the expert system technology.

The expert system technology is knowledge technology, which can be
characterized, for instance, as follows:

- It is suitable for handling large amounts of knowledge and it makes
working possible even when software does not yet contain all knowledge.

- Knowledge and software are in such a form that knowledge can be easily
updated.

- Good user interface.

Acquiring sufficient knowledge in the field of plant design takes several years. The work requires different capabilities, such as mastery of
wholes and, on the other hand, extreme attention to details; knowledge of

processes but also ability to design the equipment required by these processes. The expert system is useful in training, because all stages of
plant design can be gone through educationally by means of it.

The good experiences obtained in boiler plant design have already
proved the possibilities of the applied technique. On the other hand, it
is clear that in plant design there are so many features to be developed
in co-operation with different establishments that it is sensible to
expand the project. The intention is to pursue the development of plant
design as a technology project of Finnish Metal Industry.

This paper presents the results obtained in boiler plant design and
outlines the above-mentioned technology project.

It is through methodological development that the investment of five
per cent can be directed most profitably.

2 EXPERT SYSTEM OF BOILER PLANT DESIGN

A typical situation in plant design and manufacturing processes has been
that design data have increased only linearly. This is due to the fact
that the knowledge required for the project design is spread among different persons and even among different companies. Due to the linear increase
of data, both layout design and detail design required by manufacture are
going on at the same time. Changes in layout design cause alterations in
manufacturing drawings and vice versa, which strongly increases the iter-
ativity of design and retards the rapid progress of the project.

Figure 1 illustrates the great influence of preliminary design on the
implementation costs. Preliminary design also has a great influence on the
availability of a boiler plant. If the preliminary design of the boiler
plant has been carried out properly, the availability of the plant is high.

For the above-mentioned reasons, changing the design process has been
set as a target. Design knowledge has been collected from experts and
worked up into design rules. Preliminary design and detail design have
been separated into different stages. The customer is involved in preliminary design, but detail design is carried out by the manufacturer on
the basis of his own knowledge.

The design process leading to the above-mentioned results has been
developed on the basis of VDI 2221 using the expert system technology.

When the development of the expert system for plant design was
initiated, it was natural that well-operated data systems for technical
calculation and the conventional CAD were included in the expert system.
Figure 2 presents the computers, programs and areas of operation of
different data systems.

Design by means of the expert system is initiated by creating a
product structure description on the basis of which the model of the
actual design object is formulated. The series of figures 3 to 7 presents
the progress of design.

By the expert system, design is carried out in one sixth of the time taken by the conventional CAD design.

3 THE POSSIBILITIES OF THE IMPROVED BOILER PLANT DESIGN METHOD TO ANSWER THE DEMANDS FOR THE EDUCATION AND TRAINING OF A NEW PLANT DESIGNER

In his demanding task the plant designer must have a thorough knowledge of the power plant process including the equipment, machinery, space requirements and main solutions related to the process.

It was presupposed that a plant designer needed ten years' practice after the actual university education to acquire professional skill.

Ten years ago it was said about plant designers that they were a 'dying breed of elderly men without any followers'. So far the development of integrated plant design has not been very extensive.

Due to the small number of boiler plant designers in Finland, it was decided to analyse the current situation and the future of plant design in more detail. The questionnaire was answered by six companies.

From the answers some new emphasis for plant design was found:

The design method must be

- flexible but easy to control
- customer-oriented
- suitable for teamwork both inside and outside the company
- provide organized treatment.

To summarize the answers: The designer's attitudes and professional skill have to be developed.

At the same time it was considered that the knowledge required in plant design could be split into three broad categories:

- Continually expanding specific technological information and design rules to define the product. Direct knowledge.

- The context and influences of the fields, such as economics, politics and business in which the design activity takes place. Indirect knowledge.

- Management knowledge, which is required to coordinate the design process effectively.

As a high level of expertise is essential for the plant designer, it has become important to reduce the learning time to raise the designer's expertise to a sufficiently high level.

The new design method has concentrated on enhancing preliminary design. The method allows the amount of design data to grow exponentially in accordance with the design rules right from the start of the project.

For the preliminary design stage, methods of visualization are developed that enable evaluation of the essential matters requiring decisions.

The design stage for each task is visible (outline, preliminary plan, accepted solution, fixed solution), as also the customer's approvals.

Combining work versions as an official version is also possible.

The graph of the product structure presented in the product structure window makes it possible to direct the different design operations to components and their attributes, such as

–   Start-up of the recomputing of the design rules related to attributes
–   Study of the dependencies of the design rules between attributes
–   Changing the values of the attributes
–   Removing or adding components from/to product structure
–   Visualization of component geometry.

Using the method, young designers can be trained for plant designers increasingly effectively and quickly, because several alternatives and their effects on the entire plant can be examined rapidly.

It has been found that the method changes plant design into teamwork between experts with the main stress on product development and research.

4 DEVELOPMENT OF INDUSTRIAL PLANT DESIGN AS A GUARANTEE OF COMPETITIVENESS FOR THE METAL INDUSTRY

Finnish metal industry sells plants, such as

– Wood-processing plants
– Power generating plants
– Chemical plants
– Ships
– Oil drilling rigs
– Sub-sea technology

In selling plants a competitive edge is achieved by superior design and project control.

The above-mentioned plants forming the main part of Finnish metal industry, an idea of a technology project by Finnish Metal Industry was expressed for developing a general plant design method.

The technology project has now been initiated. In the project four stages can be separated:

(1) Development of design theory for plant design.

(2) Development and optimization of design rules for sub-areas.

(3) Development of design and project control tools based on the expert system technology.

(4) Development of systems applying to a particular company and product. (These are developed as projects according to the companies in question).

Plant design is characterized by the following features:

- Amounts of knowledge are large.
- Plants are unique.
- Major part of the design is combination of data.
- Design is rule-based; however, the rules are not chrystallized and different designers can have different rules.

4.1 VDI 2221-Methodology

In the plant design technology project, the methodology according to VDI 2221 is used.

Figure 8 presents plant design on the basis of VDI 2221. The presentation is completed by illustrating the business idea as a roof under which the operations take place. A business idea is a comprehensive description of the nature of the business operations and the reasons for the existence of the organization. The customer requirements to be met form the starting point.

4.2 Preliminary definition of the technology project for industrial plant design

Suppliers of industrial plants must have superiorities in their plants. These superiorities consist of the knowhow which is so highly valued by the customer that it is the supplier concerned that he wants to buy the plant from. Generally the process knowhow related to these superiorities is complex. The design rules have been proved in pilot plants or in full-scale plants. The number of the plant sections relating to superiorities is, however, small compared to normal technical solutions. Pumps, fans, valves, etc. are almost standard products. The technology project aims at developing the design and project knowhow of normal technology, which, in terms of the work involved, takes most of the working hours in the project.

The following target is set for the development work: CONSIDERABLY REDUCED DELIVERY TIME, HIGHER QUALITY STANDARD AND IMPROVED OUTPUT OF A PLANT.

4.2.1 Development of Process Design

Today there are computer programs which can be used to analyse flows in pipework and ductwork. These programs compute flow distribution and pressure loss. A weak point of these programs is that even if suitable for analysing the ready designed process and plant, they cannot be used as a tool for the preliminary design of a plant.

In the technology project the most essential processes and their unit operations are selected and their own process modules are made. The expert system dimensions, in accordance with the design rules, a corresponding module on the basis of the customer requirements. It also suggests connections for the module, dimensions the design values for the related

processes on the basis of the selected connections and defines the transfer process between the process modules. The mixing process is given as an
example, Fig 9.

4.2.2 Plant Layout Design

When the preliminary design of the process is completed, the preliminary
design of plant layout is started. At the preliminary design stage the
corresponding structural modules are defined for the process modules. The
structural modules are in fact space reservations but they include connection data for transfer processes. The physical counterparts of the
transfer processes in the plant are tubular, mainly consisting of

- pipework
- ductwork
- routes for maintenance, passage and transport
- cable trays

By moving modules and above-mentioned connection tubes in relation to
the basic module (which includes superiorities) space utilization is optim-
ized. The design system must be such that in design it is possible to
return easily to the process design module and optimize the operational
costs of the process.

The third element of the methodology is process control. When process
and plant design is carried out in an integrated way, it is possible to
proceed automatically to control loop diagrams and, in the long run, to
process simulation and an expert system of process control.

4.2.3 Development of Plant Modules Corresponding to Process Modules

An object-oriented approach is developed in which the different design
disciplines are integrated into the structural module itself. A pump, for
instance, is defined in the following aspects:

- as an individual part of the process
- as a process module
- as a receiver of ductwork forces
- as an object of thermal expansion and dynamic forces
- as an object of environmental stresses, such as
  - corrosion
  - temperatures
  - moisture
- as an impact on the environment
  - noise
  - pump and pipework damages
  - heat

Particularly the design of the properties pertaining to operation,
maintenance and safety of the process modules is developed.

4.2.4 Reduction of Plant Layout Design

For achieving the required delivery time, it is very important to reduce plant design into design of modules, see Fig 10.

5 CONCLUSIONS

- The use of the expert system of systematic boiler plant design will particularly clarify design and considerably reduce the delivery time of a plant.

- The expert system has proved to be an excellent tool for training engineers for diversified operations of plant design.

- In the field of plant design it is necessary to increase cooperation. The purpose of the technology project is, coordinatedly, to develop general plant design theory and make applications for various plants.

REFERENCE

RIITAHUHTA, A. Enhancement of the Boiler Design Process by the Use of Expert System Technology. Dr.Tech. thesis. Acta Polytechnica Scandinavica, Mechanical Engineering Series No. 92, 1988, pp. 122.

ACKNOWLEDGEMENTS

The authors thank Tampella Ltd for permission to publish this paper.

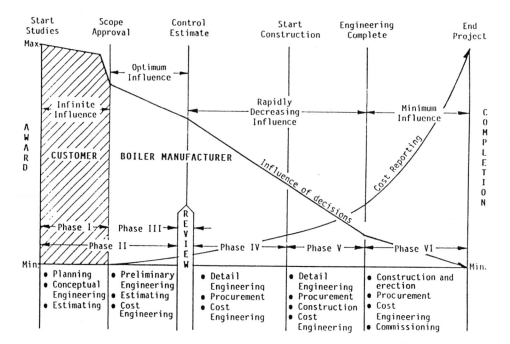

Fig. 1 Influence of the time of design decisions on the costs

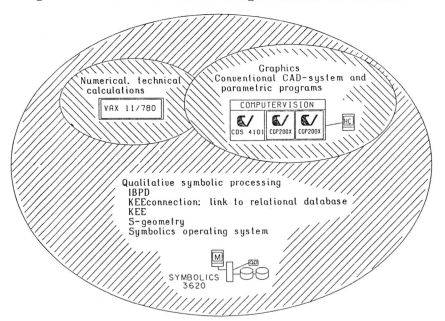

Fig. 2 Areas of operation, programs and computers of different data systems

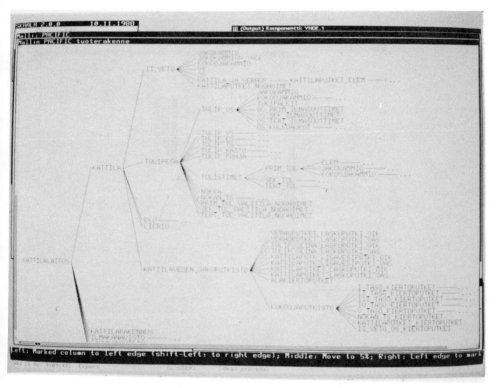

Fig 3 Hierarchical product structure

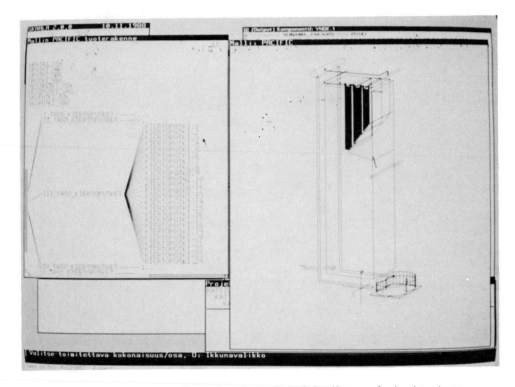

Fig 4 Circulation pipe of III level, pointed in the product structure
and visualized by the expert system

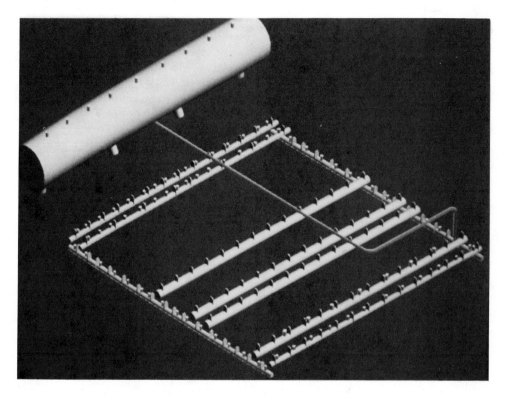

Fig 5 The same pipe, transferred into the CAD system

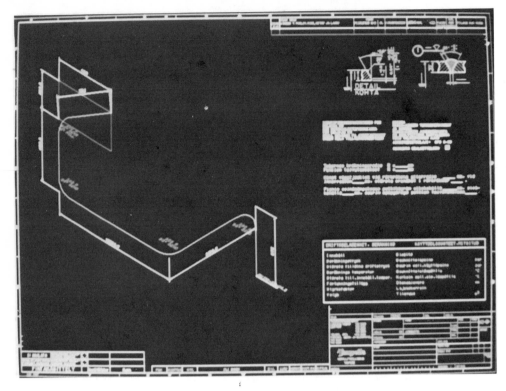

Fig 6 Manufacturing drawing made by the CAD system

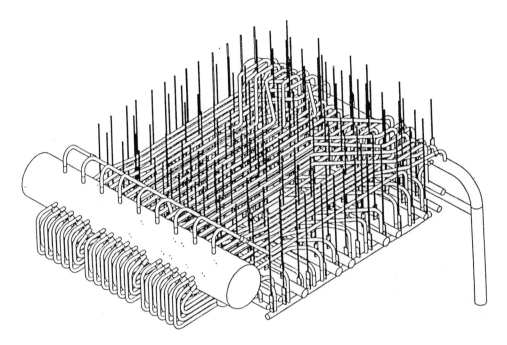

Fig 7 The entire upper circulation pipework of a recovery boiler

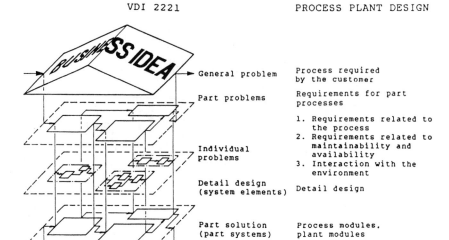

General problem   Process required
                       by the customer

Part problems   Requirements for part
                       processes

                       1. Requirements related to
                          the process
                       2. Requirements related to
Individual   maintainability and
problems   availability
                       3. Interaction with the
                          environment

Detail design
(system elements)   Detail design

Part solution   Process modules.
(part systems)   plant modules

General solution   Profitably operating
(system)   process and plant

Fig 8 Method for solving the problem and finding the system structure
(VDI 2221) in connection with plant design. Operation takes place
within the frame of the business idea.

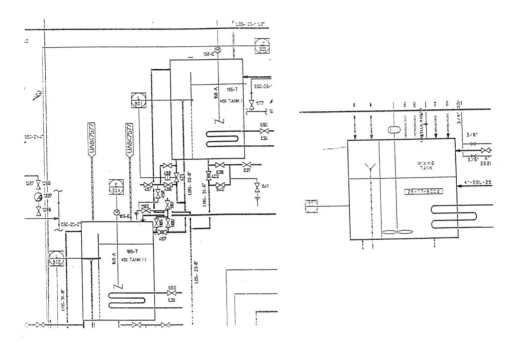

Fig 9 Two different connection alternatives for the mixing process

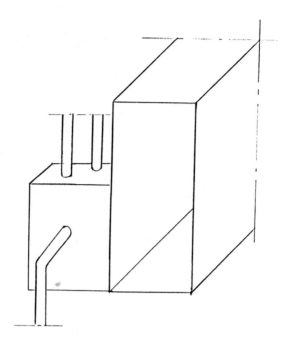

Fig 10 Reduction of plant layout design into
design of modules and connection tubes

C377/104

Initiatives in UK engineering design education through industrial involvement

G DRUCE, PhD, CEng, MIMechE
Department of Mechanical Engineering, University of Surrey
I GILCHRIST, MSc, PhD
Department of Mechanical Engineering, Brunel University, Middlesex
P H HAMILTON, BSc, MSc, CEng, MIMechE
Division of Mechanical and Aeronautical Engineering, Hatfield Polytechnic, Hertfordshire

This paper reviews some innovative developments in the
integration of industrial experience, resources and
expertise into Engineering Design Education in UK and
recommends appropriate action by industry, the
Government or its agents and the Engineering Institutions
to extend effective cooperation including the provision
of necessary resources.

Engineering departments in academic institutions of
higher education are making great efforts to work in
partnership with industry in order to equip graduates
effectively for a career in industry. Nowhere is this
more important than in the field of Engineering Design
and no medium is more appropriate than that of design
project work.

The findings of this paper are based on a survey
carried out in October 1988 which confirmed that industrial
involvement in engineering design studies is widespread
but predominantly conventional. A few departments had
attempted innovative ways of involving company personnel.
A number of these are described in some detail and from
these a number of recommendations is distilled,
requiring action from all sectors of the engineering
world if the significant advantages of such experiments
are to be realised.

1 BACKGROUND

For many years, ever since the advent of engineering studies into higher
education, there have been attempts to relate the study aspects of a
student's initial 'formation' to the world of industry. ('Formation'
is the term used in the Finniston Report [1] to define a student's
total educational experience in preparation for a career in engineering.)
As Engineering Design became a more widely accepted part of first degree
schemes the trend has become more determined. Reports have emphasised
the paramount role of Engineering Design in the engineering industry and
stressed the need to include, indeed to integrate, design project work in
engineering courses. [1] [2] [3] [4] [5] [6] [7] [8] [9] [10] [11] [12].

Partly as a result of pressure for 'accreditation' of engineering degree courses by the Engineering Council and the Engineering Institutions, almost all engineering degree schemes now include Engineering Design as a thread running through the course. Following further recommendations by Fielden [2] , Moulton [3] and the Engineering Council in their SARTOR document [6], such schemes require students to tackle a significant engineering project in the final year. Many of these projects have their roots in industry and involve Engineering Design. Often they are directly related to the student's own experience or commissioned by a sponsoring company.

Nowadays there is nothing innovative about industrially-based student project work [13] as such but there are signficant developments in several course disciplines. These merit attention by virtue of their enhancement of academic attainment as well as strengthening the mutually valuable relationships between industry and academia.

There are many opportunities for students to engage in design activities and thus for industry to be involved. Many delegates will be aware of the work undertaken by SEED* in exchanging successful engineering design projects through a Compendium [14]. This work identified two categories of design task.

> (a) *Projects* are those which involve the total design activity, total integration of topics and may last up to one academic year.

> (b) *Assignments* are those which relate to a single topic or to the limited integration of a small number of topics or skills and may last only a few hours. Assignments may be less direct use to industry than 'projects' but it is just as important that industrial partners are involved in order to ensure that some industrial realism is there.

2 VALUE OF INDUSTRIAL INVOLVEMENT

The casual observer might be tempted to ask the reason for developing such a relationship since many potential difficulties are evident. Would it not be easier to educate students using in-house academic exercises and then let them discover the industrial world when they enter employment? The answer is, no doubt, complex but a number of factors emerge.

Engineering courses have benefitted immensely from the influence of integrated industrial training in 'sandwich' courses [15] [16]. Now this benefit is extended by integrating industrial influence within the academic parts of the course. Although this may be implicit in 'Engineering Applications 1 and 2' as defined by Finniston, it is frequently attempted

*SEED is an Association of those involved with, and interested in, Engineering Design Higher Education. It was founded in 1979 in UK and now has nearly 200 members in six countries.

with paper exercises instigated   by the tutor concerned. Such exercises, whilst suitably emphasising certain aspects of a design, are liable to lack background appreciation and realistic timescale.

There is plenty of evidence that industry in general, and many firms in particular, have benefitted from student projects and that firms who have been involved have found them to provide for a real need. Benefits accrue to companies through a fresh approach to problems through the efforts of students, a means of identifying and encouraging potential recruits and an exposure to rigorous techniques of approach or analysis. Some benefit also from the enhancement of a product range through detailed product design and development.

The results of national competitions such as those sponsored by Molins/ BICC and the Worshipful Company of Turners/IMechE indicate the quality of work. However, the arrangements for these competitions do not always match the organisation of the academic institution.

The interaction of students with industry at an early stage exposes them to commercial dimensions in terms of organisational rigour, timescale, values and limitations of industry, the requirement for a complete specification and a detailed design which satisfies all the objectives (incorporating quality, costing, etc). These all help students to form a more accurate perception of industry before entering full-time employment. These graduates should have improved ability and confidence to make a constructive contribution to their employer more rapidly than those who have followed a purely academic course.

The conventional wisdom in educational circles is that 'realism' is important in education both from technical and commercial viewpoints. Realism means discipline in approach and resource accountability, discipline in dealing with detail in a practical way and discipline in not dodging issues or glossing over objectives. It means getting the right 'end result' as if the exercise were part of a commercial business.

The reducing number of 18 year-olds offering themselves for higher education in engineering in UK, along with the demand by industry for 'better qualified' graduates, implies that students who are more readily assimilated into industry are more likely to be employed and reflect well on their *alma mater*. This, in turn, may well have an influence on the marketing of courses and the attraction of students to an institution. This latter issue is particularly relevant today.

3 ESTABLISHING PARTNERSHIPS

The need for 'partnership' has been stressed for some years [17] and it is ever more vital that the theme is adopted by all influential bodies. There are many instances of valuable work being done at a local level as shown by the survey reported below, some evidence of national action of an advisory category but a large gap somewhere in the middle where specific action by national organisations could greatly ease the task at local level.

Currently, contacts are made and maintained through research and consultancy activities of academic staff, through the placement and tutoring of 'sandwich' course students in industrial periods, through special initiatives from academia such as Industrial Liaison Bureaux or as a spin-off from funded activities such as SERC* Teaching Companies, which provide an ideal framework for a 'special relationship' between the company and the college.

Many of these contacts are initiated from the educational side and there is scope for a more positive approach from industry who might see a worthwhile resource for product design and development in academia. One survey carried out with the help of the SERC at Hatfield Polytechnic has shown that many small companies, which comprise 90% of UK manufacturing industry, have a wealth of design problems and are anxious to get specialised help but cannot afford to employ an expert full-time. They are being encouraged to turn to the latent potential in the Polytechnic. The personal approach seems to be important in this work.

4 SURVEY

Such is the interest in the current state of cooperation between industry and academia that it was thought necessary to attempt to discover what innovations have been initiated recently. Accordingly, some 76 departments were asked to respond to a short questionnaire, seeking to discover the ways in which industry is involved in their design project work. The responses from 38 departments revealed many instances of successful industrial/academic partnerships, much at a local level through personal contact and some at Board level. Only two respondents did not use industrial links in their design teaching. Some of the examples of successful cooperation are reported below in order to give a flavour of current thinking. Further details can be obtained from the departments concerned if required. The major conclusions reached from the replies received covered the following points.

Currently it is in the final year 'major' individual projects in undergraduate schemes that industrial integration is most prevalent (30 respondents). In six courses it was claimed that much of this work was actually carried out in industry. For the 'sandwich' student the project frequently derives from his placement organisation which should have provided several months' experience in a 'design environment'. The project may involve the design and development of a product, service or system which is required by the employer. Any of the instigation, monitoring and assessment functions of the project may thus be a joint concern. For students on a full-time scheme, while the task may be based in industry, the link is likely to be more tenuous with the academic institution being primarily responsible for the work done.

In nine courses design work in intermediate years was arranged in conjunction with industry, mainly supervised by academic staff. Much use is made of industrial input to lectures (fifteen courses), tutorials (six

* SERC is the Science and Engineering Research Council of the Department of Education and Science. This body administers government funded research in UK

1010

courses) and case studies (seven courses). Company visits were also arranged for five of the courses.

In only four courses was the company liaison carried out at Director or Senior Management level while most corresponded at senior designer or project manager level. In two courses an industrial liaison group had been set up to promote interactions. One local Scottish IMechE Branch Committee had been instrumental in setting up contacts.

Almost all respondents recognised that 'realism' was the main advantage for students in this interaction. Some also looked for ways to increase student motivation (seven courses), maintaining relevance (three courses) and giving exposure to practising engineering situations (two courses).

Industrial partners expected a professional approach (eleven courses), well presented documentation (five courses), a useful source of ideas (four courses) and a means of assessing potential recruits (eight courses). A few companies also recognised the possibility of influencing the curriculum. Most companies required confidentiality, some expected access to expertise and equipment and a few regarded cooperation as a cheap form of consultancy.

Vital ingredients of success were seen to be an honesty of approach from both sides, a recognition of different objectives and timescales and a personal contact at a high level in the company.

The major concerns voiced by academic staff regarding the extension of industrial interaction were the danger of allowing the industrial partner to under- or over-estimate the students' capabilities, the need to rely on the goodwill of people working to short timescales, recognising the company which is only interested in cheap development work and the difficulties encountered when the 'live' project goes wrong.

Finance is rarely a severe restriction since student design projects are not 'paid for' at a commercial rate and the company has the option of accepting or rejecting the proposals. The potential for a conflict of interests is, however, more of a problem.

> Much reliance is placed on the goodwill of industrial middle management, who are consequently expected to divert their own resources to be involved in the actual liaison work;
>
> industrialists often expect students to perform as well as their own staff, in terms of output and quality. This can lead to misunderstandings;
>
> the need for confidentiality is an important reason for reluctance by industrialists to collaborate forcing dependence on obsolete data;
>
> timescales will be different both in terms of time devoted to the job, the start/finish constraints in academia and the total elapsed time available. Delays in replies from

firms can often exacerbate the student's problems despite the best efforts of supervisors to build up a suitable 'data source';

the 'natural course of events' may change the nature or requirements of a project during its progress in industry. To a student this would constitute an unnecessary and confusing situation detracting from the educational aspects;

educational aims are often inadequately explained and are not necessarily compatible with commercial expediency;

students are often unaware of the company strategy on product development and are liable to contravene the 'rules' if not guided by well constructed *Briefs* which, nevertheless, must not overconstrain the students' creative ability;

students vary greatly in abilities, background, study mode and motivation. The success of a cooperative project will depend in part on the student's approach and reaction to the task. They should be well prepared for the differing constraints between the academic approach and the commercial approach.

On the face of it there might be more to be lost than gained in a cooperative venture. However, in practice this is unusual and all parties can generally derive benefit from the exercise.

The principal benefits include:-

enhancement of the student's experience in many ways, especially from the commercial and business aspects, but also by gaining specialised technical knowledge;

they also come into close contact with staff from a particular company and understand a little of how it works. This should help them when they seek employment;

academic staff derive benefit by maintaining and extending their knowledge of engineering practice;

curriculum and teaching methods benefit from the exposure to the realities of commercial engineering.

Potential benefits for the academic institution include:-

direct funding of specific projects, financial support for academic development plans and indirect income from short courses, sponsored research and consultancy;

its reputation is enhanced within industry, thus improving the chances of student recruitment.

Advantages to the company include:-

the company's contribution is rewarded by the design proposals
which relate directly to its need for a minimum of effort
and expenditure;

it will also have the opportunity to work alongside potential
recruits;

realisation of the potential for further collaboration with
academia.

The extent of the cooperation is an important issue, emphasising the
responsibilities of all concerned. A company cannot expect a finished
product to be delivered if it bows out after initiating a project. It is
vital that the commitment to provide continuing support is recognised. It
should be encouraged to demonstrate this commitment by allocating the
necessary time and financial resources. The exercise should not be seen
as a voluntary extra on top of all the other activities of the designer
but an integral part of the company's work. Support generally required
is in the form of background information, rapid response to queries,
provision for visits by students to the company, loan of hardware, a
contribution to assessment and feedback of results.

The company staff involved should be practising Chartered Engineers
with engineering design experience, preferably at the Project Manager level
or above, who are actively involved in project work and with a
technological expertise appropriate to the student project in hand. They
should also be keen to participate.

The academic institution should consider appointments of suitable
company staff as associate academic staff with appropriate status and
facilities. Some part-time lecturing or tutoring might result, thus
cementing the relationship and developing the cooperation.

Likewise academic staff need to prepare thoroughly. This is likely
to involve a study of the company's product range and the collation of
an adequate resource file of anticipated data requirements besides
understanding the relevant theory and practice.

5 RECENT INNOVATIONS

5.1 Postgraduate

It may be supposed correctly that links with industry would be essential
to the integrity of postgraduate courses involving engineering design.
Indeed the few schemes in this sector have very strong relationships with
one or more companies.

Among those active in this field is the Engineering Department
(R Lawrence) at the University of Warwick, with their Integrated Graduate
Development Scheme [18]. It is worth noting that *companies*, rather
than individuals, join the scheme, enabling a real partnership between the
company and the University. A continuity of relationship between a

company and an academic institution is time-efficient because time spent in acquiring background information and data and in establishing relationships with the company and its suppliers is reinvested.

Recent graduates of engineering who are employed by the companies attend part-time to study on specific modules which can accumulate towards the award of a higher degree and accords with the Engineering Council's initiative in encouraging 'Continuing Education' [19]. There is thus an integration between experience in industry and knowledge gained in the modules. The scheme includes a major project of some six months' duration sponsored by their company and jointly supervised by company and industry staff.

The scheme's flexibility and 'modular' nature enable it to be large without being unwieldy. Although the scheme originated in 1981 with a 'Manufacturing Systems' label, a parallel scheme is now running in the area of Design Systems in Production.

The 'Advanced Course in Design, Manufacture and Management', run jointly by Cambridge and Lancaster Universities (J Gattiss) integrates industrial experience with an academic element. It is a full-time one-year scheme which is highly structured with lectures, visits and project work, culminating in a group project sponsored by a company and comprising some 14% of the total time. More than half the total time is devoted to tackling current problems in factories.

The rationale for the scheme is that most engineering graduates are dissuaded from careers in a design office by the apparently slow progress to a position of real influence in the design process. To address this problem the course places total responsibility for the project on the students resulting in a high degree of motivation and hence impressive results.

The other problem identified by the course leaders is that much of the realism of 'project work' is removed by bringing the work into an academic institution since the industrial culture is absent. The interaction of designers' knowledge, attitudes and opinions is a necessary environment for design work to develop. There are the twin dangers that the student is cushioned from the real world by the supervisor and the students are removed from the 'negotiating' environment of design.

Much of the course is spent in industry, small groups of students being placed full-time in a firm for a period of about three weeks to solve an identified problem, with access to company personnel, expertise, records and so on. They are expected to manage their own project.

A special relationship has developed between the University and the companies as a result of the quality of participants. The experience enables students to advance rapidly in their subsequent careers although no formal qualification results from the scheme.

One example of a project involved a pair of students being placed in a company which makes agricultural machinery. Their brief was to improve the design of a potato harvester which was subject to malfunction in adverse weather conditions. They investigated the problem, identified a suitable solution, designed and made the modification and then tested it to the company's satisfaction. All this was done in a period of six weeks in the company.

5.2 Undergraduate

In the undergraduate scene notable examples of industrial participation have been found at several universities and polytechnics.

The continued success of the Teaching Contract Scheme begun at the University of Technology, Loughborough (I Wright) in 1982, provides some effective evidence of the willingness of industry to collaborate in academic engineering design projects. Based on a Finniston recommendation [1], the scheme involves twelve major UK companies. Initial support to establish the organisation was provided by a three-year pump-priming grant from the DTI. This has been succeeded by 'consultancy fees' paid by firms to fund visits by students to their premises and incidental expenses at the University. It is equally important that they provide internal budgets allowing for the time spent by senior engineers in briefing and liaising with academic staff and the students, tutoring at the University and organising the visits.

Awareness of their elevation to 'consultant' status boosts the students' enthusiasm and encourages the preparation of high quality designs and reports which form the return to the company.

For the 1987-8 session students at four academic institutions, including the Department of Mechanical Engineering at the University of Surrey (G Druce), were invited to compete on a project promoted by The Marine Society, with the active collaboration of the Institute of Marine Engineers. This project involved the preparation of the design specification for completely re-equipping the training ship "Enterprise" of 160 tonnes with new main propulsion and auxiliary machinery, including the associated controls and instrumentation. At the inception of the project the ship was merely a serviceable hull. The open-ended customer's Brief allowed a free hand in the choice of propulsion to be used.

A significant feature of the approach adopted by the department was the allocation of all fifty-four final year students to the same project. They were divided into ten groups, one of which specialised in Project Management and another in CAD. The remaining eight groups received technical briefs.

The students' final report was written both as a paper to the learned society, being presented to a formal meeting of the Institute of Marine Engineers, and as a design proposal to the customer.

Valuable support and advice was rendered by the Marine Society and the IMarE throughout. Indeed, the role of the IMarE in this project is worthy of special note as a vital catalyst in bringing industry and academia together and a forum for presentation of the final proposal.

The approach taken by Brunel University (I Gilchrist) in its BEng in Mechanical Engineering takes advantage of the sandwich course structure in matching the industrial input to the design course to the student's developing experience [20].

The first year involves industrial support for a project which requires the students to progress the design of a consumer item from initial inception to prototype stage. Special lectures and tutorials are given by relevant industrial personnel.

The second year concentrates on a component design. It was developed in partnership with a firm of consulting engineers and attempts to reflect commercial design office practice. Third and final year design work builds on earlier experience with industrially linked projects, students working alone or in a team.

The significance of the industrial interaction at Brunel is that the course has been planned from the outset to maximise the potential for industrial interaction, involving the introduction of academic topics at appropriate stages. It benefits from a well-staffed engineering design office within the Department which has been described elsewhere [21].

At the University of Strathclyde a 5-year 'Dainton' course in Manufacturing Sciences and Engineering is offered. The course is broadly based in several engineering disciplines with a significant Management and Mathematics input.

Entrants to the course, numbering between twenty and thirty each year, are highly qualified and carefully selected by interview. Each student is sponsored by industry and works for his company during the summer vacations. A major portion of the final year is devoted to a Design-based project for teams with the cooperation of industry (R Forgie and Dr J K Patrick). The students are reported to be very competitive and security conscious.

The project starts in the fourth year with the formation of project teams, identification of the requirements, introduction to the industrial partner and the evolution of a Product Design Specification. This absorbs about 4% of the total fourth year student timetabled time. When students return for their final year they are able to make a rapid start on the conceptual design and evaluation, followed by detail design. The final year student effort is estimated to be about 200-250 hours each, equivalent to about 20% of the total timetabled for the year.

The 1988/89 project concerned the design of a means of access from the public road near Fort William, Scotland, to the restaurant and ski-lift station 600 metres up Aonach Mor (Ben Nevis range) and 2 km distant. The industrial partner was the Scottish Development Agency, a Government agency who had sub-contracted the engineering consultancy work to the Sir William Halcrow partnership.

Four teams of four students each were asked to compete in the design of a system to perform a demanding task for all ages and conditions of passenger in summer and winter. In addition to normal users it would have to carry injured skiers, elderly people and young children as well as goods. One of the main considerations was the variation in weather patterns between top and bottom. Safety, reliability and performance had to be balanced with cost, the effect on the environment, social and political implications (especially with regard to landowners and local authorities) and the comfort of passengers.

The teams were required to discuss their work formally with senior members of the industrial partners at three stages during the project. First presenting the Specification, next many different concepts for discussion, then their chosen concept after rigorous evaluation using a dynamic evaluation matrix and finally their detail proposal. Since, at the time of writing, the project was not completed, it is not possible to present the finally selected concept.

An example of a useful interaction on a smaller scale has been an assignment given to MEng students of Mechanical Engineering on entering their final year course in Engineering Design at Hatfield Polytechnic (P H Hamilton). The MEng is an extended first degree in engineering which, at Hatfield, has an emphasis on engineering design and management. The assignment concerns the topic of Project Management, so crucial to both the student's final year and to his later career in industry. It is a topic which is difficult to make exciting in the classroom and lends itself ideally to a cooperative approach. The assignment is limited to that topic and does not necessarily require students to have a depth of knowledge about the technology involved in the product under consideration, nor does it require them to **do** any design work. However, it enables them to learn a great deal about the way projects are carried out and what can, and does, go wrong.

Students will have returned from a six months placement with their sponsoring company and will have had some experience in a design office. This is used as a basis for the first assignment for which they are taken to a local company which produces an appropriate product and given a detailed account of a recently completed project involving the design and development of a new product. Their task is to critically appraise the **project**, as distinct from the product, and during the full day visit they must find out everything they can about the company, its strategy and methods, the project methods and management, the people involved, the difficulties encountered and reasons for failure to meet the requirements in any way.

This approach requires a special relationship with the company because confidential matters are bound to be discussed. It requires a very frank presentation by senior company staff, as well as by those who have been actively involved in the project. It requires students to reach a significant level of understanding of a new situation in a very short time. Students are required to present honest and constructively critical conclusions within two weeks of the visit, addressing their reports to the Technical Director of the company. They find this a challenging task but normally rise to it admirably.

The company staff are given copies of the reports for their information and comment. In all cases the company have expressed their satisfaction with the result, despite the fact that three or four valuable staff have given up more than a day to the exercise. Often students' perceptions have been corrected by company staff at this stage, providing a valuable feedback from the 'sponsor'.

A feature of the BEng degree course offered by the School of Mechanical Aeronautical and Production Engineering at Kingston Polytechnic (D Alcock) is the final year industrially linked group design project. The student group, supervised by a member of academic staff, is presented with the Brief at the beginning of the session and remains in close contact with the firm throughout. The group meets its supervisor for two hour sessions each week and presents two formal seminars to the company.

The industrial involvement demands a professional approach from the students and the resulting long-term collaborative relationships have generated substantial financial support for the Polytechnic.

Another course having a strong thread of industrial involvement is that offered by the University of Hull's Department of Engineering Design and Manufacture (K Hurst), where four members of academic staff are sponsored by industry. The industrial component of this initiative came from the advisory group which has actively supported the course since its inception in 1982. Originally consisting of ten companies, the number involved has now doubled. In addition to the continuing commitment to providing facilities for this project, about 50% of the final year projects are industry-based. Additionally, individual firms sponsor one 'Chair' and three Readerships within the Department. From 1989 every student on the course is eligible to receive a bursary.

Industrial interaction includes visits to companies in the first year, design projects in the second and third years, one of which lasts three weeks and takes place within the firm, and a final year design project sponsored by, and carried out largely within, a company. Regular meetings with company personnel are a feature of all these projects. This cooperation results in a course closely tailored to industry's needs. As knowledge of the successful interaction reaches industry, more opportunities are created to collaborate and the process is self-sustaining. A pre-requisite for this success is the presence of well-qualified practical engineers as academic staff who can create a climate of confidence in prospective industrial partners.

Historical case studies with the benefit of hindsight are well established. A notable extension of this approach at the University of Glasgow (A Fairlie-Clarke) involved a project in parallel with the actual development of a product by the collaborating company. The analyses and recommendations made by the students followed periodic briefings and visits to examine progress and resulted in obtaining rapid feedback on their work.

Liaison at Board level facilitated cooperation. The project was arranged so that the company was not dependent on the progress of students

but could benefit from any positive ideas. It is also noteworthy that
academic staff not usually concerned with design gave specialised lectures
needed to provide appropriate theoretical support. Other lectures were
given by engineers from suppliers to the collaborating company of components
and materials. The topicality and appreciation of the timescale involved
enhanced the benefit derived by the students. The company derived benefit
from the greater breadth of thought brought to bear on the problems and from
their assessment of the students as potential recruits.

During their industrial placements, sandwich course students acquire
a thorough understanding of the sponsor's products and/or processes. Hence
they are capable of more specialised design work than can be expected from
students following a purely college-based course. The associated frequent
contact between their academic tutors and the company management provides
good opportunity for identifying and defining projects based on genuine
needs.

The MEng degree course in Electrical and Electronic Engineering at
the Polytechnic of Wales (S R Brown and M Charnley) has responded to a
succession of projects from industry. It culminates in the students
manufacturing and testing their designs, some of which are now in industrial
use.

Similar arrangements apply in the Chemical and Process Engineering
course at the University of Surrey (A Millington) where all the design
projects in the second and final years come from industry. These years
are separated by an industrial (sandwich) year. Supervision of final year
projects is the responsibility of a part-time lecturer who took early
retirement from the post of Principal Process Engineer in a premier company.

The problem of confidentiality restricts the students to published (and
hence) obsolete data, whilst the timescales involved preclude manufacture.
These constraints have no adverse effect on the educational merit, the best
solutions replicating those used by the collaborating industry.

6 OPPORTUNITIES

There is a vital role, also, for national bodies in the encouragement of
these initiatives more widely among their members. The Engineering
Institutions, for instance, could become actively involved at both national
and regional levels through both their Divisional and Regional structures.
The Engineering Council should be using its influence more overtly,
nationally, regionally and locally, through the ECRO* network. The Council
and Institutions have great influence in the engineering industry which
could be harnessed to encourage the involvement of industry in education.
Student papers, prizes and competitions could be used regionally and locally
to enhance a competitive spirit and raise the profile, placing principal
emphasis on the quality of engineering design.

*ECRO – Engineering Council Regional Organisations. There are 19 of these
local associations of professional engineers organised under the aegis of
the Engineering Council.

Indeed, the interaction envisaged could be integrated neatly with current initiatives in the continuing education and training of young engineers. Recent graduates might be involved in advising and supporting the students, deputising for the Chartered Engineer when necessary.

The problems encountered most frequently relate to resources and conflicts of interest. Successful industrial involvement with student projects and assignments always depends on good relationships between the company and an academic institution, usually at a personal level. It follows that links must precede the organisation of the work and there must be a commitment by both parties to get involved. Arrangements always take a significant time to initiate, often involving visits to both sites and lengthy meetings. Mutual expectations and objectives must be explored so that:

> industrialists understand the academic context, including facilities, course requirements, timescale, expectations and assessment;

> the commitment expected from industry is clearly stated;

> a satisfactory set of instructions is prepared for the students. This *Brief* is of vital importance to the success of the project and it must be approved by all parties involved. A proforma for such a 'Brief' is included in the Compendium produced by SEED [14] and reproduced, with permission, in Appendix 1 to this paper.

Normally, the greatest resource constraint is time. Educational requirements are related to an academic year, schemes of study, timetables, and have to take into account the students' learning process. Assignments may be linked to a particular curriculum and teaching scheme and time allowed for design coursework must be 'sufficient' without being out of proportion with the rest of the course. Almost inevitably the timescale will differ from that of the industrial partner, who will normally expect an answer 'as soon as possible' if the work is to be 'real'.

In addition, the academic tutor's time is limited by the academic system of 'recognition' which generally continues to rely more on published papers than on 'innovation' as criteria for promotion. Thus, priority for developing novel approaches to student design work is low. This is unacceptable if engineering design is to contribute its full potential to courses. Promotion criteria need to be developed which recognise 'product innovation' activities. Academic supervision of design projects, for instance, is notoriously time-consuming and demanding but even more so when the extra dimension of an industrial link is involved.

Conventional academic assessment may require many students to work individually on the same task, an arrangement which would be quite unrealistic from the industrial point of view. Therefore, assessment methods must be carefully selected and may include an input from industrial sponsors, the use of a competitive element, peer assessment and so on. Where industrialists are involved the right balance between industrial and academic criteria must be found, made clear to the student and applied in a fair and uniform way.

Time spent on these projects is usually only a part of the students' study programme and progress is liable to appear slow to the industrialist. At the same time, student confidence varies widely with age, experience and personality. All students may not 'see the point' of design assignments nor wish to be designers when qualified. Motivation is thus a potential problem on both sides.

Equally important, but less appreciated by some industrialists, is the comparative lack of background knowledge of students. They often have to go through a steep learning curve in order to come to terms with the project and this constraint may limit the scope for real projects in the early years. As the supply of school leavers seeking engineering education in UK tends to diminish, the problem is liable to increase in severity.

Industrial project leaders, having a number of projects and many staff for whom they are responsible, may be reluctant to give more than token attention to what they may perceive as a 'training' function. The advantages of cooperation are not immediately obvious to the Project Leader. For this reason, personal contact and discussion in preparing the ground and covering the objectives and advantages to both sides is important.

7 CONCLUSIONS AND RECOMMENDATIONS

The vast majority of collaborative design projects are successful, their value to all increasing as individual experience is gained. As the number of academic institutions realising the potential of collaboration increases, so coordination becomes more important, providing a vital role for the Engineering Institutions, ECROs, etc.

Students following a 'sandwich' scheme are more likely to have the opportunity of undertaking design coursework assignments involving industry and academia working together within the academic part of the scheme as a result of the inherently closer relationship which exists through the placement of students. This relationship usually improves the company's appreciation of academic requirements and the company's knowledge of, and commitment to, students. However, industrial involvement in, and commitment to, full-time courses can also be beneficial to both sides.

Quite novel engineering design projects have been successfully carried out at some academic institutions, though many others follow the traditional pattern of industry-based major projects and some rely heavily on 'college' generated projects.

Significant difficulties have been encountered in setting up, monitoring and assessing industry-based assignments which seek to involve company personnel to any great extent. It is rare to find companies who deliberately set aside time for this purpose.

Some difficulties have been found in reconciling the inherent conflict of interests between industrial and academic objectives. Confidentiality and time constraints, including starting and finishing dates, are paramount.

Where innovative projects have been successful, success has often depended on relationships between individuals from each partner. However, the mobility of personnel can quickly destroy painstaking foundation work and the attitude of senior management to such links can profoundly affect progress.

Poor communications have sometimes been the cause of abortive attempts to set up a working relationship; this may involve a failure of the 'mechanism' of communication or a lack of understanding of the demands and responses of correspondents. Personal links and a high level of commitment on all sides is vital. It is vital for industrial partners to understand the educational objectives and timescales.

It may be that some companies get involved as 'third parties', supplying information, design data, etc. without a visible return. They may regard this as an imposition since no sales income is likely to derive. It is important, however, to take the long term view that the engineering industry as a whole will benefit from these activities, and with it so will individual companies through the improvement of the quality of graduate recruits. It is also a very cheap form of advertising. Thus it is important to encourage these indirect helpers to continue their sympathetic attitude.

There are clear messages to all concerned:

> Collaboration between industry and academia in engineering design education in the classroom is good for all concerned. It must be encouraged by all who have influence in order to ensure that all first degree schemes in engineering benefit and to invite development of even more innovative approaches. Commercial competitiveness dictates enlightened self interest. The twin problems of a lack of skilled manpower and the under-use of academic resources must receive coordinated attention in an attempt to speed up the learning process.

> The successful innovations revealed by this survey should encourage other academic institutions to develop industrial collaboration. This requires that the initiative and time expended should be officially recognised as a legitimate academic activity and rewarded appropriately. Staff who are involved must accept the need for a rigorous approach to the definition of the educational and industrial objectives, in the initial planning and in the preparation of a clear *Brief* which is agreed by all parties.

> The return to the company takes the form of design proposals, reports and the recruitment of suitable staff. Students gain and respond to increased responsibility knowing that

their work is to be used in practice.

The most effective innovations develop from decisions made
at Board level for the long-term commitment to support a
specific course in a specific manner. Continuity is a
great benefit in academic planning and in minimising the
time involved in setting up projects.

More engineering companies should be made aware of the
benefits to them of collaboration, be more willing to be
actively involved and should set aside, say, 1% of
designer effort per year for these activities and build
that into project plans. That might typically be one
hundred man-hours per year for a medium sized company.
This amount could be linked to a budget for training of
the designers themselves. Support at Board level and
allocation of a specific budget are critical to the success
of collaborative projects.

The Engineering Institutions, SEED and the agents of
Government concerned, such as DTI, DES and the Design
Council, should actively collaborate to promote the
approach as expected company policy for engineering
companies. Special use must be made of Divisional and
regional organisations such as ECRO and Divisional Boards.

Competitions between academic institutions could be used to
encourage involvement. The initiative of the IMarE illustrates
the potential for coordinating an imaginative proposal with
appropriate academic courses. This is an inspiring example
for the bigger institutions to follow.

It is not necessary to confine the work to large scale projects,
since some very successful assignments are quite small and can
be very resource efficient.

8 REFERENCES

1 Finniston Sir M F
   (Chairman)
'Engineering Our Future'
HMSO London 1980

2 Fielden G B R
   (Chairman)
'Engineering Design'
HMSO London 1963

3 Moulton A E
'Engineering Design
Education'
The Design Council 1976

4 Dawson J G
'Towards a Qualifications
Policy for the Future
Institution of
Mechanical Engineers
1978

5 Corfield K G
'Product Design'
NEDO London 1979

6
'Standard and Routes to
Registration' (SARTOR)
Engineering Council
Policy Statement 1984

7 Grant Report
'The Formation of Mechanical
Engineers'
Institution of
Mechanical Engineers
1985

8 Kimber M
'Degree Course in
Engineering Design'
Engg Designer V 14,
n5   Sept/Oct 1988

9 Sheldon D
'How to Teach Engineering
and Industrial Design –
a UK Experience'
European Journal Eng
Education 13/2/1988

10 Blows L G
'Objectives of Design
Education'
Engineering Design
Autumn 1987

11
Curriculum for Design –
Engineering Undergraduate
Courses
SEED 1985

12
The Initial Formation of
Mechanical Engineers
IMechE 1988

13 Hayes S V and
    Tobias S A
'The Project Method of
Teaching Creative
Mechanical Engineering'
Proc I MechE V 179,
P1, n4   1964/5

14
Compendium of Engineering
Design Projects
SEED 1987

15 Pollard D and
    Druce G
'Surrey Makes Superior
Sandwiches'
Engineering Design
Education Autumn 1982

16 Pullman W A
'Teaching design to
sandwich course students'
Proc I MechE V 179
n4   1964/5

17	Hamilton P H	'Partnership in Design'	IMechE Workshop on Design Education, University of Bristol, May 1984
18	Lawrence R	'IDGS An Industry – Education Partnership'	Engineering Design Education & Training Spring 1988
19		A Call to Action – Continuing Education and Training for Engineers and Technicians	The Engineering Council 1986
20	Gilchrist I	'Design Education and Project Work – a Mechanical Engineering Approach'	10th Annual SEED Seminar, Brunel University, June 1988
21	Large J H	'A Design Office on the Campus'	Chartered Mechanical Engineer V 31, n 9 1984

C377/091

Competition develops organizational skills

R P WELLINGTON, BSc, DipEd, MEd, ARACI
Department of Mechanical Engineering, Chisholm Institute of Technology,
Caulfield East, Australia

The development of a solar powered car for
the World Solar Challenge provided an
unparallelled opportunity for first degree
engineering students at Chisholm Institute
of Technology to design and build a totally
original vehicle. The chance to work with
industrial experts and to compete against
the world's best is a chance few
undergraduates have. Finishing in the top 6
in such a field, is an appropriate
reflection of the high level of technical
competence, organisational skills and
personal dedication which were developed.

1   INTRODUCTION

One of the realisations of the 1980's throughout the world is
the importance of good design to the health of local manufacturing industry. Governments in both Australia and Britain have
been promoting attention to design by industry and to design
education as a major emphasis for universities and polytechnics. However if real design skills are to be developed and
proven, the production of drawings on a drawing board or
computer screen are only a partial answer. Where possible,
projects which allow a student to participate in all aspects of
product development including specification, R and D, design,
construction,testing and finally competition with others, must
provide the ultimate design experience.

    The purpose of this paper is to describe some aspects of
the development of the "Desert Cat", a solar powered car
entered in the 1987 World Solar Challenge by Chisholm Institute
of Technology. (Figures 1, 2). The range of skill development

by the students involved, the essential role of industry to the project's success, the benefits of such projects to the students, staff, the college and employers of those students upon graduation shall be discussed. The current development of a new vehicle for the 1990 race improving upon the shortcomings of the earlier vehicle will also be considered.

2   THE MILEAGE MARATHON EXPERIENCE

The confidence to attempt such a project was largely based on prior ex- perience over the years 1980 to '85 with the fuel economy Shell Mileage Marathon competition in which Chisholm vehicles had achieved over 1000 kpl and were 2nd in Australia only to the then world record holder, Ford. This programme had been successful educationally as well as technically with over 50 students being involved between 1980 and '88.

Industrial support for this philosophy, has been most tangibly demonstrated by ICI Australia who has sponsored the development of a new vehicle during 1987 and '88. A further example of the success of this programme, is the continuing involvement of a number of graduates who are now tutoring current students and are taking increased technical and organisational responsibility for the whole project.

This Mileage Marathon experience provided the confidence that the technical and organisational skills necessary to see the solar project to a satisfactory conclusion were available at Chisholm and helped convince a number of potential sponsors to provide support.

3   THE WORLD SOLAR CHALLENGE

The World Solar Challenge was a race for vehicles which used photovoltaic cells to convert the sun's energy into electrical power. The race was held in November 1987 between Darwin and Adelaide, a distance of over 3000km. The cars travelled between 8:00am and 5:00pm each day with 2 hours prior to and following the days journey for exposing the cells to the sun and recharg- ing batteries.   The regulations required the vehicles to fit within a volume no larger than 6 by 2 by 2 metres with all cells limited to a 4 by 2 by 2 space.

The staging of the World Solar Challenge was announced in July, 1985 and Chisholm's Division of Engineering immediately established a research and design team of staff and students. It was apparent that competitive vehicles would need a comb- ination of high available power and efficiency, low aerodynamic drag and rolling resistance, plus good stability, reliability, competent drivers, well considered strategy and good organisation.

An executive committee was established with one academic, one technician and one student from both Mechanical and Electrical Engineering to ensure equal representation in decision making.

3.1 Mechanical and Electrical Student Projects

In 1985 and '86,final year mechanical students undertook projects including investigation of aerodynamic drag and stability; cell cooling; effects of panel tilting; body, cockpit, steering and suspension design; wheel,tyre and brake selection. Electrical projects included cell selection and array design, motor specification, battery evaluation, motor speed control, development of a digital radio link and design of a microprocessor control. As part of the preparation, four team members undertook a reconnaisance trip along the route 12 months prior to the event, so as to produce detailed data on road gradients and conditions, location of cattle grids, traffic density, availability of supplies, and to determine actual power availability and likely cabin temperatures. This information was essential to the computer simulation which was developed as part of the strategy process. The survey trip also lead to a change in selection of tyres and wheels due to the failure of the adhesive holding the tyres on to the rims at the high temperatures and speeds experienced.

The Chisholm "catamaran" design differred from all other designs in the race. This design was believed to be the most stable option as side winds would be unable to generate lift forces under the top collecting panel, possibly causing it to overturn. With the route used by 50 metre long road trains, this was a real concern.

The catamaran was also capable of generating more power over the day by having cells attached to the sides of the body as well as the top. Figure 3. compares power available from 4 designs which were used by different entrants.(2). An asymmetric design was chosen to minimise frontal area. The right hand side was designed to hold a 50th percentile female driver (500mm wide including some room for movement), but the left hand side was only 300mm wide having only to fit batteries and a motor. This asymmetry added complexity in the design of suspension units, brakes and motors.

3.2 Industrial Contributions

From the very beginning of the project, management had to address issues such as financial and product sponsorship as well as developing industrial contacts with necessary expertise and equipment unavailable within the college. CIBA-GEIGY, provided fibreglass-aluminium honeycomb sandwich panels, F-board, which was an ideal stiff, lightweight structural body material, particularly appropriate for the straight sided monocoque design chosen. With CIBA assistance, cut and fold techniques were developed for not only the right angle corners but also for more complex curves.

Telecom Australia made available their research laboratory facilities and staff expertise on solar cell selection and evaluation, motor speed control, power maximization and battery specification. With high efficiency (and hence high cost) solar cells essential to making the vehicle competitive, Telecom, Australia's largest user of photovoltaics, provided contacts which lead to Hoxan, Japan's largest manufacturer, donating the 1200 solar cells required. Following suggestions by Chisholm/ Telecom, 9 by 4 cell modules using 100mm square monocrystalline silicon cells sandwiched in a plastic laminate were supplied. The normal 3mm glass front face was replaced by a 0.1mm plastic film, Tedlar resulting in weight saving of approximately 200kg. A similar cooperative effort, developed a system of peripheral bonding of the fragile and flexible modules onto an aluminium backing sheet for lamination and transport, but which allowed easy removal for adhesive bonding direct to the vehicle body. These modules were evaluated upon arrival for resistance to temperatures between -40 and 90'C, humidity, hailstone impact and abrasion resistance; all of which exceded specification.

Telecom personnel were also involved in developing the power control system. High frequency switching systems were developed to maximise the power delivered to the batteries from the array and control power delivered to the motors. The control process was complicated by the use of 2 motors, a 1HP motor driving the right hand rear wheel and a 3/4 HP motor driving the narrower and lighter left hand side. The vehicle was fitted with a JED microprocessor which was designed to monitor array current and battery voltage, temperature of vital components and speed as well as control the power switching devices. The data fed into the JED was to be trans-mitted via a digital radio link to a Terran computer carried by a control vehicle travelling about 100 metres ahead of the Desert Cat. The data was to be used in conjunction with a route simulation to determine optimum speed but due to the noise generated by the switching devices and the aluminium honeycomb core of the sandwich panel acting as an antenna, radio interference made this impossible and the automatic speed control did not operate.

Pacific Dunlop donated the Pulsar batteries which had been selected as offering the best energy density lead acid batteries available, and along with Telecom, devised and implemented a schedule of deep discharge testing. These tests involved a deep discharge cycle over a 9 hour period followed by a 4 hour recharge period. Although the battery performance deteriorated with time, their performance was considered satisfactory for the anticipated 7 to 12 days of the race.

A motor development programme to make Neodymium permanent magnet motors was undertaken in conjunction with CSIRO and Preslite Australia but manufacturing problems enabled only one motor to be developed just prior to the race and the decision was made to use conventional (although heavy and less efficient) ferrite magnet motors. This motor was the first neodymium magnet motor to be developed in Australia, and has lead to an ongoing development programme which should see an improved and lighter motor being developed for the 1990 race.

The Victorian Solar Energy Commission provided the other essential contributions - cash sponsorship and industrial contacts which, in 1987 has included an entrepreneur interested in joint development of a commercial alternative energy vehicle.

3.3 Industrial Engineering Student Contribution

Having involved over 30 Mechanical and Electrical Engineering students in the initial design and development phases, twelve 3rd and 4th year Industrial Engineering students undertook projects in 1987 to raise further cash sponsorship, publicity , plan the logistics of the trip, design race strategy and select and train prospective drivers. The driver selection criteria required drivers to be small enough to fit comfortably in the cabin, and to perform well in an accelerated race simulation and a driving test. Each driver was required to sit for 1 hour in a 480mm wide simulated cabin, steering a cursor along a road on a computer screen while the temperature was maintained at 55'C. Heart rates were continually monitored during and after the test which all prospective drivers passed without complaint.

3.4 Educational Implications

To ensure that the project progressed satis- factorily and met its deadlines, frequent interaction between all team members was essential. During the research and design periods, the 20 to 30 students and their supervisors met for 1 hour a fortnight to disseminate external information, hear brief reports back on the progress of individual projects, and make decisions central to the project as a whole. Students soon developed the confidence to put and argue a case for a preferred design option. Two of the most heated debates concerned the catamaran or tilting panel concept and the size of driver who should be chosen for cabin design to procede. Decisions were usually made democrat-ically although many detailed design decisions were made by students with their supervisors.

Following the end of formal meetings, several small subgroups would usually form from those present to resolve conflict between people working on interdependent projects. Rarely did project management need to become involved in resolving these problems although some students had difficulty in meeting deadlines. Many of these informal meetings took place over a meal or a drink and this social interaction became stronger and more important as the pressures to meet critical deadlines increased.

Most students performed to or exceeded expectations but 1 or 2 projects did not meet their deadlines and had to be completed by supervisors which contributed significantly to the lack of available testing time.

As the project developed, students learnt to discuss their designs in detail with technical staff to ensure ease of manufacture and many students found this interaction one of the highlights of the project.

3.5 Development difficulties

The vehicle sufferred numerous setbacks during development including damage to several cells during bonding which required a time consuming replacement process, rain penetrated under the cells short circuiting some of the arrays, the motor speed controllers had to be redesigned due to frquent destruction of fets due to voltage spikes, the aluminium suspension mounts distorted during welding, the left hand motor sprocket could only be aligned with the wheel sprocket by machining slots in the motor housing and placing the drive sprocket between the windings and the bearing. These and other setbacks, typical of all real projects, provided the basis of a series of invaluable learning experiences for staff as well as students. However these problems did minimise the time available for vehicle testing and most of the minor delays which occurred during the race resulted from this lack of testing.

The finished vehicle was substantially heavier than the early weight estimates which resulted in higher rolling resistance, more punctures and high power consumption when climbing hills. The failure of the radio link made strategic decision making difficult and the objective of maintaining a constant speed throughout the day was not achieved. The chain drive came off the sprockets on several occasions and the DC-DC converter overheated on the last day, requiring the driver to use a crude but adequate power off/power on backup switching procedure to be used. Several cells failed due to hot spots which may have resulted from microscopic cracking caused during attachment of the modules. Some modules also absorbed moisture after the rain storms encountered because the encapsulant had been trimmed too close to the edge of the cell.

4.   THE RACE

The Desert Cat was presented for scrutineering in Darwin on October 29th but was initially rejected on the basis of driver visibility caused by the left hand side "outrigger". However when resubmitted the next day with additional mirrors giving enhanced left hand side vision, the vehicle was passed. A stability test was also required in which the entrants were driven as fast as possible past a 50metre long road train travelling at 80kph. in the opposite direction without veering off course. In the speed test to determine grid position, the Desert Cat achieved 67kph which was 6th fastest.

The entrants left Darwin at 9:00 am on Sunday, November 1st with the multi million dollar GM Sunraycer winning after 44 hours 54 min. at an average speed of 67 kph. with a Ford Australia entrant second. Chester Kyle (3) analysed the first 6 vehicles to finish and identified the critical performance factors in order of significance as

1.   Reliability
2.   Panel power
3.   Aerodynamics
4.   Weight
5.   Rolling resistance

A comparison of the first 6 vehicles is shown in figure 4. (4).

The Desert Cat finished 6th in a field of 23 starters from 7 countries finishing the 3004km. on the 12th day with an average speed of slightly over 30kph. While driving in bright sunlight, speeds of 50-60kph were easily achieved on flat road. The weather proved to be disappointing with several days of continuous cloud and heavy rain which restricted the vehicle to speeds of 10 kph. The first 5 placed entries used silver zinc batteries with substantially higher energy density than Chisholm's lead acid batteries, which did however perform well despite the harsh treatment.

5   WORLD SOLAR CHALLENGE 1990

A second World Solar Challenge has been announced for 1990 and sufficient enthusiasm exists among former team members and existing students to start developing a new vehicle. Accordingly, three 1/5th scale models have been wind tunnel tested and the best design chosen. Most former sponsors have expressed the desire to be involved in the next project but increased sponsorship is being sought with the help of Chisholm Marketing students who have undertaken projects to develop a marketing strategy and identify industry sectors who will be approached for sponsorship. Industrial design expertise has also been enlisted to ensure the next vehicle is more aesthetically designed and better attention is paid to detail.

International co-operation is also being developed between Chisholm and The Institute for Solar Electrification, Worcester Polytechnic Institute, Massachussets with the visit of a team from WPI in January '89.

6   CONCLUSIONS

A project such as the development of a competitive vehicle requires a massive commitment of time and resources but the benefits listed below would appear to significantly outweigh the disadvantages. Students benefit from:

1. Applying theoretical knowledge to a real problem.
2. Developing enhanced communication and interpersonal skills
3. Following through the whole development process
4. Comparing their design to the world's best
5. Working in a team with academic and technical staff and graduates
6. Working with industry based engineers and experts in other fields
7. Using equipment and technology usually unavailable
8. Being responsible for part of a high profile, successful project.

Dr. Paul MacCready, president of Aerovironment, identified the greatest value in the project when he said (5): "Because of the effort going into the low powered vehicles in this race across Australia, gasoline-powered cars will someday be achieving better fuel economy, and battery and hybrid-powered cars will become more practical. Overall, this race moves us toward placing lesser demands on the earth's environment and resources while handling our transportation needs."

7 ACKNOWLEDGEMENTS

The author wishes to again thank all sponsors especially Hoxan, Ciba-Geigy, Telecom, Victorian Solar Energy Council, Terran Computers, Jed Microprocessors, Pacific Dunlop, Greater Shepparton Development Council, Aspect Computing, Aerospace Technologies Australia, CSIRO, SKF Bearings, Craftek, John Hart,   Preslite and the staff and students of Chisholm Institute whose continuous cooperation and dedication made the "Desert Cat" a reality.

REFERENCES

(1)   WELLINGTON, R. P.   The Mileage Marathon Competition; An
Exercise in Engineering Reality. Research and Development in
Higher   Education ,1984, Vol 5179-185

(2)   PATTERSON, D.J.   Design Philosophy for a Solar Powered
Vehicle   for a Cross Continental Race. Darwin Institute
Internal Publication,   1988

(3)   KYLE C. Analysis of Solar Cars, World Solar Challenge,
Australia. Private Communication, 1988

(4)   ROBBINS, T.,HAND, R.,HINCKLEY, S.,MURFETT, A. Design and
Evaluation of Solar Powered Car, Telecom Australia Technology
Branch Paper, 1988

(5)   MacCREADY, P. Pentax World Solar Challenge Official
Programme, 1987

Fig 1   Chisholm's 'Desert Cat' with driver's side foremost

Fig 2   Rear view of the 'Desert Cat' showing the different width pods
of the catamaran

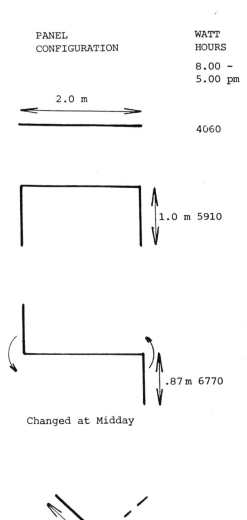

PANEL CONFIGURATION	WATT HOURS 8.00 – 5.00 pm
2.0 m	4060
1.0 m	5910
.87 m (Changed at Midday)	6770
2.0m (Tracking)	5110

Fig 3   Power output for alternative designs. (Courtesy Patterson D. J.)

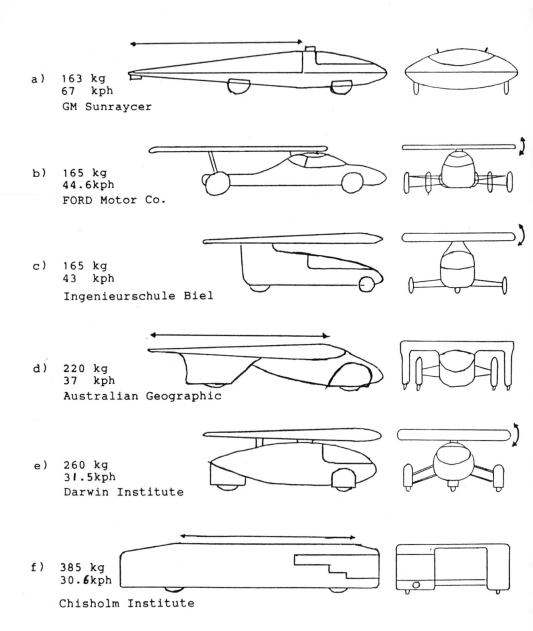

a) 163 kg
   67 kph
   GM Sunraycer

b) 165 kg
   44.6kph
   FORD Motor Co.

c) 165 kg
   43 kph
   Ingenieurschule Biel

d) 220 kg
   37 kph
   Australian Geographic

e) 260 kg
   31.5kph
   Darwin Institute

f) 385 kg
   30.6kph
   Chisholm Institute

Fig 4   The first six finishing solar cars — kerb weight, average race speed, name, and front and side projections. (Courtesy Robbins T. et al)

Commercial software—a medium for teaching 'life-long learning'

W S VENABLE, EdD, PE, MASME, MASEE
Department of Mechanical and Aerospace Engineering, West Virginia University,
Morgantown, USA

In the United States, teaching 'design' includes conveying professional attitudes. One goal is teaching students to continue their technical education in an informal setting. The author uses a 'continuing professional education' model when introducing software to students. The behaviorally based model uses minimal formal instruction with commercial software and documentation. The author and his colleagues have employed this technique in large classes with 'spreadsheet' analysis and graphics, word-processing, and kinematics and dynamics of linkages. Some undergraduate students now use computer-aided drafting and finite element analysis in a self-taught context. The paper contains a discussion of the results to date.

1 INTRODUCTION

In the United States, assignment to the teaching of 'design' includes the assignment of conveying a wide variety of professional skills and attitudes, as well as specific design skills and practices. (1,2) Among these more or less hidden items on the instructional agenda is the responsibility of teaching students how to continue their professional and technical education in an informal setting after graduation.

How can we meet the responsibility for 'teaching' an ever increasing number of peripheral professional skills at the same time as we are forced to accept responsibility for seeing that students gain experiences with modern computer-aided design methods *and* still devote sufficient time to fundamental principles of design? One answer lies in integrating multiple objectives into our assignments, and one way to do this is by changing the way we introduce computer methods into our courses.

Let us consider, for a moment, how practicing engineers acquire new skills, computer or otherwise. In a few cases, engineers may return to a university for extensive formal instruction. More frequently, but still rather rarely, engineers may attend a training session lasting several weeks. Most often, engineers will have no more than a day or two of formal instruction on a new technique or a new design tool, followed by an extended period of learning from self-study, on-the-job practice, and informal interactions with colleagues. Both individuals and institutions who provide 'continuing professional education' place great emphasis on formal offerings and instruction, but, in fact, the 'life-long learning' for which students should be prepared is effective primarily when it is based on individual efforts.

In the computer area, it appears that access to any classroom instruction in new software may be the exception, rather than the rule. A computer magazine conducted a survey of over a thousand users of its on-line information service on how they learn to use new software. They found no reliance on classes among their respondents, with most learning coming from reading manuals and experimentation. (3) While the population sampled was a select one consisting primarily of advanced users of personal computers, and might not be representative of engineers assigned to routine tasks and tied to net-worked corporate computer systems, it does reflect the experience of many engineers in a position to initiate the use of new software.

The author is now in his third year of incorporating student use of packaged computer products into both analytical and practice oriented courses. (4) His experiences have convinced him that there are several reasons to use a 'contin-uing professional education' model when introducing software to students. That is to use minimal formal instruction and to make maximum use of high quality commercial software, and the associated documentation and tutorials. The method relies, in large part, in creating an atmosphere in which the software is merely a tool needed to accomplish an engineering task, rather than something to be mastered for its own sake.

2 THE MODEL

2.1 A theoretical base

The author approaches instruction from a behavioral point of view. (5) His approach begins with the assumption that the primary responsibility of the instructor is to structure circumstances so that students engage in a series of activities which will make them practice the skills to be learned.

Under the behavioral approach, a plan of instruction should exist primarily as a set of student exercises, rather than as a series of topics. Similarly, a text should begin as a set of problems, rather than as a series of lectures. Information and supervision are provided to insure successful completion of the experiences, rather than assigning problems to assure students' attention to the instructor's activities. 'Teaching' may often assist in the learning process, but it is not always required, and should not be the point of departure for the design of a course.

To apply this approach to education to the introduction of computer techniques within existing engineering courses requires that the instructional designer develop a series of assignments which the students must complete with the help of computer software. The first in the series must be of the sort which require only a limited set of the computer package's capabilities, and the assignments must increase in complexity as the course progresses. At each stage, the student is expected to view the program as a tool to make the technical operation more efficient or more accurate, not as an end in itself. Students are given a few hours lecture or demonstration as an introductory 'short course,' then are expected to refer to the documentation for the software, and to informal interaction for most learning. The formal instruction is focused on the capabilities of the software, not on the actual methods of use.

Students are required to become productive almost immediately. They must do so in order to do the analysis in textbook problems, and to complete design calculations or prepare reports.

Reducing the apparent involvement of the teacher in students' acquisition of computer skills is not to belittle the importance of skilled instructional work. In fact, the success of a strategy of 'self-taught' computer use is highly dependent on the proper structuring of the demands placed on students for computer generated results. Thorough mastery of a specific software package will often depend on structuring a series of demands over a period of several years, with the degree of prompting and assistance being reduced as the sophistication of output required is increased. Traditional teacher activities such as consulting with students and marking assignments are frequently important aids in monitoring the process.

The behavioral approach also requires very careful review of the software employed. Students require success for reinforcement of the behaviors being acquired. Assignments must be designed as much with an eye on the probability of successful accomplishment as with concern for technical interest. Design for success must consider the strengths and weaknesses of the documentation and 'user friendliness' as for the power of the program itself.

The 'continuing education' approach also requires that software be carefully reviewed for utility. Within a design sequence, for example, only software which processes data applicable to the designs under study should be aggressively introduced, although, of course, much additional learning about hardware and systems operating software will naturally be involved.

When software needs can be defined in terms of general requirements which can be met with more than one package, students can, and should be given the responsibility of selecting to particular package which they will use.

Under this approach, mastery of specific software procedures, and development of general computer skills are fundamental instructional objectives, but not necessarily announced ones. Students' attention is focused on the 'scientific' or 'engineering' learning, i.e. dynamics or heat transfer or gear selection or cost reduction, rather than on computers or programming.

There is an obvious danger in the behavioral approach that what is taught may be chosen because it is easily taught, rather than because it is important or useful. The same, or similar, argument can also be made about other approaches. For instance, there appear to be cases where lecturers choose their topics because they create a strong impression, rather than for importance or utility. No approach to instruction obviates the need for professional maturity in those setting curricular goals.

2.2 Applied Model

2.2.1 The educational context

West Virginia University is a major state owned and financed university offering both undergraduate and graduate degrees in nine engineering fields. The College of Engineering receives a relatively low level of financial support, and the expenditures for computer hardware and software are quite modest.

At the present time, we have roughly 1600 undergraduate students. These students share about sixty terminals without graphics capabilities which are linked to a VAX system and used primarily for FORTRAN programming. In addition, they have access to about fifty IBM PC's and PC/AT compatible machines. Most students use PC's in a stand-alone mode, although a sizable number can access the VAX network. In our third year of PC use, the ratio of over thirty students per personal computer seems to be adequate except at mid-term and semester end when a large number of reports become due.

The College of Engineering makes a large quantity of public domain and 'shareware' software available to students on an optical disk. While this provides students with many useful utilities and recreational programs, only a little has been found which is suitable for serious engineering use.

Students are neither required nor expected to purchase computers. Fewer than ten percent of our students own computers suitable for engineering use. Students are required to purchase computer supplies such as floppy disks, as well as selected programs and/or manuals.

2.2.2 Practical application

In practice at West Virginia University, the instructor begins by identifying those topics within a course on which computer software can help students to make a more rapid or more accurate solution, and the places in which the use of computers can assist students in making better presentations of their results.

The instructor then identifies specific software products, and makes arrangements for providing the products to students. In our university, when the programs are of a general purpose nature, word-processors or spreadsheets for example, this is generally done by requesting that the bookstore stock copies of commercially available 'student versions' of the software and manuals. When the programs are of an expensive and specialized nature, such as CAD or finite element programs, the software is installed on hard disks in university owned computer labs, and manuals are provided on a short-term loan basis.

The instructor then develops a project assignment which includes a clear statement of the requirements for computer use which are associated with the project. This is generally made in writing. When new software is introduced, the instructor usually provides one class (about fifty minutes) in which the general features and operations of the package are outlined in a lecture or lecture-demonstration. Students are the expected to begin working with the package. Student work is most often done on university owned IBM PC/AT compatibles, although a minority have their own computers at their residences.

Instructors also function as consultants. In this regard, they generally adopt the role of a colleague of their students who happens to be an advanced user and who will help find answers to questions, rather than the role of tutor. Usually a class of students will quickly develop their own network for the exchange of such information, so this instructional responsibility is generally of limited duration and it is often focussed on problems which are of genuine interest to the faculty involved. As is often true in industrial settings, computer problems develop into a two way exchange of information beneficial to both users.

Student achievement of required competency is usually assessed solely on the basis of the project assignments submitted, and is not marked separately. In rare cases, a few questions about software procedures have been included on examinations.

3 RESULTS

At West Virginia University the author, and several of his colleagues, have been using this model to introduce personal computers and software in second, third, and fourth year courses. The results described are based primarily on the author's personal experiences in the Mechanical Engineering curriculum, but they reflect the experiences of other faculty in aerospace, civil, electrical, and industrial engineering with several products.

3.1 Courses and software involved

At the sophomore level, our experience has been primarily with the introduction of spread-sheet analysis and word-processing, along with basic PC/MS-DOS operations. These second year students are required to prepare project and laboratory reports in two mechanical engineering courses. They are permitted a free choice their word-processing package, but are expected to use a 'Lotus 1-2-3 compatible' spread-sheet. In one course they use the spread-sheet to reduce laboratory data from tests of material specimens and to plot graphs such as stress-strain curves and logarithmic fatigue life diagrams. In the other, they organize and present such things as monthly and annual energy costs and economic paybacks using line and bar charts. During this first year, few students go beyond the use of formulas involving addition, subtraction, multiplication, and division, although the graphic capabilities are rather completely explored.

In the third year the author has required all students to individually develop spread-sheets for the analysis of the kinematics and dynamics of machine linkages during each of the past two years. These exercises require students to use the extended mathematical functions available in the spread-sheets. This work has lead to the introduction of spreadsheet analysis of such things as powerscrews, and for the plotting of involute gear tooth profiles.

Already several colleagues within the university have had a few undergraduate students successfully use computer-aided drafting and finite element analysis on special projects in a self-taught context. In most of these cases, the level of program familiarity and skill in usage which the students have developed has far exceeded that of the faculty members directing the project.

During the fall of 1988 the author began the development of a fourth year required course on the design of machine elements in which all students will use computer-aided methods within the context of a traditional course. In the first offering of the course, all students successfully completed an exercise requiring them to analyze stresses in a thin shell with the SuperSAP finite elements analysis package. Future work will involve assignments requiring computer-aided drafting and additional analysis packages, even though the students involved will have had little or no previous experience with the procedures. The use of their previously acquired skills in word and data processing will be required, in addition.

In addition to their use of 'applications packages,' students have routinely developed high skill levels in the use of several programming languages not taught within the formal curriculum. Many engineering students (with faculty encouragement) have made QuickBASIC their programming language of choice for senior level projects in analysis and controls, despite an almost exclusive focus on FORTRAN with the formal curriculum.

3.2 Effects on students' attitudes

Our students appear to have a positive attitude towards both the use of computers and toward this method of introducing it. Criticism is more often directed at the producers of specific products than at the faculty members who have required their use. While many students initially pro-offer such excuses as 'I can't type,' the majority quickly attack the keyboards. Only a very few attempt to 'get George to do it' in team situations.

Students are proud of the professional appearance of the work they produce. Their reports on most industrial projects are given directly to the sponsors without additional services from our clerical staff, other than duplication and binding.

It is perhaps surprising to find that students are willing to acquire skills on their own in the face of a general feeling on their part that it is unfair to require learning of information, whether conceptual or factual, without formal classroom presentation. Whether or not this is due to some American perception that any man can master simple tools on his own has not been determined. No particular problem has been observed, however, with our limited foreign undergraduate student population.

Students do not object to any significant extent to being required to purchase software. A survey of third year students conducted last year indicated that many are willing to spend as much as fifty dollars on good software, if it is of general use. Student editions, and clones of many of the most popular and powerful software packages are available at prices ranging from twenty to fifty dollars. Many student purchase additional software which is not directly required for courses.

It is true that there is a significant amount of software exchanged between students in violation of copyright laws, but this is not a situation unique to educational institutions. Faculty using the method described here actively attempt to discourage such practices by declaring the importance of access to full documentation, and by referring many students to their manuals for the answers to operating questions.

4. RELEVANCE TO INDUSTRY

It is the author's belief that the model described is related to industrial practice in two ways. One is related to preparing students for industrial reality. The other is guidance for managers concerned with industrial computing.

First, this method of instruction places students in a situation similar to that which many will face in industry. That is, each individual has a direct responsibility for making productive use of computer products. This includes responsibility for software and hardware selection, installation, training, and data interchange. They are practiced in decision-making regarding computation and data processing. Such persons should be much more effective in industrial work than students with 'classical' computer training limited to writing simple high level language programs in a highly structured environment.

Second, the method places students in a position where informal learning, much of it self-directed, increases job productivity and may increase payoffs, in the form of grades. For many students in America, this may be their first experience with such a situation. For some, it may also be the first situation in which they have been responsible for 'professional' learning based primarily on textual references without the guidance of lectures or the stimulation of examinations. Such experiences should help them to prepare for the professional responsibility of maintaining general technical competency.

Third, the experiences of educators with such an approach to computer use should help assure corporate and institutional managers of computing services that flexibility in selection of computers and software need not necessarily result in explosions in training costs. In fact, adoption of a wide variety of products may result in increased productivity if each individual user can select and use those best suited to his individual objectives.

5. CONCLUSION

Personal computer use and applications software can be practicably introduced to undergraduate students in a 'continuing education' context requiring minimal expenditure of class time, or direct teaching. Such learning experiences are accepted by students, and prepare them to accept personal responsibility for continuing their technical education informally after graduation.

REFERENCES

(1) Venable, W.S., Developing a Design Attitudes Structure to Support a Capstone Design Course, *1987 International Conference on Engineering Design Proceedings (ICED 87)*, 1987, 1089-1095 (American Soc.of Mech. Engineers)

(2) Venable, W.S., Developing Objectives for an Engineering Design Course, *Proceedings of the Fourteenth Conference on Frontiers in Education*, 1984, 696-699. (Inst. of Elec. & Elec. Engineers)

(3) Anon., Survey, *PC Magazine*, March 28, 1988, 40-41.

(4) Venable, W.S., Experiences in Introducing PC's in Undergraduate Courses, *1987 Annual Conference Proceedings*, 1987, 1986-1989. (Amer. Soc. Engineering Education)

(5) Venable, W.S., Developing a Behavioral Approach to Teaching Design, *International Journal of Applied Engineering Education*, 1988, 4, 237-242.

C377/087

Prototype design—the missing link

P E R MUCCI, DIC, CEng, MIMechE
Department of Mechanical Engineering, University of Southampton

Pre-production prototype work done in the University
environment for an industrial customer permits a
one-off working product to be designed, made, and
tested under controlled conditions, which are strongly
influenced by the needs of the marketplace and the
constraints of production methods and costs.

The organization and running of similar groups in
other universities and polytechnics is proposed as an
important way of transferring advanced technology and
design innovations to industry and at the same time
provide the missing link between research and production.

1 INTRODUCTION

Less than 2% of the 80,000 manufacturing companies in the UK invest in
research and development (1), and although this figure does not separate
out the design function, it can be fairly assumed that a large number of
these companies choose not even to pursue an active design policy for
creating new products from existing lines.

Research, Design and Development is an expensive business and whilst
money (or lack of it) may not be the only reason for this worrying
statistic, manufactured goods selling into competitive markets do not make
the sort of profit margins that would encourage modest sized companies to
invest large sums of money into what can often be a bottomless pit with no
visible payback in sight.

Increasing technical complexity of manufactured products, recent laws
on product liability and the growth in competition for international
markets has, however, made the process of thorough product development
absolutely essential for the survival of manufacturing industry.

2 THE NEED FOR PROTOTYPE DESIGN

Engineering research may be a well established function of higher education
institutions (HEI's) but the subsequent production of the results of
research is left to industry. The very low numbers of R & D staff in

industry previously mentioned are, perhaps, a reflection of financial stringency or even a lack of capability to improve product design.

There has been a tendency in recent years for companies to devolve the responsibility for supply of semi-finished and finished components made to their specification to outside suppliers whereas before they may have done it in-house with their own design teams. Improvements in material quality, international standardization and communications have made it attractive financially to move toward the creation of a product by the assembly of purpose-designed parts manufactured elsewhere.

Creating products by assembly or 'catalogue engineering' as this is unflatteringly called, can be justified by the argument that this is merely a natural extension of, for instance, using standard fasteners, electrical components or bearings.

There is, however, the danger of making a product which, for reasons of cost-cutting and ease of production, begins to drift away from its original purpose and starts to lose market appeal. Perhaps worse still, if all the competition is doing the same thing, then the customer is left with the choice of several products looking and behaving in the same way but which all have the same faults.

The alternative is for industry to do original design and development work on all the important aspects of its products, and to bear the cost of permanent high quality engineering staff and facilities needed to push new ideas into production. The financial risks are high, though, because it is common practice to start spending on costly tooling as soon as it looks as if a design will work on paper. Well before the production phase is reached an intermediate 'pre-production prototype' (PPP) is needed which is operational in the laboratory environment and satisfies shape, size and technical specifications but which has not yet been taken to the jig and tool stage for final production. Table 1.

Although this stage is costly it is nowhere near as expensive as having to go straight from design drawings to production, only to find serious faults which force the scrapping of special tools. The PPP stage has another advantage - it is quite portable in that it does not necessarily have to be done on company premises as long as there are enough pairs of experienced and qualified hands and the right facilities available.

Every major city centre in the UK has a college of technology, poly-technic or university in close proximity. Most of these establishments run degree level courses in engineering and have research equipment and facilities. The 'opening-up' of these establishments to industry in recent years should have provided a cost-effective solution for prototype design projects, but the numbers of design-related contracts placed has been disappointing despite strenuous efforts from the SERC and the Design Council and encouragement from the government. There are a number of arguments which must be familiar and frustrating to both sides which may explain this, and they include: the need for universities to publish results of research and issue higher degrees; the need for industry to have work done to strict deadlines, and the ownership and transfer of intellectual property rights.

Southampton has tackled some of these problems by allowing much more flexibility in the duration and educational aims of research contracts, at the same time offering a businesslike, objective approach to project management.

3   SETTING UP A PROTOTYPE FACILITY

Over the past ten years, Southampton has developed an approach to student project organization which has become an essential feature of the Master of Engineering course (2).   Since 1982, students have won ten Design Council prizes for their work, the great majority of these being awarded to inderdepartmental teams working on industrial projects.   There has been a strong trend toward innovative design in the field of electro-mechanical devices (3) and in most cases sponsors have asked the University to carry the work onto the next stage of prototype design so that it can be properly assessed for a potential production run.

Gradually, from the early 1980's onwards, ideas were developed for breaking the impasse caused by conflicting industry/university needs described earlier.   By 1987 the concept of a prototype design activity had been evolved with the aim of providing the missing link between research and production for industrial customers.   The main criteria for its operation were:

(1)   Efficient and full transfer of design work from institution to industry

(2)   Full financial support for the project from each sponsor

(3)   Proper business contracts and close financial control

(4)   High status and genuine potential for future career development for staff employed by the group

(5)   Maximum feasible use of information technology systems and computers in all areas related to the projects and the running of the group

(6)   The widest possible educational advantages without putting confidentiality at risk.

Recent success in another sponsored design project accelerated the demand for a viable activity within the University.   It was decided to propose to the latest satisfied customers in industry that work on their product design could carry on to a pre-production prototype stage without loss of momentum and at modest cost to themselves.

A medium sized company agreed to place a contract with the University which permitted the formation of a Prototype Design and Development Group (PDDG) on the campus.   The financial backing of the sponsor company supports two design and development engineers (both M.Eng. graduates from the original design team), all test rigs, instrumentation and control equipment, CAD and administrative computing power, office equipment, day-to-day purchases and expenses.

Two members of academic staff (one mechanical, one electrical engineer) are retained as paid consultants, and there is an academic supervisor for the group.   There is additional funding to cover loss of intellectual property rights (IPR) and this can be used as seed corn funding for attracting more industrial projects.

4   PROTOTYPE DESIGN STRATEGY

The fact that pre-prototype design can be done in a university does not automatically permit the designers the same freedom to pursue their objectives as those working in pure or applied research.   Engineering design is always product specific and very often company specific.   Before work can start, therefore, there must be a full understanding of the sponsor's needs for the product, especially if there is an existing product line into which the new design must fit.

The strategy used in the successful completion of industrially-sponsored student projects at Southampton has been developed into a working method for handling product design in the PDDG.   Rather than attempt to squeeze the needs of the customer into a pre-ordained methodology, the design approach first and foremost serves the needs of the customer (4).   Some stated principles can, however, help the designer to establish priorities. Table 2.

5   INTELLECTUAL PROPERTY RIGHTS AND DESIGN 'CONTRACTS

Design work done for industry in an educational establishment is often the subject of complex negotiations not normally found with more traditional research projects.   The reason for this is the possibility that not only will something new come from the work but something saleable as well. Although the same could be said of more fundamental research, this is usually over a longer time scale and without specific product objectives.

Innovative design, however, can make a fairly quick impact on a new product and improve its sales over the competition, but if the designer has no protection against ideas being simply put into production with no financial gain or other advantage, then there is little incentive for creative work to start in the first place.   An added difficulty is when the sponsor prohibits publication on the grounds that it will reveal all to competitors. If an academic can neither publish nor gain financial or other rewards, then it is not surprising that there has been a limited flow of good design from university to industry.

The first step toward agreement on IPR is to offer the client a study of the problem over a fixed period of time - possibly even as a student project for which payment is made for the final model or rig, but the ideas contained within it remain the property of the institution.   Should the client wish to buy the rights to this intellectual property, then it is almost certainly more favourable to the parties concerned to accept a single one-off cash payment.   This can prove to be more satisfactory than trying to negotiate royalties from future sales.   Even if it were easy to obtain sales figures (the company might change hands, for instance), the administration of an educational establishment is not usually set up to interpret them (for example, how to charge royalties on large discount sales).   There are also circumstances where the client could claim that Mk I of the product may be subject to royalties, but Mk II, which bears no resemblance to the original prototype, is not.

Most higher education institutions (HEI's) have now got a clear policy on the gains to academics and students from sales of IPR's.   A common approach is for small sums, under £10,000, to be shared equally amongst staff and students involved, say, in a student project, with 20% payment being made to the institution.   As the sum to be paid increases, there can be a 50/50 split between the institution and the designers.

If a sponsor wishes to collaborate with the institution in taking a study or a student project further toward a prototype for production, then a formal contract is essential.   Unless the institution has a department able to handle the day-to-day administration of a commercial design and development group, then partnerships, shareholdings and limited liability companies should be avoided.   Instead, the project may be set up within the terms of reference of an ordinary research contract but with the following provisos:

Duration of contract	Allow flexibility; the standard three years is often too long for product design
Equipment and materials	Limit to day-to-day expenditure only - large items to be paid for by the sponsors as demand arises
Running costs	A specific amount agreed to be spent on the running costs of the whole project, eg. secretarial, office and general expenses
Conflict of interest	The group should contract not to work on a similar product for a competitor for an agreed period - 3 to 5 years is sensible
Confidentiality	The institution should agree to keep the work secure from prying eyes and practice a 'need to know' policy for visitors etc. In return for restrictions on publication the institution should be given recognition when the customer launches the product and publishes information about it. It should also be free to use some of the work for teaching purposes
Supervision and consultancy	Whereas it is accepted that supervision of projects is at no cost to the sponsor, it should be agreed that as soon as institution staff (other than those employed on the project) do work for the client then a charge must be made
Intellectual property rights and patents	Lump sum, once and for all, payments can be the best way of dealing with IPR's unless the individual or institution has put in considerable research effort before the sponsor came on the scene. Patents must benefit the individual concerned and either the institution or the sponsor may wish to foot the bill, which can be considerable. For this reason the inventor should avoid investing personal cash.

Ownership of prototypes and associated materials	The customer will normally own the prototype(s) at the end of the project duration, but the institution should retain examples for teaching purposes
Health and Safety at Work Act (HSAW)	When any working prototype device is handed over to customers then they should agree to sign an exemption letter within the terms of the Act. This then imposes on them the responsibility for safe use of the prototype

6   CREATING THE RIGHT ENVIRONMENT

Successful design depends on people being given the freedom to be creative and inventive with the minimum bureaucracy and the maximum encouragement. The decision to improve an existing product or design a new one should cause companies to review design management methods so that the right atmosphere is created for product change.

So often the administrative system of a manufacturing company which was set up for sales and production is expected to serve the design function as well. The handling of prototype work which needs specialist people and facilities becomes a difficult task if the technology used by the graduate engineer goes beyond the knowledge and experience of existing managers.

Minimum constraints should be placed on young design engineers with the accent on trust and respect for their abilities to handle modern technology. Although careful supervision may be needed where external affairs are concerned, such as dealings with suppliers and contracts, the learning curve with enthusiastic graduates is steep and clear guidelines at the outset on good business practice will usually suffice.

Apart from offering sensible salaries to encourage good staff, positive steps can be made to develop their careers. Membership of professional institutions should be encouraged from the outset. If there are noticeable gaps in knowledge needed to do the wider tasks demanded of those involved in new product development - such as experience in marketing or sales - then effort put in at an early stage will reap future rewards.

Successful approaches to teamwork in previous student work and industrial project management (5) helped set the scene for the PDDG. A modern difficulty for project managers in any organization is deciding how much freedom staff should have to use information technology and communications systems without prior reference to immediate superiors. Computer links and Fax machines are quite cheap compared with total project costs and information can now pass directly between a design team and a supplier or customer without getting lost in a paperwork system.

It was decided that the PDDG should give freedom coupled with high personal responsibility for all engineers involved in design work. Fortunately the sponsor is also operating in this enlightened way in his own organization and there was therefore no clash of ideologies.

The real decision making on design steps, however, is firmly held as a group rather than individual activity, and regular working meetings are held where ideas are sounded out and developed. The method of operation is such that, once a design step is agreed upon, the engineer is given full use of the system to implement it within the allowed budget. If the decision has major financial implications then this will be discussed at a meeting with the sponsor, but the day-to-day business management of the project is left entirely to the group staff. This is a growing trend with industry/ university work at Southampton and elsewhere (6).

The low level of responsibility often complained about by new grad- uates in industry and their disenchantment with old-fashioned company systems could largely be eliminated if new product development could operate in a unit within which staff have the freedom to use the most appropriate methods and technologies available to achieve their aims.

7   THE ROLE OF THE COMPUTER

The Mechanical Engineering Department at the University had considerable experience of developments in CAD and the application of desk-top computers in general so it was decided that, as far as was possible, the product design work and its associated administration should be done entirely on a single 'engineers workstation'.

An Apple Mac II was selected, following an extensive survey, for its ability to handle alpha/graphics and operate very advanced storage and retrieval of data. It was an essential requirement of the system chosen that it should operate all required software efficiently and quickly with the minimum of staff training. The Apple is clearly not the only machine capable of doing this, and there are many systems available which are equally powerful. Apple were, however, one of the few companies which offered a total commitment to hardware and software support. The workstation regularly performs the following functions:

    Letters and memos
    Reports
    Accounts
    Calculations
    Database
    CAD

Fax is used extensively to communicate drawings, specifications, and letters to sponsors and suppliers. By this method the use of office typing and postal services is minimised. More important than the administrative advantages are the gains for design staff in terms of total project control and freedom of action.

The effect of giving highly qualified staff routine office work to do as well as major design tasks could have overloaded the project from the start and threatened its development. The real effect, however, has been to cut out bureaucracy, speed up communications, and reduce wasteful paper- work. Furthermore, the use of the computer to record all financial transactions has proved to be vital for control of the various monies passing between the sponsor and the University and from the Group to its suppliers.

The ability of the workstation to process both text and drawings and mix and match between them creates a new dimension in technical communications.

The 'Versacad' CAD software on a high resolution screen performs at least as well as more established packages and has the advantage of being 'bug-free' at the beginning of its development. It is likely therefore that it will get even better as new versions are introduced.

8   HOW IT IS WORKING

The sponsored design work which brought the PDDG into being is one of many electro-mechanical designs completed at Southampton.   Table 3, Figs. 1 and 2.   It requires an extensive pre-production design, build and test programme before jigs and tools can be commissioned for the first production run.

The sponsor liaises closely with the PDDG but takes no part in its day-to-day supervision which is left to University staff. He does, however, get directly involved in design decisions which may affect future marketing and sales.   Meetings are held with him monthly and there is close contact with the financial director with regular visits to various company premises. This two-way involvement with the sponsor is considered the key to future success as others have found (7).

Apart from some small modifications to a laboratory to assist with the test programme, bringing a highly commercial project of this kind into the University has been reasonably trouble-free.   There has naturally been a need to inform technical and administrative staff of what is going on - and particular emphasis has been placed on safety and a level of security higher than might be normal with an ordinary research programme.

Unlike applied research, there is also a direct design input to what will become a saleable product.   Even at the early stages, original design work was being done on what will become mass-produced plastic mouldings, diecastings and electro-mechanical assemblies.   A surprising development has been the amount of time that needs to be spent on ergonomic and industrial design considerations, even though the product does not fit into the category of 'consumer goods'.

Constant effort is spent on reducing assembly costs and simplifying various manufacturing operations.   Suppliers are already being asked to agree to contractual terms and orders being placed for long lead items which will form part of the final product.

The good and bad points of similar products are proving to be an important input into the design process and the sponsor is supplying information about the market from both the end user and the distributors.   It is planned to put the first pre-production prototype into the field under controlled conditions at an early date so that real operating problems can be ironed out prior to the first production tooling being commissioned.

9   CONCLUSIONS

The concept of pre-production prototype design and development can be a vital link between design on paper and the first production run.   For it to work properly it must allow engineers the freedom to achieve their goals with a high level of responsibility for making decisions within a team uncluttered by bureaucracy and arbitrary company systems.

True industry/university collaboration on commercially viable engineering design projects can be made to work.   For this to happen changes are needed to pre-conceived ideas about the aims of research in higher education, and manufacturing industry needs to reassess the time-honoured view that academics can play no part in short-term product development.

Whereas the educational institution stands to learn a lot about the business of engineering from such projects, industry may understand better how to manage and keep young, well qualified engineers.

There are clear signs that modern information technology and computer systems could be used to better advantage in product design, especially if engineers are allowed to use these systems to their maximum potential.

If only a fraction of the large number of HEI's were linked to industry through prototype design projects, there could be a major improvement in the level and quality of original design generated in the United Kingdom.

REFERENCES

(1)   Duckworth, E. 'Research Worry', Statement by the Assoc. of Independent Research and Technology Orgnisations, report in Daily Telegraph, 22nd Sept. 1988

(2)   Mucci, P.E.R. Role of Projects in Engineering Design Education, IMechE Conf. Paper C313/85 1985

(3)   King, D., Luk, R. et al   The Design and Development of a Non-Contact Printer for Product coding, Proc. Instn. Mech Engrs, Vol 202, No B2, 1988

(4)   Winstanley, D., Francis, A.   Fast Forward for Design Managers, Engineering, March 1988

(5)   Mucci, P.E.R.   Teaching Engineering Design from Paperwork to Prototype, IMechE Conf. Paper C226/83 1983

(6)   Tacey, E. 'Scaling New Heights', report in The Engineer, 18/25 August 1988

(7) Davies, P. University/Industry Links - An Integrated Approach, <u>Engineering Designer</u>, May/June 1987

STAGE	ACTIVITY
Applied research	Potentially valuable idea proves repeatable under laboratory conditions
Research and development	Reliable application of the idea to a purpose under rig test conditions. First estimate made of value and cost.
First prototype design	Design-build-test iteration to guarantee product value and performance. Shape and size unconfirmed. Major cost elements known.
Pre-production prototype	Shape, size, performance, market and quality aims confirmed by design, build, test and field trials. All specifications and manufacturing methods confirmed. Synthetic estimate made for final cost and selling price.
First production prototype	All production tooling commissioned, assembly methods confirmed. Materials and component suppliers approved. Production costings made. Market feedback obtained on price, quality and reliability needs.

Quality	Build in high quality and low cost at all stages of design
Customers	Deal direct with customers and discuss design with them at all stages
Suppliers	Create an early dialogue with specialist suppliers and enlist their technical expertise
Competition	Analyse competitive products to fully understand their strengths and weaknesses
Standards	Use materials and components which conform where possible to recognised standards
Assembly	Aim for minimum parts, ease of assembly and maintenance
Testing	Get an accurate picture of the working conditions for the product and test as close as possible to them

PRODUCT	CUSTOMER
Hydraulic laminating press for the automatic production of Melamine and Urea based samples for quality control on production lines Supervised by: P.E.R. Mucci	Wiggins Teape Orchard Ltd. Christchurch
Computer-controlled non-contacting printing head for coding of products passing through high speed packing machines Supervised by: R. Holmes and P.E.R. Mucci	Jacob White Packing Ltd. Dartford
Automatic resin impregnation plant for the coating of paper with water based resins without the need for operator intervention Supervised by: P.E.R. Mucci	Wiggins Teape Ltd. Hele, Exeter
Warm air vehicle heater for the independent heating of enclosed spaces without the need for continuous running of engines Supervised by: P.E.R. Mucci and R.M. Crowder	Beta Ltd. London

All winners of Design Council Awards, 1982 to 1989.

Fig 1   Non-contacting printing machine (foreground) fitted to a high
speed packaging machine

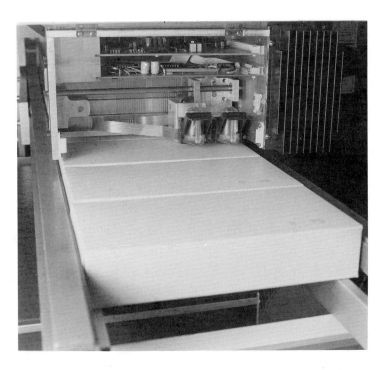

Fig 2   Close up of head of machine in Fig 1 showing cardboard cartons
         (foreground) being printed with product coding

C377/066

Design education: industrial input through the use of video

J E A BROADBENT, BSc, CEng, MIMechE, ATI, CDipFA
Department of Engineering, University of Manchester

SYNOPSIS.   In design courses, direct contact with industrialists is valuable, but logistic problems are significant. The Engineering Department at the University of Manchester has used video (interviews and edited programmes) to provide industrial input. Even at this early stage, the teaching benefits are apparent and the production of videos is shown to be a convenient and mutually beneficial form of cooperation between industry and universities

1. INTRODUCTION

The final year Design course in the Department of Engineering is an optional course. The course builds on the foundation of practically orientated design studies established in the previous two years and on the increasing knowledge of engineering science acquired by the students at this stage.

The course has two main objectives: to continue to develop the design skills of the undergraduates by studying the design process itself (i.e. design methodology) and to make students aware of design practice in industry. In particular, they must appreciate the magnitude and variety of skills involved in the launch of a new product, the role and importance of the designer within this process and the importance of good design to the success of a business. The management aspects of design, marketing and production are covered in the final year Management course. The Design course had three elements: lectures, seminars and design exercises.

For several years the Department has used professional engineers from industry to give the seminars. In general they discussed individual projects in which they had been involved, as well as the wider aspects of design. Typical of the companies who have given us assistance are Fords, ICI, British Rail, Mullards and the National Centre for Tribology, Risley. These seminars are popular with the students and effective as a teaching technique. The direct contact with real industrial problems and industrial personnel brings an awareness of the situations which the students will face upon graduation. The use of external lecturers extends the range of experience and greatly enriches the course.

Despite the undoubted advantages of involving industrial personnel in the seminars there are several problems. For industrial personnel to prepare and give a seminar involves considerable commitment and effort. This can only be justified if it is a major contribution to the course; thus, the number must be limited (in our course to four or five a year). The timetabling must be mutually convenient; a far more difficult problem than you would imagine. On several occasions this resulted in seminars not being in the ideal sequence. Because we wished to introduce at least some new seminars each year, we have the ongoing problem of identifying suitable people, securing their agreement and briefing them for the seminars.

In 1987, therefore, we began to search for some method of retaining and increasing the educational advantages of the industrial contacts without the accompanying disadvantages. With the Department's extensive experience in the use of the video, it was natural that we should investigate this medium.

2. PREVIOUS EXPERIENCE IN USE OF VIDEO

Since the Department made the first video for use in the teaching of Design during 1981, the growth in the use of video has been dramatic. Over thirtyfive programmes are in regular use, mainly in the first year course. In general the programmes are quite short (10-15 minutes is typical) so they can be fitted quite easily into a lecture. There are three main series of programmes, one on production engineering, one on engineering drawing and one on components.

The programmes on production techniques were the first venture and are continually being extended and updated[1]. In the main, the programmes support the first year Design and Manufacture course. Each programme describes a process (for example, the hand lay up of glass fibre or powder metallurgy) and also shows the types of components which can be produced.

In this application the use of video gave a far more realistic idea of the process than is possible with any other lecture room technique. The undergraduates see the process, the role of the operator and machinery, the pace of work and the environment. The use of a camera combined with television production techniques shows aspects of the process which would not be apparent even on a visit to a works. For example; the use of slow motion and the ability to enlarge shots are techniques which are easily available on video. In addition, the camera can give views of areas inaccessible because of safety or space restrictions.

Experience with the use of video showed several other advantages. Video is an extremely fast way of communicating information. Visual information is being supplemented by commentary. The introduction of video certainly makes lecturing easier. Television holds the attention, even of students!

It is necessary to adapt teaching to accommodate the characteristics of video. At Manchester short videos (10-15 mins) are being used to enhance the lecture. This method of using video allows the lecturer to gauge the student response, adding explanations, or even rerunning sections of tape if some aspect seems to be causing difficulty. This avoids the one-way communication aspect of television, where there is some evidence that recall of material is impaired[2].

Handouts are necessary to summarise the contents of the programmes. It is difficult, if not impossible, to take notes during programmes. Facilities need to be available for revision, e.g. an audio-visual section in the library.

The costs of making programmes is significant, both in technical effort (typically £3,000) and staff time. At the University of Manchester we are fortunate to have a well equipped and professional T.V. unit [MUTV]. Their expertise and facilities make it possible to produce good quality videos, a necessity for an audience used to the standards of UK public broadcasting. Using these resources can only be justified when the programmes are to have considerable use, although some costs have been recouped by selling tapes to other educational institutions and to industry[3].

Industry has almost invariably given the University considerable support in producing the programmes. These contacts have formed the basis of several enduring relationships.

The success of the production engineering programmes gave us sufficient confidence to develop a second series on engineering drawing. Teaching engineering drawing involves imparting;

a) intellectual knowledge of the principles of drawing, and

b) the development of the physical skills necessary to produce a good drawing.

(At Manchester we teach students to draw freehand, with instruments and with CAD).

On video it is particularly easy to show the relationship between an object and its representation on a drawing. This enables the basic concepts of orthographic projection to be developed quite rapidly.

Video is particularly useful for the introduction of symbols and conventional representations. Many of the students come directly from school and have little general knowledge of engineering. Using video, they can be shown the object and then the symbol.

While the advantages of video in the teaching of the intellectual knowledge required for engineering drawing is evident, it also has advantages in the teaching of the physical skills. On video students can see an experienced draughtsman at work. They can study the way he produces a drawing, how he handles his instruments and the rhythm and pace of his work. Effectively it provides them with a model to copy. Our experience shows that this is an efficient way of teaching, although we chose to exploit the improvement by reducing the time devoted to teaching traditional engineering drawing rather than aiming at higher standards. Further details of the series are given in reference 4.

On CAD our main problem is too many students for the available terminals. This forces the adoption of small group teaching; this is an effective and pleasant teaching method but it absorbs considerable staff effort. We experimented with trying to develop a video self-teaching package. The initial work was promising but, before the programme could be revised to overcome the problems that had been identified, the CAD system was changed. This made the video package obsolete and there has been insufficient staff effort available to remake the programme.

The third series is on common engineering components (bearings, seals, etc.). Again, the visual nature of the material makes it easy to video. Diagrams (animated if necessary) can be added to clarify points and it is possible to show applications. However, our experience is that by themselves the programmes are insufficient to develop an adequate understanding of the topic. Therefore, at present, we are using the video as part of a seminar. For example, in the bearing seminar, students handle the bearings, fit bearings, and study damaged bearings, as well as completing exercises on applications. This is an area of active development in our teaching. Industry has been most helpful in the production of these programmes. The commercial advantages of familiarising large numbers of undergraduates with a particular brand of product are evident. From the educational viewpoint it is essential to be factually correct and avoid blatant advertising. (It would be counter-productive anyway). Naturally, credit is always given for the assistance of firms in the production of programmes.

The series of programmes discussed above (production techniques, drawing and components) are all essentially descriptive. They exploit the ability of video to give visual information. This is a straightforward application of television to fill an obvious educational requirement.

3. PROGRAMMES FOR FINAL YEAR DESIGN COURSE

The objectives of the programmes for the final year Design course are more subtle. The course concentrates on design methodology and design practice in industry. Consequently, the discussion centres on abstract concepts rather than physical objects or processes; although it is essential to provide practical examples to illustrate the points. Secondly, the aim is to introduce industrial experience, a knowledge of typical situations and an understanding of the relative importance and range of factors involved.

The initial intention was to make a video tracing the development of a component or sub-assembly. However, during the preliminary discussions with the Ford Motor Company, who were approached for assistance with the programme, it became evident that there were two additional areas of mutual interest. These were a programme outlining the role of an individual design engineer (a component engineer) and a programme giving an overview of the design, development and launch of a car.

The Ford Motor Company could see a use for these programmes in their recruitment and training; in particular, the role of the component engineer and his importance in car design is difficult to explain. Since all these topics provide useful examples to include in the final year Design course, it was obviously in the University's interest to make the programme. The chance to record many aspects of car production was valuable and this material will be used in other programmes. In addition, it would enable the Department to assess the educational benefits, in the final year design course, of three distinctly different types of video. The main disadvantage was the increase in the size of the project and the staff effort involved.

The response of the Ford Motor Company to our proposal has been enthusiastic and slightly overwhelming. Not only did they cooperate fully by giving access to their personnel and sites, but they agreed to support the project financially. In addition, they provided considerable help with the research for the script and also with the graphics.

The first programme, entitled 'Concept to Reality', is an overview of the design, development and launch of a car. It traces the process from the business development plan (which determines the appropriate time and type of vehicle to launch) through the various stages of development until the launch.

The programme emphasises the size and complexity of the operation and the wide range of business skills required. In addition, it aims to make students aware of the necessity to operate to a strict timetable and the risks and costs associated with short timescales.

This programme involved recording at both Halewood and the Research Centre at Dunton. Extensive use was made of library material provided by the Ford Motor Company. The programme was completed in Autumn 1988 and shown to students during the Michaelmas term. Although finding the programme complex, initial reaction of the students was favourable. It is too early to assess the effectiveness of this video, which may well be revised.

The second of the programmes, entitled "The Component Engineer", is an interview with a designer who specialises in fuel rail components. The recording was made in April 1988 and the editing completed during the summer. In the programme the component engineer, as well as outlining his job, describes two projects (the redesign of an airfeed hose and the replacement of a brazed fuel rail with a forged design).

This programme has been used both in the final year Design course and in first year tutoring sessions and has proved very effective. The extensive testing necessary before a new design can be introduced and the timescale involved are two of the points which surprise students.

The programme provides an interesting comparison with an interview given by Colin Ledsome of the Design Council. In this programme he discusses the design of the sole bar of the power car in the Advanced Passenger Train. In this interview the range of factors and the logic which determined the specification are particularly interesting. [The design problem is described in reference (5)].

Both interviews introduce the students to design in practice. Experienced and articulate professional engineers are discussing their work. The students begin to appreciate the factors, often non-technical. which dominate design decisions. Television is a revealing medium. The problems are described in the interviewee's own words; this is direct and personal. The students respond to this far more than to information filtered via a lecture.

The dangers of this approach are to be found in the rationalisation and simplification which inevitably occurs when the discussion of a complex topic has to be condensed into twenty to thirty minutes.

The third programme which we made with Ford Motor Company was entitled '*Car Seat Development*'. The programme begins with the Anglia seat, and traces the evolution in the design through to the seat of the B13 model (the Fiesta replacement). The programme is essentially descriptive. It is used to introduce the concept of continuous development in products to meet changes in market requirements and to exploit changes in technology. Although car seats are commonplace, few students have considered them as examples of engineering. As a design problem it involves balancing the marketing, ergonomic and technological requirements. The programme caught the interest of the students but it is evident that it will have to be supplemented by a handout on related topics (principles of ergonomic seat design, manufacturing processes and polyurethane chemistry) to exploit the full potential of this topic.

DISCUSSION

Overall, these programmes have made a useful contribution to the final year Design course. All will be supplemented with handouts for next year to increase their educational value. The only programme which may require revision is the overview '*Concept to Reality*'. At this time it is difficult to assess whether too many new ideas were introduced too rapidly.

However, although there is some fine tuning required, the programmes have enriched the course, enabling a wider range of industrial experience to be introduced. Because this is now on video tape, it can be introduced when it is required for maximum educational benefit.

As the programmes were developed it was realised that this is an effective way to study the design process. It is hoped to build up a library of case studies which can be used to analyse the general models widely quoted in the literature.

Other educational possibilities are to introduce contrasting approaches to design into the course, or to get experts to talk on their specialities. The probability that some of the views will be controversial will make the courses more challenging and interesting. It will dispel the idea that engineering education is the passive absorption of knowledge until the answers to all likely problems have been covered.

This paper describes the development of a teaching aid. However, case studies of this type also improve our understanding of design by showing how designers operate in practice. This work is of considerable research interest as this is a difficult area to investigate. The production of video programmes is an exceptionally suitable area for cooperation between industry and universities. In general, industrial personnel and academics have different timescales, working patterns and objectives. This can often cause problems. However, the commitment and workload required to produce a video is more convenient. Script research can be done in several short sessions over a period of weeks or months. The actual recording is a heavy commitment but usually requires only one to three days. With active cooperation, this is usually not difficult. Most industrial personnel find this a pleasant break in routine. The studio shots, editing and voice over can be done at leisure, although our experience is that to leave this more than two to three weeks is inefficient as it becomes necessary to refamiliarise yourself with the material.

In our experience, video programmes are an effective and convenient method of teaching. In many cases industry will be able to make direct use of the University programmes for training. Ford Motor Company intends to use the programmes '*Concept to Reality*' and '*The Component Engineer*' for training and graduate recruitment, thus both parties benefit. However, even if this is not possible, it is easy to re-edit the original recordings to suit another script. For example; when aluminium extrusion was recorded at H.D.A.E. (Workington), the material was used to make two programmes, one for the University and a longer training programme for H.D.A.E.

However, even without the direct benefit of training videos, most firms are only too willing to assist fully in this type of project. They are aware that universities depend upon the active cooperation of industry to keep courses up-to-date and that a supply of well-educated engineers is essential to their long-term prosperity.

ACKNOWLEDGEMENTS

The author wishes to acknowledge the assistance of the Ford Motor Company Limited and, in particular, Mr.G.McCabe, in the production of the programmes. Without the active cooperation of Mr.K.Wrench and the entire M.U.T.V. unit, these programmes could not have been completed.

REFERENCES

1. BROADBENT, J.E.A. *'Engineering Production Techniques'*,
   MUTV 1988
2. KEEBLE, M. and WEININAN, J. *'Immediate and delayed recall of information presented in a live and a televised lecture'*,
   Med.Educ 20(4) 1986
3. BROADBENT, J.E.A. *'The role of video in the teaching of computer-aided production engineering'*, CAPE Conference, Edinburgh, 1986 419-421
4. BROADBENT, J.E.A. , *'Engineering Drawing'*, MUTV 1988
5. LEDSOME, C. *Engineering Design Teaching Aids 1987*, 143-150,
   The Design Council

C377/035

A human-powered aircraft project as part of an undergraduate course

R S J PALMER, BSc, MSc, CEng, MIMechE, MASME
School of Mechanical and Production Engineering, Nanyang Technological Institute,
Singapore
K SHERWIN, BSc, PhD, CEng, MRAeS
Department of Mechanical and Production Engineering, Huddersfield Polytechnic

SYNOPSIS   A human-powered aircraft provided the ideal
basis for a creative design-build-test project for
second-year students in the School of Mechanical and
Production Engineering at Nanyang Technological
Institute, Singapore. The students completed the project
in sixteen weeks; this period represents two in-house
training programmes during 1984 and 1985 of ten weeks
and six weeks duration respectively.

The students learnt that the translation of any
idea into engineering practice requires sound
engineering design and construction.   They worked
enthusiastically throughout especially as the project
represented a pioneering activity within the ASEAN
region.

1   INTRODUCTION

This paper describes the design-build-test of a human-powered aircraft (HPA)
by second-year undergraduate students at the Nanyang Technological Institute
(NTI) in Singapore.   The decision to build such an aircraft was based on the
previous experience of one of the author's with the 'Liverpuffin' HPA (1).

Liverpuffin was started in 1969 as a result of a search for a major creative
project for an undergraduate course in engineering design.   It was argued that
there was no standard solution existing for a HPA and that such a device could
be built by students using simple hand tools.

Since 1969 the development of human-powered flight has been rapid,
culminating in the Daedalus aircraft of 1988, which achieved a flight distance
of 70 miles, 112 km, from the island of Crete over the Aegean sea (2).   Like
most engineering projects, Daedalus, was not a unique 'one-off' but the result
of continuing development by staff and students at MIT over the previous two
decades.   Their efforts had resulted in four previous HPA, namely 'BURD I',
'BURD II', 'Chrysallis' and 'Monarch', of which the last two achieved successful
flights.

In this context it is interesting to note the link between universities and human-powered flight. Of more direct relevance, it is considered that the arguements for adopting a HPA as a student project are still valid today. This was reinforced in the NTI project by the large student numbers available to build such a machine.

2   BACKGROUND

NTI was established in August 1981 to conduct engineering degree courses with an emphasis on engineering applications. This was a direct result of recommendations to increase the enrolment numbers in engineering courses at university level to meet the demands of the Singapore economy.

In order to provide practice-orientated education the degree courses at NTI consist of periods of academic study, in-house practical training and industrial attachment. All students undergo a six-month industrial attachment at the end of the third academic year. Initial practical experience is provided at NTI by a one term, ten week, in-house practical training period at the end of the second academic year of study.

The ten-week in-house practical training period provides an opportunity to gain knowledge, skills and hands-on experience within an organised and familiar environment. In addition the practical training helps the student engineers to appreciate the relevance of their academic studies.

Within such a programme a HPA appeared to provide the ideal basis for a creative design-build-test project. Students in the School of Mechanical and Production Engineering participate in five different in-house training modules. The design and construction of an HPA represented just one of the modules. With a second-year student population, in 1984, of 202, this meant a group size of 40/41 students working on the project during each two-week period.

3   ORGANISATION

It was considered group sizes of 40/41 were too large for one project, so a decision was taken to build two HPA. Consequently, the second-year class was divided into two teams, each consisting of 101 students. This decision had the bonus of engendering friendly competition between the teams.

The specification for each team was to design, construct and test its own HPA. The design and construction was the task of 20/21 students working on the project for each two-week period. There were obvious disadvantages with such a system, particularly the lack of continuity, but with only four members of staff involved in the supervision it was considered to be the most workable that could be arranged. To ensure liaison between groups and help improve continuity, student group leaders were appointed (by the students) for each two-week period, together with an overall leader for each aircraft.

The two teams managed to design and build an aircraft within the ten-week period. However it was inevitable that one team would perform better on the project. It was found that the Aslam team established an early lead over its rivals and this lead was extended over the course of the project. As a consequence only Aslam, the most successful of the two aircraft, is discussed within this paper.

With NTI being a new institution no large workshop was available for the construction of the HPA. Permission was given to use a corridor 6m wide and approximately 50m long situated on the third floor of the Auditorium building. This posed some interesting problems in logistics since large items, such as the wings and fuselage, had to be carried into the circle area of the Auditorium, lowered by rope into the stalls and then carried out through the main entrance.

Similarly the aircraft had to be transported from the Auditorium building to the sports field for flight testing, a distance of some 2km by road. As there were no suitable storage facilities at the sports field, it meant that the aircraft had to be transported to and from the Auditorium for each flight attempt.

This posed unique problems in that the HPA had to be capable of being broken down into units sufficiently small enough to be easily moved through the campus and yet sufficiently robust to be transported without damage.

It was decided that certain materials would be required for the aircraft, therefore because of long delivery times, these were ordered prior to the start of the project. Aluminium alloy tubing with a specification 6061-T6 was difficult to obtain at short notice in Singapore, and had to be ordered at an early stage. Similarly it was recognized that expanded polystyrene sheets would be required, these were also ordered prior to the commencement of the project in a range of thicknesses; 6, 12 and 25 mm.

4   SPECIFICATION

In order to simplify the design of the aircraft a limit was placed on the required performance, a flight length of 100m in a straight line. With such a limited performance it was assumed that the aircraft would be compact and, therefore, easier to design and construct.

Compact in the context of a HPA was defined as a single seat aircraft, in which the pilot both propelled and controlled the aircraft, with a maximum wing span of 18m.

Apart from these constraints the design specification was open-ended.

5   DESIGN

During the first two weeks the design of Aslam was completed. Design of a HPA represents a fascinating challenge because of the low power available. For a given wing section Power can be related to the mass of the aircraft and the wing area;

$$\text{Power} \propto (\text{mass})^{3/2}(\text{wing area})^{-1/2}$$

so that the design must be a compromise between minimum mass and the largest possible wing area.

The group carried out initial studies into the design and construction of previous HPA (3,4). They also performed tests to check the power output of potential pilots. Choice of pilot raised no major problem as there were a large number of very fit young men on campus, all with a mass of 55 kg or below. Tests revealed that the majority could achieve outputs of 300W for the required duration.

Design was divided into the main areas of aerodynamics, structural, propulsion and stability/control.

At the low Reynolds numbers at which HPA operate there are very few purpose designed aerofoil sections. The students decided to use the Lissaman 7769 section, developed for the 'Gossamer Condor' aircraft in 1977. It had the advantage of having a flat undersurface making it easier to cover during the construction stage. However, quantitative details of its performance were unobtainable. Its creator, Peter Lissaman, considered it had a 'good' performance compared to equivalent sections. More importantly, he was convinced it was a very 'forgiving' section, defined as constructional errors not causing a noticeable deterioration in performance. Extrapolating from the little data available gave an assumed operating point with a profile lift/drag ratio, L/D = 100 at a lift coefficient of 1. The performance of this aerofoil section has since been reviewed in more detail (5).

Structural considerations quickly convinced the students that the most important structural member of the aircraft was the primary structure of the wing. The weight of this member can be reduced by external bracing which led the group into the realm of statically indeterminate structures, a topic they had not previously discussed during class. Eventually practical considerations of transportation and simplicity of construction caused them to choose a cantilevered wing using aluminium tubing as the primary structure. The secondary structure was provided by ribs of foam plastic covered by 12 micrometre polyester sheeting.

There are advantages in both the tractor and pusher positions for the propellor. The latter has a better performance but at the expense of a more complex drive system. On this basis the students chose a tractor position in front of the pilot. The propellor was designed with a constant chord using the 7769 aerofoil section. This decision appeared sensible at the time but in retrospect a model-aircraft section would have given an improved performance.

The drive system was designed to be as simple as possible utilizing bicycle parts. Power input was by pedal crank and chain wheel through a chain to an idler shaft and then from the idler shaft by means of a crossed chain to the propeller shaft. The aim was to achieve a drive/propulsion efficiency of 80%. With a distance of 1.4m between the centre of the idler shaft and the centre of the propeller shaft, the use of a crossed chain to change direction through 90° proved to be satisfactory. However, the loads on the chains proved to be greater than expected and the bracket holding the idler shaft had to be redesigned and built several times during the course of the project, in order to withstand the loads.

Previous experience with 'Liverpuffin' had shown that by incorporating sufficient dihedral into the wing, it can be inherently stable without the need for ailerons. In passing it may be noted that the 'Daedalus' aircraft employed a similar technique.

With a limited performance objective it was decided that Aslam could fly with just one form of control, the elevator.

Although utilizing some ideas from previous aircraft the final design of Aslam was highly original, see Figure 1. Details were as follows:

Wing span	18m
Wing chord	1.64m
Aircraft mass	49.5 kg
Flying mass	105 kg
Flying velocity	7.5 ms^{-1}
Power input	300 W

6 CONSTRUCTION

Following the initial design stage which was completed by the end of the first two-week period, the remainder of the project was devoted to construction and solving detail design problems.

The students had to use their ingenuity to construct something that was well outside their previous experience. For example, certain students developed specialist techniques such as how to mould expanded polystyrene sheets into the leading edge of the wing, using warm air from a hair dryer.

Throughout the project, the emphasis was on testing parts to ensure structural integrity. A wing test section was constructed and tested by mounting the unit on a lorry which was then driven at the appropriate flight speed. Useful feedback emerged from the test relating to the buckling of structural members including information on the second moment of area and slenderness ratio. Slenderness ratio in this context means the ratio of the length to the radius of gyration of the section.

Construction of the Aslam wing raised a problem in how to prevent distortion of the outer wing ribs and the trailing edge under the tension loads induced by the covering material. The 12 micrometre polyester sheet was held in position with double-sided adhesive tape attached to the rib sections. The covering being finally tensioned by applying heat with a smoothing iron.

Care had to be exercised in solving the distortion problems; increasing the rigidity of the structure by adding too much material would result in increasing the weight by an excessive amount. In the event, attaching pieces of expanded polystyrene sheet between the outer wing rib and the adjacent inboard rib solved one problem. Buckling or distortion of the trailing edge was not excessive and accepted as having little detrimental effect on the aerodynamic performance of the wing. Figure 2, Aslam wing.

7   TESTING

By the end of the penultimate week of in-house training, Aslam was complete apart from the pilot fairing.   In reality shortage of time meant that the fairing never was incorporated in the aircraft.   However, with the low frontal area of the pilot it was considered that the drag penalty was very small and would not affect the success of the aircraft.

On completion the aircraft was transported to the sports field to be dynamically tested for the first time with the lift loads on the wing.   To check that the structure was sound the authors ran with the pilotless aircraft at speeds varying from zero to flying speed.   As a result several of the rib/aluminium tube glued joints on the starboard wing failed.   Investigation showed that the joints had been made incorrectly with a poor mix of epoxy resin, pointing to the need for strict quality control on a project of this nature where the structural integrity is of vital importance.

Repairs required the first few days of the final week and the only other time that the aircraft could be taken to the flying field was on the final Thursday of the project.

A runway of 6mm thick plywood was laid out on the soccer field.   Wind conditions were ideal with strength of force 2.   During the take-off run the aircraft accelerated until after 20m the rear wheel lifted off the ground.   If the pilot had then actuated the elevator control so that the tail came down the aircraft would have lifted off.

In the event, with the rear wheel off the ground, the angle of incidence of the wing was reduced, reducing lift from the wing and preventing take-off. The aircraft ran to the end of the runway and the front wheel embedded itself in the soft ground of the soccer pitch causing the whole aircraft to overturn, what has since become known in HPA terminology as a 'nose over'.

The resulting damage could not be repaired in the time remaining to the project and had to await a resumption in 1985.

8   DEVELOPMENT

The Aslam project was continued during the in-house training programme in 1985, but as one of the several projects within the School of Mechanical and Production Engineering.   As a consequencce the project ran for six weeks only, with a smaller group of eight students working on the aircraft during each two-week period.

The size of group allowed for improved supervision relative to the much larger group of the previous year.   The opportunity was taken to build an entirely new wing, and to increase the span to 18.5m.   Although the basic configuration was fixed the students still used their ingenuity in both the detail design and improved production techniques.   For example the sheet polystyrene leading edge was changed from an open section to a D section, making for a much more rigid structure.   Inevitably construction took far longer than anticipated so that it was not until the Monday of the sixth and final week of the project that Aslam was fully assembled.   The remainder of the week was spent awaiting suitable opportunities to carry out taxiing/flight trails.   In

C377/280

Design education for a developing country

L DROSZCZ, MSc, MNOT(Poland) and **I HERNANDEZ**, PhD, MASME, MANIAC
Department of Mechanical Engineering, Instituto Technologico de Celaya, Mexico
J VALDES-NERI, MSc, MASME, MSME
ACEROS, SA, La Loma, Tlalnepantla, Mexico

This article explores the educational causes for the low output of qualified designers in a developing country. The general characteristics -
of Mechanical Engineering Teaching in Mexico are presented and the causes for unsatisfactory results are sought. The time devoted to mechanical design teaching is compared between several universities in Mexico.
The attributes of design education are also compared in several coun---
tries and finally an adecuate teaching model, according to the authors,
is presented for Mexico. Although it is based in the Mexican experience
the diagnosis can be applied to other developing countries.

INTRODUCTION

The causes for underdevelopment are varied. Some third world countries try to
develop "high technology" products to enter the international market. However-
the success is almost nil. One of the main reasons is a poor basic (traditio--
nal or mature) industry. It is believed that if a country invests in mechanical engineering related industries (among other disciplines) the net output --
will be greater as compared with other areas. The point in this approach is --
the ability that mechanical engineers can develop in design and manufacture as
a key factor for innovation and quality of products in the precision, medium -
and heavy engineering fields.

DESIGN IN THE CONTEXT OF ENGINEERING

To date, in public and private universities, the traditional careers in engi--
neering are offered: civil, chemical, mechanical, electrical, electronical and
industrial. From this base, there are ramifications: e.g. bio and nuclear engi_
neering. The present work is concerned with mechanical engineering for which -
85 institutions offered the BS degree in 1986 (1).
The National Bureau of Science and Technology (CONACYT) defines ME (2) as: ---
(the goal of ME is) "...the design (or project) of machinery and all the con--
ducting phases to the manufacturing and operation of the same in the more effi_
cient, safe and economical way. Although the center of these activities is the
mechanical designer...".
The Technological Institute of Monterrey (3) requires from its graduates to --
have "abilities that will enable him to approach easily the problems present -
when designing or selecting equipment for electromechanical systems".
Other universities (4,5) reffer to design as a main characteristic of a mechanical enginner. However, only one or two stress the importance of design with
a specialty module at the undergraduate level. There are graduate programs in
machine design in several universities.

The output of this educational system is the formation of mechanical engineers which are supposed to be well versed on mechanical, electrical, manufacture, - energy and management engineering, all at the same time, or at least, the ---- actual curriculum points to that direction. An ME graduate student will likely pursue a graduate program in machine design, industrial engineering or management.

The curriculum in ME has not changed very much in the last decade. In some a--reas, the changes in the state of the art have been incorporated: e.g. finite element method and CAD. However not in design as a discipline. This state of - facts is probably the result of a development model chosen 30 or 40 years ago which permitted internal growth at the expense of technology dependance . There fore, the curriculum available at present is adecuate for the mechanical engineering demand: to run and preserve equipment bought from other countries.

MECHANICAL ENGINEERING: WHY AND WHAT FOR

The development of a country requires efforts in different areas, with limited financial and human resources. The results must be evaluated technologically - and productively in limited time units. Which are priority areas?, is design among them?. One methodology for development planning (6), uses the concept of technological front to measure the desired level of development. A possible -- set of technological fronts is shown in table 1. One step is to assign priorities. The next will be to decide the desired level of development for the next unit of time. Finally, a crossed effects analysis can reveal the influence --- each front has over the others. The result shows a great dependance on mecha--nical engineering technology. That is, if a country invests in the development of the mechanical disciplines, other areas will receive benefits. This proposi tion has been discussed -surprisingly enough- in developed countries (7,8).

This type of programs are government promoted (9). For the so called metal-me-chanical area (see table 1) the following objectives are sought in the pre----sent:

1. Development of technological capacity on design
2. Development of standards
3. Quality assurance
4. Development of manufacturing and metallurgical technology
5. Assimilation of technology
6. Development and training of human resources
7. Rational use of national resources
8. Development of technology enterprises
9. Development of technological capacity on automatization systems

Documents like those cited above (6 & 9), depict clearly a main goal: deve--lopment of technological capability on design. On the other hand, it would be virtually imposible (with the teaching resources available) to train effective mechanical engineers for each and all the technological fronts. A division by branches could be used. One can be: 1) general design, 2) energy, 3) manufacture, 4) heavy machinery, 5) precision engineering. For this or any other proposal it must be remembered that design is the basic component of any engi---neering activity. Also, depending on the country's interest on developing --- shipyards, the naval engineering specialty could be created later on. There is a solid foundation to develop areas 1,2 and 3 readily. 4 and 5 can be offered as specific graduate programs as a start.

THE ACTUAL CURRICULUM

Table 2 shows a comparison of the relative weight of different disciplines in the mechanical engineering curriculum from the average of six important engi-- neering schools in Mexico.

All universities and institutes include courses like machine elements and/or - machine design. These can be classified in one of three labels: as an exten--- sion of the strenght of materials courses, as an extension of the solid mecha- nics courses or some sort of engineering science combining both of these as--- pects. Drawing courses (also included in table 2 as design courses) are offe-- red in one or two terms and sometimes are imparted by architects and therefore drawing courses are mechanical drawing courses. The industrial exposure during a five year period is close to zero except for some visits to local factories. The result is a high distrust towards engineers with a BS which the industry - hires with low initial salary and usually think in terms of a ten year period for an in-house training before the engineer is considered to be technically - productive.

A recent study (10) calls for an urgent need for mechanical engineers at the - graduate level. This is a response from industry which finds no qualified engi neers and leaves the cost of further training to the state and private univer- sities. Unfortunatelly, most of the time, at the graduate level, the students get more engineering science. The fundamental causes according to this study - are:

1 The main problem is the lack of technological research and development in ME
2 Few personnel with MS or PhD degree
3 Although there is a high academic level, professors show very little knowledge of the problems that the industry confronts.
4 Poor facilities (tool shop, library and computational capacity)
5 Low graduation output (69% BS, 35% MS)

It is very sugestive that the requirement to enter a graduate program in all - universities is either an examination or one semester of review courses. Most of the students take the second alternative.

DESIGN TEACHING: THE REST OF THE WORLD

A comparison to the time devoted to design teaching in ME in other countries - is presented in table 3. The paper by Hongo (11) has been used as a basis. The data presented in table 3 has been sumarized in averages for Europe, Canada, - USA, and Japan. A column has been added to include Mexico as a developing coun try representative. Table 3 shows the amount of hours for each subject of de-- sign teaching and the total amount of hours dedicated to design. Mexico has -- the least amount of hours dedicated to the "practical side of design". Mean--- while, the total time dedicated to design is comparable to other universities except for those in Europe. The analysis presented above can be applied to all engineering disciplines with design as a common base. Then, design becomes a - major priority as a discipline (12,13). Table 3 contains a very importantmessa ge: design must be practice oriented in Mexico. This conclusion is coincident with the concept of design (14).

MODELS OF DESIGN EDUCATION

In general, it can be said that there are two different models of design tea-- ching which are illustrated in Figure 1. Model 1 considers that the university gives only theoretical background and the practical training is obtained -----

through professional work. Model 2 is that in which the student is transformed into a designer with great influence of the university.
There are variations arround these two very general and rough models and it -- can be said that their development corresponds to social and economical fac--- tors within the countries where they first appeared. As an example: the "engi- neering science" (15) tendency developed in USA. This approach will match well with model 1 and could be convenient for that country's industry, where the -- companies hire new engineers well developed in theory. Then, training will be- gin in R&D, design and manufacture departments that have high tradition and ex perience. So, by model 1 education, the engineers are trained in comparatively long periods of time. By the education directed to the "design engineering" -- approach of model 2, the recently graduated engineers are formed in few years as designers or specialists.

DESIGN: THE CURRICULUM

"The designer has to be a man of many parts" (16). "(One characteristic of the designer is...) the specialized, exigent technical and comercial work devoted to the elaboration of non existing structures" (17).
If one is to implement an educational program on design, the first thing to do is to find out what design is and the profile required for a designer, from -- the practicing designers. A study of this sort (18) concludes that a designer must be formed on four basic areas: methods, knowledge, practice and group dy- namics. Based on these areas, the next step is to determine wether an existing educational model can be used to fit the goals of a developing country. The -- answer is no. Mexico, like many developing countries must determine its own -- educational model without reproducing irreflexively foreing models: what works in a developed country, not necessarily is adecuate for countries in a develop ment process. To reduce the non equilibrium between the offer from the acade-- mic institutions and the demand of qualified personnel from industry, it is -- reasonable to create a model based on practice and adapt it to national condi- tions. The goals of this educational model of design for a developing country are to:

1 develop design engineers in a 5 year school period to enter industry in a productive way.
2 teach design based on a developed and specific practical base
3 prepare design engineers to enter not only large, established corporations but mainly the medium and small industry
4 adapt education to the computational facilities available locally
5 make the potential designer aware of the limitations of its context and at the same time of the exiting career future and responsabilities.

An educational model based on the four areas mentioned before should be pla--- nned on the following premises:

KNOWLEDGE: Increase the number of design oriented courses and distribute them evenly during the five year period.

METHODS: Teach the design process gradually and in an iterative fashion, clo sely related with the courses in progress and with illustration of an artifact product or machine currently in the market (or failure!). Besides the general process of design, specific methods should be thaught according to the stage of of the process in study, and again, closely related to a product.

GROUP
DYNAMICS: Encourage group work and change these groups at least each semester

Give feedback and theory in one or two humanities courses.

PRACTICE: The substitute for case studies is the analysis of existing products (a Japanese practice, ref. 13). The meaning of practice must be taken to the optimum, therefore the study of a mechanical device is proposed to be covered in every semester. This means having the device phisically and in a reasonable number to be disassembled, measured, tested, drafted, etc.

ACTIONS FOR ACADEMY AND INDUSTRY

This last section is the case of the egg or the henn. Who is responsible for - the disadvantage, industry or academy?, non or both?. Anyway, a solution must be sought.

For the academy are:

1 Implement programs that emphasize "practical skills".
2 Make the drafting, manufacturing and model shop areas, the core of the tea-- ching facilities to work as a modern design center.
3 Arrange programs with industry to receive students during at least two su--- mmer vacational periods before their terminal project.
4 Separate the engineering design career from the theoretical mechanical engi- neering (which is also needed!).
5 Establish at least two institutions offering this specialty with a higher -- concentration of specialists.
6 Open specific graduate programs on heavy and precision engineering.
7 Select, translate and promote more adecuate texts at accesible prices.
8 Open continuing education courses for engineers in industry to be trained in methods and knowledge and share their experiences.
9 Bring in practicing specialists as part time professors. One for each area of specialty (gears, bearings, couplings,etc.) to ease the shortage of de--- sign academicians.
10 Work with industry to work out specific problems of the later.

The consecuences for the lack of practice when teaching design for the new gra duate are:

1 Low performance in applying theoretical knowledge to technological problems.
2 Low contact with reality: information sources, decision making, human rela- tions and creative and motivating work.
3 Low income for new graduates.

The actions for industry are: to

1 Open design departments even if they are small firms (usual practice is to - allocate 2% of sales to a venture of this sort).
2 Contact universitiy's design centers (or departments) to work together in - specific projects.
3 Install parallel career paths for technical positions as compared with mana- gement ones, to promote technical excellence (equal remuneration)
4 Take a radical decision about the practice of standards (e.g. to adopt fully DIN, ANSI or ISO standards).
5 Prepare an adecuate 'introduction program' for newly graduates in design.
6 Allocate funds for continuing education

The consecuences for industry for receiving engineers with a lack of practice are:

1 Long training periods after graduation.

2 High cost of design mistakes in professional work.
3 Slow and limited development
4 Continuing dependance on foreing technology.

CLOSURE

Not everything is bad in developing countries. Many of the forementioned ac---
tions are being taken (19,20). There is a lot to be done in this respect due -
to the great need from industry of qualified designers. That is why proposals
like the ones from Bonsiepe (21) are trully valid: "...to leave the didactic -
methods which award memorization in detriment of a productive assimilation of
what has been learned..." and "...to leave the honors of academic excellence".
"What is needed are not more academicians but more industrial designers who --
know how to carry out industrial projects".

(Note: In this paper, the word HE is equivalent to SHE with all due respect).

REFERENCES

1 CATALOGO DE CARRERAS, LICENCIATURA 1986
   (Región norte, centro occidente y área metropolitana)
   ANUIES, México 1986

2 BOLETIN DE ESTUDIOS DE ESPECIALIZACION Y GRADO EN INGENIERIA MECANICA
   CONACYT 1986, México

3 BOLETIN INFORMATIVO PARA LA CARRERA DE INGENIERO MECANICO ELECTRICISTA
   Instituto Tecnológico y de estudios superiores de Monterrey, 1987 México

4 GUIA DE CARRERAS EN INGENIERIA
   UNAM 1987, México

5 BOLETIN INFORMATIVO PARA LA LICENCIATURA DE INGENIERO MECANICO ELECTRICISTA
   Universidad anáhuac, 1987 México

6 HACIA UNA METODOLOGIA DE PLANEACION DE DESARROLLO TECNOLOGICO Y PRODUCTIVO
   Mario Waissbluth, Articulación Tecnológica y Productiva
   Centro para la innovación tecnológica
   UNAM, 1986 México

7 THE MYTH OF A POST INDUSTRIAL ECONOMY
   Stephen S. Cohen & John Zysman
   Technology Review, March 1987

8 HOW TO KEEP MATURE INDUSTRIES INNOVATIVE
   C. F. Sabel et. al.
   Technology Review, April 1987

9 PROGRAMA NACIONAL DE DESARROLLO TECNOLOGICO Y CIENTIFICO 84-88
   Poder Ejecutivo Federal
   CONACYT 1984, México

10 NECESIDAD APREMIANTE: LA FORMACION DE INGENIEROS MECANICOS A NIVEL POSTGRADO
   Comité Técnico asesor para la formación de recursos humanos en Ingeniería mecánica.
   Ciencia y Desarrollo, Abril 1987.

11 A SURVEY OF MECHANICAL ENGINEERING DESIGN EDUCATION IN UNIVERSITIES IN JAPAN IN
   COMPARISON WITH EXAMPLES FROM EUROPE AND NORTH AMERICA
   Kaoru Mongo
   ICED 1987, Boston MA

12 GOALS AND PRIORITIES FOR RESEARCH ON DESIGN THEORY AND METHODOLOGY
   ASME, 1987

13 SYSTEMATIC APPROACH TO THE DESIGN OF TECHNICAL SYSTEMS AND PRODUCTS
   VDI Guideline 2221
   Verein Deutscher Ingenieure 1987

14 LA IMPORTANCIA DEL DISEÑO EN LA EDUCACION DE INGENIEROS Y EN EL DESARROLLO TEC---
   NOLOGICO DEL PAIS
   Cyril Semenov
   2º Congreso de la Academia Nacional de Ingeniería
   Monterrey NL 1976, México

15 WHY WE NEED HANDS ON ENGINEERING EDUCATION
   Arnold D. Kerr & R. Byron Pipes
   Technology Review, October 1987

16 ENGINEERING DESIGN
   G. Pahl & W. Beitz
   Springer Verlag 1984

17 THEORIE DER KONSTRUKTIONSPROZESSE
   V. Hubka
   Springer Verlag 1976

18 MEMORIA DEL PROGRAMA EN DISEÑO MECANICO
   A. Pita et al.
   Centro de investigación y desarrollo VITRO
   MONTERREY NL 1981, México

19 PROYECTO DE ESTANCIAS INDUSTRIALES EN DISEÑO MECANICO
   CONACYT 1986, México

20 PROGRAMA DE MAESTRIA EN DISEÑO MECANICO
   Instituto Tecnológico de Celaya
   México 1986

21 EL DISEÑO DE LA PERIFERIA
   G. Bonsiepe
   G. Gilly, Barcelona 1985.

TABLE 1   POSSIBLE TECHNOLOGICAL FRONTS

1 Basic sciences & engineering	14 Hydraulic resources
2 Social sciences & education	15 Agriculture
3 Geophysical sciences	16 Cattle technology
4 Industrial engineering	17 Forest technology
5 Programming	18 Fishing technology
6 Health	19 Nuclear energy
7 Environmental & sanitary engineering	20 Thermomechanical energy
8 Costruction materials	21 Chemical industry
9 Management	22 Agro industry
10 Housing & construction technology	23 Electronical components
11 Consumer goods	24 Electronical equip. & components
12 Mineral resources extraction	25 Metal-mechanical products
13 Metallurgy	26 Machinery and equipment

TABLE 2   RELATIVE WEIIGHT OF DIFFERENT DISCIPLINES IN MECHANICAL ENGINEERING
          FROM THE AVERAGES OF SIX IMPORTANT UNIVERSITIES IN MEXICO

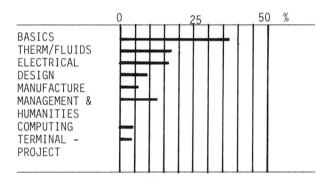

FIGURE 1   GENERAL MODELS OF DESIGN EDUCATION

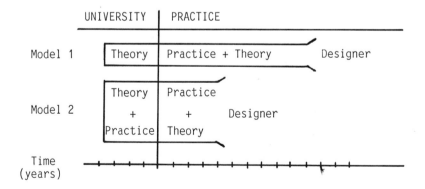

TABLE 3

COMPARISON OF TIME DEDICATED TO --- DESIGN TEACHING IN DIFFERENT UNIVER_SITIES. BASED ON (13).

SUBJECT	HOURS				
	E[1]	C[2]	A[3]	J[4]	M[5]
MECHANICAL DRAWING	70	120	192		
MACHINE ELEMENTS	104	74	36	198[6]	170[8]
DESIGN EXCERCISE	115	0	0	67	30
DESIGN PROJECT	337	63	96	0	0
DESIGN THEORY	133	67	0	0	0
TOTAL DESIGN	767	387	324	265	210[10]
CAD	64	56	48	25	11

1) Europe: average of 9 universi--ties
2) Canada: average of 2 universi--ties
3) USA: Kansas State University. Dept. of mechanical engineering
4) Japan: average of 19 technolo--gical institutes
5) México: average of 19 technolo-gical institutes
6) Both subjects
7) Technical drawing is given in a theoretical fashion with no me-chanical drawing.
8) Named "design theory" based on texts like Shigley's*. Practica lly an extension of strenght of materials courses.
9) Same as 8
10) Almost 100% of teaching is the-ory.
11) Carried out appart from the u--sual activities.

* MECHANICAL ENGINEERING DESIGN
  J.E. Shigley
  McGraw-Hill 1972

C377/101

Design engineers from industry as university tutors in design

J E L SIMMONS, BSc, PhD, CEng, MIMechE, MIProdE
School of Engineering and Applied Science, University of Durham

SYNOPSIS   This paper describes the direct industrial involvement in student design activities which has been taking place at Durham University for more than twenty years. Practising Design Engineers from industry serve as part-time members of University staff, supervising and marking student projects. The approach adopted is seen as having considerable benefits for all those involved; the students, the Design Tutors and their employers, and the University. A key element in the success of the system is that the Design Tutors receive a fee for the work they do. They thus have a formal relationship with the University and a professional commitment to the design programme which is described.

1.   INTRODUCTION

British Polytechnics and Universities involved in engineering education are being encouraged to enhance the quality of the design teaching they offer by forming stronger industrial links. The purpose of this paper is to describe one model for direct industrial involvement in student design activities which has been in use at Durham University for more than twenty years. In this model practising design engineers from industry serve as part- time members of university staff, supervising and marking student projects. This approach is seen as having benefits for all three parties involved; the students, the industrial design tutors and the full-time university staff.

The Durham system is just one way of linking engineering design education with industrial input and perhaps it will not be appropriate or even possible in other situations. Nevertheless there may be courses where an adaptation of the procedure described here is a useful way forward. In any event it is hoped that the issues considered will be of interest to those concerned with teaching design to student engineers.

2.   ENGINEERING IN DURHAM

The present course at Durham was established in 1965 as a unified programme for the first five terms covering the fundamentals of engineering science in mechanical, electrical and civil engineering, followed by a degree of specialisation in the remaining four terms.   As the course has evolved over the years the number of options in the second part of the course have increased to include for example electronics and manufacturing engineering. Nevertheless the fundamental structure has remained unaltered with an emphasis on broad based education and the need for future engineers to be able to transcend conventional discipline boundaries.

The Durham engineering course attracts an annual entry of 75 (shortly to rise to 95) generally in the 18 to 20 years old age bracket.   All are high "A" level achievers and up to a third will have spent time in industry after leaving school and before coming to university.

3.   DESIGN IN THE DURHAM COURSE

Right from the start, design has been seen to be central to engineering in Durham.   Specific design activities, listed in Table 1, take place in each part of the course.   A first year design, build and test (usually to destruc- tion) project is followed by design projects in both second and third years. In the final, third year of the course engineering students also carry out a major engineering project which often includes a substantial amount of design.

Table 1   Design Content in Engineering and Engineering/Management Degrees

TIMETABLED SESSIONS	
YEAR 1	
* DRAWING	18 hours practical
* CAD	4 hours lectures + 6 hours practical
* DESIGN, BUILD & TEST PROJECT	18 hours practical
YEAR 2	
* INTRODUCTION TO DESIGN	18 hours lectures and workshops
* DESIGN PROJECT	18 hours
* PROJECT PRESENTATIONS	2 hours
YEAR 3	
* DESIGN PROJECT	18 hours

Project work in design is supplemented by the taught courses indicated and by the design parts of other courses which range from the design of micro- processor systems, to the design parts of the materials engineering courses, to parts of courses in structural engineering.

First year work and final year major engineering projects are supervised by permanent members of the academic staff although in the latter cases there is frequently considerable industrial contact. The second and third year design projects, by contrast, are supervised principally by engineers from industry and it is this arrangement which is the subject of further discussion here.

4. DESIGN PROJECTS

The design projects follow a similar pattern in both second and third years. The students work in pairs although each must submit an independent final report. Drawings accompanying the reports may be shared. Projects are selected from a broad list of titles such as the recent examples given by Table 2. The list of projects is compiled from briefs provided by the various industrial design tutors and by a smaller number of full-time members of the university teaching staff who also act as tutors in design.

The sole restriction for students in project choice is that each industrial tutor can only accept a quota of five student pairs. Internal members of staff normally take on two student pairs at any one time. More than one pair of students may opt for a particular project although each pair must then operate completely independently.

The course arrangements are such that second and third year projects are carried out in different terms of the academic year, thus avoiding a clash in tutoring commitment. The third year project takes place in the first, Autumn term of the year, whereas the second year project is done in the term after Christmas. In both cases two hour design sessions are scheduled for each week of the term during which the project teams are seen individually by their design tutors and guided through the design process. Each pair of students needs to put in substantial time outside the scheduled sessions. The amount of this extra time is difficult to estimate. However, it would not normally be less than the total timetabled hours given in Table 1 and in many cases is considerably more. Reports are required by the end of the term in which the work is done and are marked during the following vacation.

In the case of the second year students, design sessions continue in the third term of the year, following the term in which they have done their project work. These third term sessions are given over to detailed feedback on the project reports which the tutors provide for their students and to live presentations to groups of fellow students and design tutors.

Table 2

<u>SECOND YEAR 2H DESIGN PROJECTS - SPRING TERM 1988</u>

<u>Tutor A</u>

1. Scanning System for Isolation Amplifiers
2. Non-Intrusive Flow Meter
3. A Dynamic Simulator System

<u>Tutor B</u>

4. Gantry System for an Overhead Robot
5. Fail "Safe" Pneumatically Operated Gripper System
6. Software for a Pick and Place Robot used in Electronic Assemblies
7. Redesign of an Existing Industrial Robot to Reduce the Manufacturing Costs by 50%

<u>Tutor C</u>

8. Warp Patterning Mechanism for Intermediate Technology
9. Automatic Sample Making Machine for Carpets
10. Mechanical Engineering Product of Your Own Choice

<u>Tutor D</u>

11. Highway Design
12. Golf Club Footbridge

<u>Tutor E</u>

13. Directional Antenna Controller
14. 1 M bit PROM Programmer
15. Telephone Switching System

<u>Tutor F</u>

16. Cantilevered Canopy
17. Factory Unit on Filled Site
18. New Pitched Roof over Sports Hall
19. Temporary Bridge

<u>Tutor G</u>

20. Guidance System for Automated Guided Vehicles (AGV's)
21. Vehicles for the Elderly and Disabled

<u>Tutor H</u>

22. Spin Table for Metrology

<u>Tutor I</u>

23. Reinforced Concrete Silo
24. Reinforced Concrete Chimney

5. DESIGN TUTORS

The design tutors from industry are all practising engineering designers working in North East England. They are employed by a range of international, national and local companies and cover a wide range of engineering disciplines. The tutors are released by their employers for the weekly two hour design sessions plus travelling time between their normal place of work and the University. These student design sessions take place twenty-two weeks in each year.

In the University the design tutors are officially part-time members of staff and they receive a fee for the work they do. This fact of payment is important because it underlines the seriousness with which the University views the industrial tutors' role. The tutors, for their part, have a formal relationship with the University and a professional commitment to the design project programme described above.

6. ADVANTAGES, PROBLEMS, DEVELOPMENTS

In order to be successful a system such as that described here needs to have something in it for all the parties involved; the tutors and their employers, the students and the University. The overwhelming perceived advantage from an educational point of view is that the students are exposed to real and often current design problems from industry and they have the opportunity to meet and work with engineers for whom design is their principal work. Thus design projects have at best a credibility which it is difficult for members of academic staff to replicate. Moreover, precisely because the design tutors are not the students' usual teaching staff, they provide a change which many students find stimulating and a professional model which many others find helpful.

The tutors themselves enjoy the contact with fresh minds and the opportunity to share some current and recent problems outside their usual work context. Some definitely see personal contact with students during their degree courses as an opportunity to identify potential recruits for their organisations and claim benefits also from their informal contacts which arise naturally with the full-time academic staff.

From the point of view of the University, having external design tutors is a highly effective way of including a rich experience of current engineering applications within the overall educational programme. It is also the case that full-time members of staff enjoy and are stimulated by the presence among them of their part-time colleagues from industry.

Although the advantages of having external design tutors in the way described here are considerable, it is important to note that there are some areas of possible difficulty which may require special attention. One problem is the restricted amount of time tutors can give each student pair. At present this is twenty to twenty-five minutes per week. In some cases and in some parts of the design cycle this may be enough, but in others it may well not be sufficient to allow for development and proper discussion of the students' own ideas. Recruitment of more external design tutors would allow a reduction of individual loadings and steps to do this are in hand. Durham experience over the years is that finding suitable tutors in local companies which are themselves prepared to release the people in question to attend at the University on a regular basis is possible. In no sense, however, is it easy and it always takes a considerable amount of time to arrange.

Another potential problem is that of ensuring a sufficiently uniform standard of project specification and design task when these are compiled by a group of people outside the University. There is a similar difficulty concerned with providing a uniform standard of marking. At present Durham design reports are marked by the project supervisor alone on working to a reasonably well defined scale. So long as the design mark only contributes a small proportion to a student's final total for degree awarding purposes this is probably a reasonable way of doing things. Planned course developments envisage more importance being given to the role of design, to the extent that in the majority of course options the third year design project will be equivalent to half a written finals exam. This change may make it necessary to move to second marking of project reports.

A development to be followed in the 1989/1990 academic year by students taking civil engineering options in the final year is an enhanced design lasting two terms and equivalent to a complete finals paper. Students will do these projects on an individual basis supervised by a full- time member of staff but with weekly consultative access to one of the external tutors. It will be interesting to see how this change works out in practise although the consequential increases in staff loading make it unlikely that such two term projects will be available across the range of final year options.

7. CONCLUSION

A recent survey of opinion from all those concerned including students in the Durham arrangements for design teaching has revealed widespread support to the system and a conviction that the advantages all round far outweigh any associated difficulties. Nevertheless, like any system, there is room for improvement and changes possibly in ways which have been described here.

8. ACKNOWLEDGEMENT

The author is very grateful to his fellow members of academic staff and to the external design tutors for their advice in preparing this paper. Greatest credit should go to the founding head of the then Department of Engineering Science in Durham, Professor R Hoyle, who established the system of external design tutors more than twenty years ago.

An integrated and networked approach to design, technology transfer and engineering design education

D O ANDERSON, PhD, M R CORLEY, PhD, C R FRIEDRICH, PhD and
R O WARRINGTON, PhD
Department of Mechanical and Industrial Engineering, Louisiana Tech University, Ruston, USA

The mechanical engineering departments of two major technical universities have established digital and video links to strengthen their educational, research and industrial development missions.

1 INTRODUCTION

This paper describes recent activities in the mechanical engineering departments of the two leading technical universities in the state of Louisiana (USA). Louisiana Tech University (LTU) in the north central region of the state and Louisiana State University (LSU) in the southeastern region have set out to develop a fully integrated and networked environment to enhance the design and manufacturing education of our students. An equally important goal is to increase our involvement with industries located in the state.

It is widely recognized that the challenge of greater productivity growth is of critical importance to the future of the United States. At the White House Conference on Productivity (1), held September 22-23, 1983, former President Reagan urged the American people to find strategies and methods to regain a competitive position in world markets. At the state level, Louisiana's over dependence on oil and gas, coupled with mismanagement by previous administrations, has left the state with a one billion dollar deficit at the beginning of the 1988 fiscal year. The new Louisiana state administration, led by Governor Roemer, has made a commitment to the strengthening of the state's economy, with improved education and development in the manufacturing sector given high priorities. Fig. 1 shows the strategic locations of the two universities in the natural resource-rich USA Deep South. Also, some of the major industries of the area served by LTU and LSU are shown in the figure.

The current integrated efforts include an innovative (and possibly unique) mechanical engineering departmental digital/video network linking two major state universities, a manufacturing research center at LTU, an Energy Audit and Diagnostic Center at LTU, and the Center for Advanced Microstructures and Devices (CAMD) at LSU. The relationships between the two universities and the various industry interactions are depicted schematically in Fig. 2. The remain-der of the paper discusses how each of these areas influences our design education and our relationships with industry.

2 THE DEPARTMENTAL NETWORK BETWEEN LTU AND LSU

Louisiana Tech University (LTU) and Louisiana State University (LSU) have proposed to establish an Ethernet network within each of the institutions and connect them using a high speed data link (2). These two networks are termed MEnet.LTU and MEnet.LSU, and the communication path between the two institutions is via a leased land telecommunications line operating at 9.6 Kilobaud and initially running DECnet network communications software. This will be upgraded to the full TCP/IP protocol or to the full OSI standard as soon as possible.

This computing infrastructure, combined with the new Ethernet campus-wide network and other networks and gateways into the new supercomputer systems being installed in SNCC, will assist LSU and LTU in their efforts to be competitive in current fields of endeavor and to venture into new areas. Academics will improve and grantsmanship will prosper with the new publishing tools proposed here.

A central file and compute server (VAX) and its software, when combined with a high speed Ethernet network connecting faculty, staff, and student offices, provides a commonality in computing that allows faculty from each University to share ideas, software, proposals, and other material with the greatest ease. One of the major advantages of the proposed system lies in the commonality of the software installed on each system. Because the operating system (VAX/VMS) and network system (DECnet/Ethernet) chosen are so well integrated, the resulting network immediately lends itself to faculty and students commu-nicating with peers at other universities which leads directly to national and international collaboration. The transfer of technology between LTU and LSU will benefit industries in both the southern and northern regions of the state.

The LTU network will employ DECnet and TCP/IP (and OSI as it becomes available) protocols on an Ethernet network which will support the VMS, UNIX, MS-DOS, OS-2 and Apple Macintosh operating systems. It will be developed as a multivendor network with intelligent workstations, shared file storage and shared computa-tional resources. DEC VAX computers, running the VMS operation system, will provide file server and remote computing capabilities as well as a common environment for interaction with LSU. Additional professional workstations will be acquired so as to provide a range of capabilities from computer animation to desktop publishing.

A critical decision was made to use distributed computing instead of centralized computing. There are, of course, advantages to both types of computing and the advantages of both will be discussed. A centralized system has the following advantages:

(a) A single point of software maintenance.

(b) A single location for hardware maintenance.

(c) Large disk space for BIG programs.

(d) A single software license for each installed product is sufficient to provide every user with access to all software.

(e) The ability to run large compute-bound BATCH jobs at off-hours which cannot be effectively run on smaller systems.

(f) Absolutely common environment for all concerned.

The advantages of distributed computing are also many.

(a) Performance is independent of the number of users.

(b) Each user's system can (if desired) be enhanced to meet special needs. If one user absolutely MUST have a large disk, one can be added to the specific system.

(c) A small centralized computer can still serve as the common computing environment.

(d) Additional computing power can be acquired as needed in smaller monetary increments to meet the demands arising from added users and from the need for more powerful processors.

(e) Each user maintains his own disks and provides whatever security is deemed necessary. Software upgrades can be accomplished remotely over the Ethernet network.

(f) As faculty members replace their systems, their older units may be moved into research or academic laboratories for use by students.

The challenge at both institutions is to meet diverse needs and to use existing resources at each institution in a common manner. At LSU, the existing student workstations include eighteen VAXstations for high performance interactive graphics instruction, sixteen Zenith PC's and five terminals produced by various manufacturers. The faculty and staff are currently using three VAXstations, ten IBM or IBM-compatible PC's and five Apple Macintoshes. LTU has no VAX-station systems, but does have 52 IBM-compatible microcomputers directly available to students for instruction, sixteen microcomputers in faculty offices and twelve microcomputers dedicated to research.

One of the prime considerations in the formulation of the proposed plan is the goal of providing students with extensive access to interactive computing. This goal includes the following interactive environments:

(a) Providing the quality and performance of workstations the students will use in the workplace when they graduate.

(b) Providing access to workstation, minicomputer, and mainframe computing resources and to library information services for both undergraduate and graduate students at both institutions.

(c) Supporting a centralized, electronic MAIL facility allowing students to submit questions and assignments electronically, and providing the faculty with a mechanism to review, correct, comment, and return the student's work in the same manner. System administration will be developed to allow each registered student to have a permanent account on the system. This will allow them to retain any utilities they develop and any special tools they learn for semester after semester, thereby improving their efficiency and effectiveness.

Under the proposed project, the personal computers at LTU can be easily attached to the departmental network by installing a network adapter board in each machine. With the proper network adapter, these personal computers could serve as "diskless" workstations where all software is loaded from the network (including the operating system), or as disk-based workstations which can function independent of the network. The machines could also serve as terminals to the host system, allowing access to much more powerful computer resources. The proposed link will facilitate technology transfer to industries throughout the region.

3 THE LTU-LSU VIDEO LINK AND GRADUATE OFFERINGS

The need for statewide cooperation and minimal program duplication among the major educational institutions is the driving force behind a video link between Louisiana Tech University and Louisiana State University. This link would allow course offerings not only between the two schools but also to the public and industries. The industrial base of northern Louisiana is primarily the wood products industries. To have access to the natural resources, these industries have located in rural areas which may be some distance from any metropolitan area or university. These islands of technology are continually requesting in-house graduate programs. Louisiana Tech currently has a graduate program at Barksdale Air Force Base and through Tech-Bossier. Both of these programs are located near Shreveport, 97 km west of the main campus. Faculty must commute the 194 km round trip to teach at these facilities. An alternate telecommunications system has been proposed which would join Tech with LSU and would allow graduate offerings without the need for faculty travel.

The two possible modes of video and audio communications are point to point microwave transmission and satellite relay. A nearby state, Oklahoma, has been using microwave transmission since 1971. The system consists of a network of State owned and operated microwave relay facilities. The network has three switching centers which use personal computers to route the various transmissions into and out of the centers. In essence, Oklahoma has established a digital microwave switching system similar to a telephone system. At present the video system broadcasts approximately 76 hours of instructional programming weekly. The cost of the original system was seven million dollars less operating expenses. Oklahoma is currently upgrading portions of the system with fiber optic cable to allow both video and digital data communication. This upgrade will cost four million dollars. Administrators and technicians from the Oklahoma system have suggested that it would be more economical and flexible to begin a new system by taking advantage of satellite relay technology.

Louisiana Tech University has investigated the cost of operating a similar system which will be satellite based (Fig. 3). A three-camera studio, video uplink transmitter, and associated technical hardware would now cost about $400,000. The cost of such equipment is continually decreasing while the capability is increasing. A three credit hour graduate course would cost about $12,000 in satellite transponder rental per academic quarter. As more satellites become available the cost may drop from the current $400 per hour rate to the rate which prevailed in the earlier 1980's, that of $150 per hour. A typical downlink site would need to invest approximately $5,000 for a monitor based system or $15,000 for a video projection based system.

If two graduate courses are offered per quarter and enrollment is 20 students per class offering, with a 10% interest rate, the system will break even in approximately eleven years. If the graduate offerings are raised to three courses per quarter with twenty students per class, the break-even point is reduced to three years. This assumes that both LTU and LSU install similar uplink facilities. These course offerings would not be just in engineering but also in business, management, education, and general services. One paper manufacturer, Boise-Cascade in DeRidder, Louisiana, has expressed the capability to provide forty graduate students to a master's level business and management program. Louisiana Tech has recently added the Master's level program in Manufacturing Systems Engineering and will introduce a program in Engineering Management. In the past, nearly every major manufacturer in Louisiana has expressed interest in participating in graduate programs.

The region has fourteen paper-related industries which employ 100 or more people, four paper industries with 250 or more employees, five paper industries with 500 or more employees, and two paper industries with 1000 or more employees. Other manufacturing concerns in the region include AT&T (communications, manufacturing), Riley-Beaird, Inc. (pressure vessels, heat exchangers), Frymaster (consumer appliances), Vought (aerospace, defense), and other plywood, particle board, and hard-goods manufacturers. The southern part of Louisiana contains industries associated with petroleum and chemicals. The current Louisiana administration has made a commitment to attract new industry and to significantly improve the state's educational system. The linking of the two universities would allow an industry, located anywhere in Louisiana, to have direct access to needed educational programs. In addition, such a system would have a significant impact in attracting new high-technology industries to the state.

4 THE MANUFACTURING SYSTEMS ENGINEERING PROGRAM AND MANUFACTURING ENGINEERING RESEARCH CENTER AT LTU

In the summer of 1985, the AT&T Foundation invited the College of Engineering at Louisiana Tech University to submit a proposal for a planning grant to establish the nature of enhancements in manufacturing education that would be appropriate for LTU and manufacturing industries in this region (Fig. 1). This planning proposal (3) was funded and a unique relationship was established among LTU, AT&T and other manufacturing industries within the region.

Dr. R.O. Warrington (Head of Mechanical and Industrial Engineering at LTU) and Dr. R.M. Harnett (Associate Dean of Engineering at LTU) began the seven month planning phase by visiting the following universities, industries and laboratories:

In USA:
AT&T, Shreveport Works
Georgia Tech, CIM program
Texas Instruments, Dallas-Ft. Worth
Vought Aeronautics, Dallas-Ft. Worth
Boeing Military Airplane Company, Wichita, Kansas
Utah State University, Manufacturing Engineering Dept.
General Motors, Shreveport
General Electric, Shreveport
National Institute of Standards and Technology (NIST, formerly NBS), the Automated Manufacturing Research Facility

In Japan:
NEC, Ricoh, Toshiba, Hitachi, Sharp, University of Tokyo, Tokyo Institute of Technology, Waseda, Keio, and Kyoto University

This planning effort resulted in a three year Phase I grant from the AT&T Foundation to develop a Manufacturing Systems Engineering Program at LTU. This is an interdisciplinary MS degree program which consists of 36 semester credit hours distributed over five program segments (core, concentration, broadening, electives, and practicum). The program starts with an Advanced Manufacturing Technologies course and culminates in an industrial associates program with industry. A strong link with industry is made through the student's practicum. Throughout the MSE program the student is in contact with his "parent" company and during his final quarters the student selects an interdisciplinary project within that company. The student (or students), company engineers and an MSE faculty member form an interdisciplinary team to work on the project. The practicum ranges from one-sixth to one-third of the total degree program. A recent report from the National Academy of Engineering (4) reinforces this approach to manufacturing education -(excellence in manufacturing can be achieved) "... through capstone experiences in product and process design that require knowledge and participation in a number of disciplines. A growing number of universities are taking this approach, effectively teaching systems integration and teamwork through various manufacturing systems curricular sequences and interdisciplinary research center structures."

Initial success of the MSE program has resulted in a request from the AT&T Foundation and the AT&T Shreveport Works to develop a proposal for Phase II of the program. The objectives for Phase II are:

(a) Development of an innovative materials management component to the MSE program that would be taught as a capstone course and coordinate the production facility with engineering, finance, marketing (including the international perspective), and other related functions.

(b) Broadening of the MSE program to include the manufacturing aspects of chemical engineering (the current program is primarily electrical, mechanical and industrial engineering).

(c) Development of an automated inspection and metrology laboratory in support of the program's concentration in precision engineering.

Phase II, if funded, will start in the Fall of 1989.

An associated research center (Manufacturing Systems Engineering Research Center-- MANSER) has been established within the College of Engineering at LTU to coordinate the manufacturing research activities of the college and to form research links with the MSE program, industry and the CAMD center (discussed below) at LSU. Attached to this center is an industrial advisory board that provides input to both the MSE program and the research center.

5 THE CENTER FOR ADVANCED MICROSTRUCTURES AND DEVICES (CAMD) AT LSU

The major tool of this new 25 million dollar center is a synchrotron which will be one of only five in the country, and the only one designed for the "soft" X-ray region. Two of the initial areas of application for the center will be ultra-large-scale integration of microcircuitry and micromechanical engineering (5). The networking between LSU and LTU, described above, will allow researchers at LTU to take full advantage of the facility to develop applications to attract new industries to the state and enhance the competitive posture of some of the existing industries. Construction of this facility started in the Summer of 1988, however, both departments are developing precision engineering laboratories and faculty to take full advantage of this facility.

6 THE DEPARTMENTAL SENIOR DESIGN INTERACTIONS WITH INDUSTRY

Current curriculum accreditation requirements as specified by ABET call for comprehensive design experiences for engineering students. Both LTU and LSU include senior year capstone design courses which feature design projects having technical, economic and social facets. Both schools actively seek industrial partners to suggest design projects for these courses. Students work in four-man teams under close faculty supervision to develop design requirements, alternative solutions, analyses of alternatives, and detailed designs for the industrial sponsor. The teams periodically report to their project manager who works for the industrial sponsor. They prepare a final detailed written and oral report of their work and the sponsor is allowed to provide input to the instructor for determining the grade in the course. Project sponsors contribute toward the cost of the projects. Students must develop financial plans for the project and are subject to standard accounting procedures. The object is to develop an environment that is as close as possible to that of the practicing engineer while blending in the traditional academic environment.

Most of these projects achieve creditable success in the eyes of both the instructor and the industrial sponsor. Some projects have been particularly noteworthy. One recent team developed an entirely new technique for grouping beverage containers for a manufacturer of packaging machinery. Another group developed a statistical process control system for a major electronic printed circuit board manufacturer. These and other projects generated designs that were actually adopted by the sponsor and incorporated into their products or processes.

The Energy Analysis and Diagnostic Center (EADC) at LTU also provides opportunities for selected engineering students to interact with area industries. The EADC performs energy conservation audits for industrial and governmental facilities. Each audit encompasses an interview and site visit followed by a written report which includes recommendations to improve the energy utilization efficiency of the facility. Audit teams typically consist of an experienced project staff member and one or more students. Student team members participate in all phases of the audit, and have been largely responsible for automating the engineering calculations required for typical building audits.

7 SUMMARY

Louisiana Tech University and Louisiana State University are moving ahead into a beneficial and complementary role in higher education. Funding for higher education in Louisiana over the past eight years has dropped more than any other state in the southern USA. This has been accompanied by an increase in faculty salaries which is less than one-half of the national average. Funding for higher education in Louisiana is based on the number of students enrolled in the various levels of instruction. In June of 1988, the average level of funding for higher education in Louisiana was less than two-thirds of the level set by the State. In spite of these hardships, the faculty of LTU and LSU are aggressively seeking to become more directly involved with industry at both the undergraduate and graduate levels. The universities are striving to implement the most recent technology to improve educational programs, expand industrial relations, and compete more vigorously in the national research marketplace. The programs outlined in this paper will lay the groundwork for the revitalization of higher education in Louisiana into the next century.

REFERENCES

(1) Filley, R.D. "The White House Conference on Productivity." *Industrial Engineering*, Nov. 1983, pp. 30-32.

(2) "A Mechanical Engineering Departmental Network Between LSU and LTU." Proposal to Louisiana State Board of Regents, 1988.

(3) Harnett, R.M. and Warrington, R.O. "Trends in Manufacturing Engineering Education and a Proposal for Louisiana Tech University." *Proceedings of UCAEDM-VPI*, 1986, pp. 114-117.

(4) "The Challenge to Manufacturing: A Proposal for a National Forum." National Academy of Engineering, Oct. 1988, pp. 19-20.

(5) "CAMD To Engineer Advances in Microstructures." *LSU Engineering*, 1988, 19, no. 1.

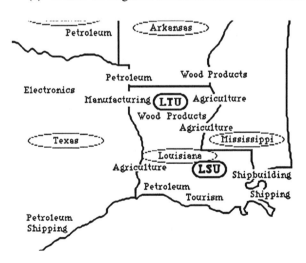

Fig 1 Major Industries in South Central USA

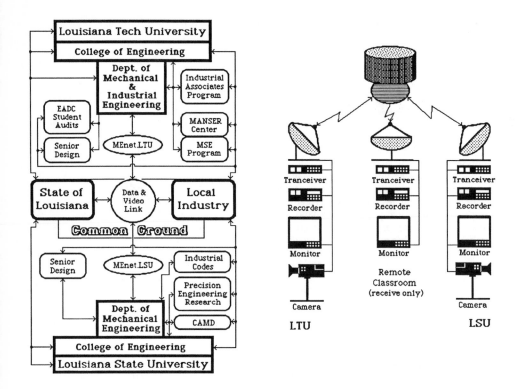

Fig 2 Major Industry-University Interactions

Fig3 LTU-LSU Video Link

Design principles: a key to improving engineering education

D G JANSSON, ScD
The Institute for Innovation and Design in Engineering, Texas A and M University, USA

Texas A&M University is undertaking a major engineering science core curriculum experiment which includes the redefinition of the core subjects and the inculcation of design principles throughout the engineering science core. The characteristics of the design process are such that the learning of fundamental engineering science principles can be greatly enhanced when these two major elements of the engineering curriculum are approached cooperatively. This paper focuses primarily on this interaction, but also includes a brief description of the proposed redefinition of the engineering science core.

1 INTRODUCTION

Considerable attention has been given in recent years to the state of engineering education.[1] In the United States, dramatic changes occurred in engineering education following the successful launching of the first Sputnik satellite by the Soviet Union. This shift is best described as a move to engineering science, or scientifically-oriented engineering education. Although not necessarily intended, one effect of the successful shift to engineering science in the post-Sputnik period has been reduced emphasis on engineering synthesis and execution.[2] Students now accumulate vast amounts of information which they may not really learn how to use. In order for this material to be usable in forms other than that in which it has been presented, a deeper level of understanding is required. Thus, students need to learn concepts in a form in which the information can be utilized; that is, it must be "enabled".[3,4]

As technology changes even more rapidly, and products and processes more frequently span the traditional disciplinary boundaries, it becomes more important that the engineering education we deliver, at the foundational level, be less and less dependent on the discipline chosen by the student. For example, no longer is it acceptable for mechanical engineers not to have usable familiarity with certain elements of electrical engineering. Also, there is a rapid merging of the life sciences with many of the engineering disciplines.

Due in part to treatment of students as receptacles for information, they seem to lose their creativity as they move through the four or so years of engineering education. By the time they enter a capstone design subject, the pedagogical system has educated them away from being creative. A change in curricular approach is required to make design a continuous element of engineering education, not something that is studied after all of the other material is presented.
These requirements for change are not simply academic matters. As technology is being utilized in more regions of the world and in more areas of human experience, there is heightened awareness of the dependence of industry on engineering practice. Competitive pressures continue to develop in the U.S. and around the world to maintain and, indeed, to increase shares of the world's technological market.

1.1 The Curriculum Experiment

The list which follows summarizes the motivation for this curriculum experiment:

- There is a general national concern for the state of engineering education.

- Technology is changing faster than education.

- The market demand for the engineers is changing as technology changes.

- More efficiency is needed in engineering education as the body of knowledge grows.

- Creative design content needs to be strengthened within the engineering curriculum.

Our conclusion is that we need to consider radical changes in the conceptualization of the engineering science core.

In response to this statement of motivation, Texas A&M University submitted a successful proposal to the National Science Foundation containing the research objectives of developing a stronger, principle-oriented engineering foundation in engineering undergraduates which will:

- be applicable to all or most of the engineering departments;

- give students a better ability to cross disciplinary lines;

- relieve pressure on the four-year curriculum; and

- strengthen undergraduate design education.

The approach to be taken in the curriculum experiment is two-fold. We will perform a "coordinate transformation" on the engineering science core subjects and infuse creative design principles throughout the curriculum. These two are not unrelated. The main reason for infusing creative design education throughout the engineering science core and not some other strategy is that creative design education enhances the teaching of engineering fundamentals. This symbiosis is the main topic of this paper.

1.2 A Coordinate Transformation

Before the design-related aspects of this experiment are addressed, it will be useful to describe in a general way the coordinate transformation being explored for the engineering science core. Our approach is based on the assumption that the engineering science core must move from a reference system which defines subjects on the basis of traditional disciplinary subject matter to a reference system based on concepts which provide boundary-spanning by virtue of their commonality to all engineering disciplines. Figures 1 and 2 exemplify this approach. It must be pointed out that the new set of core subjects in these figures are only one candidate set. As this paper is being prepared, we are in the process of performing the coordinate transformation and do not expect this set to be our final choice.

1.3 The Engineering Foundation

In order to make a proper coordinate transformation and to understand the role of design principles in the teaching of the engineering science core, it will be very helpful to clarify what the objective of the engineering science core is. Figure 3 is a schematic showing the various types of information which are presented in the traditional engineering science core subjects and the interrelationships among these types of information. It should be noted that the figure is not intended to be a diagram of the engineering process; thus, it does not contain feedback loops which describe the iterative nature of engineering problem-solving or the design process. It is only intended to help us understand the nature of what we teach and to discover just how much the body of information we present contributes to the educational objective of the engineering science core.

We believe that the objective of the engineering science core is to create in our students a solid engineering foundation. This foundation includes a usable or working knowledge of the physical laws (e.g. conservation of energy), a large number of fundamental concepts and definitions (e.g. temperature), constitutive relationships (e.g. Hooke's law), and knowledge of the properties structure and behavior of materials (e.g. thermal expansion). This body of information, coupled with what we hope is a generous supply of experiential understanding of the physical world, constitutes the engineering foundation, that which we hope the engineering science core will impart. All of the rest of the schematic represents a large body of material which is presented in the engineering science core subjects but is, or should be, only included in order that it might contribute to strengthening the engineering foundation.

This taxonomy is very revealing. A close examination of the content of many engineering science core subjects will indicate the large fraction of time spent delivering this "downstream" information while we often lose sight of our primary objective. It must be clearly pointed out that this discussion is not advocating the removal of this downstream information but is merely encouraging a conscious recognition of its role in the engineering science core. As technology changes ever more rapidly, much of the downstream information may change but the invariant in the process is the engineering foundation. Engineers who have been well educated in the engineering foundation have been educated for life; engineers who have been educated largely with application-oriented information are in danger of obsolescence early in their careers. In addition however, the application-oriented information can be much more effectively taught in the upper-level courses if a solid engineering foundation has been established.

1.4 The Role of Design

With this objective clearly in mind, we are then faced with the pedagogical question of how to impart this knowledge more effectively to the students. What makes information usable? How is information retained or cataloged by students as they learn? What makes this information come alive for the students? All too often, the educational process presents information in the form of equations which the students then attempt to retain at the level of equation. Without the physical understanding of each quantity and term in an equation, it is only a tool used to reach a numerical result. The author and others[3,4] have observed that students who apply fundamental laws and concepts in creative design problems learn this material much more effectively. This concept is very simple yet quite profound. Design educators who become frustrated because design is relegated to be almost an afterthought in many engineering curricula, should recognize that they hold one of the keys to significant improvement in engineering education.

It must be pointed out here that this is just the beginning. The implication is not that engineering science core subjects should merely become design subjects. There is much too much to be taught for that sort of irresponsible approach. The research question for us now is to learn how to use the principles of creative design in a very efficient manner in order to be continuously enabling that which is taught within the engineering foundation. A happy by-product, of course, is that design education will start much earlier in the curriculum. With respect to the ever-present question of how to balance the teaching of fundamental disciplinary material and the more application-oriented design subjects, the approach outlined here is an "everybody wins" situation.

A very simplified example will help to illustrate this point. Imagine the teaching of certain fundamentals of heat transfer, in particular, those which relate to simple heat exchanger configurations. We have several choices. Typically, however, the physical configurations of heat exchangers are presented as givens in this discussion. The new approach should allow for the creation of new heat exchanger configurations or, at a minimum, the discovery of existing configurations, based on the fundamental principles. The rate at which heat is exchanged between two fluids is proportional to the area of the heat exchanger, the temperature difference between the fluids, and the heat transfer coefficient which applies in any particular situation. Traditional approaches might examine the improvement of performance by the selection of parameters within a prescribed configuration. The "design" approach would allow for the discovery of many different possible configurations based on the fundamental relationship. For example, heat exchangers which take advantage of maximizing the temperature difference along the flow path are so-called counterflow heat exchangers. Heat exchangers which attempt to take advantage of large increases in area are exemplified by so-called direct-contact heat exchangers, applicable when one of the fluids is a liquid, the other is a gas, and when direct contact does not have any negative implications. Finally, configurations which attempt to enhance the heat transfer coefficient are exemplified by so-called stirred heat exchangers, in which the heat transfer coefficient is increased through the addition of velocity within the boundary layer.

Many pedagogical tools which can take advantage of design principles are possible. These may include liberal use of examples as motivation for new configurational concepts, the use of conceptual design exercises and projects (as opposed to design problems which are merely parameter selection), and the assignment of problems which require the application of fundamentals in areas completely outside of the traditional domains of the various departments in which the students are enrolled. It should be noted that many of these approaches are probably already being used on a small scale. The real challenge is to establish their utility through controlled experiments and to develop additional mechanisms which take advantage of the cognitive processes involved in design and in learning, such that they can be transferred to many other teaching faculty and adopted on a large scale, in a sense less dependent on the skill of a few unusual teachers.

In summary, creative design principles should permeate the entire engineering curriculum. Design process and experience enhance in-depth understanding of the fundamental disciplinary knowledge of the entire engineering curriculum. The design process provides the avenue for "enabling" the information that we force feed in ever-increasing doses to our students. The benefits of using design in the teaching of the engineering science core are thus two-fold: better understanding of fundamental technical concepts through the enabling effects of the design process, and more effective engineering design education with better nurturing of the creative spark.

The curriculum experiment at Texas A&M University has just begun. The first group of students in this five-year experiment will begin the new subjects in the fall of 1989. Over the next four years, we hope to generate significant curriculum research results in addition to the development of a new engineering science core curriculum.

Acknowledgments

The author wishes to acknowledge the support of the National Science Foundation and the contributions of co-principal investigator Carl Erdman and colleagues John Fleming, Charles Glover, Harry Jones, Cy Ostowari, and Mike Rabins in this on-going curriculum project.

References

1. Proceedings of the National Congress on Engineering Education, Washington, D.C., November 1986, see p. 18 & 19 of Curriculum Issues section of Appendix H.

2. "Report of the Committee on Evaluation of Engineering Education", Journal of Engineering, September 1955.

3. Jansson, D.G., "Creativity in Engineering Design: The Partnership of Analysis and Synthesis", 1987 ASEE Conference Proceedings, Reno, June 1987.

4. Flowers, W.C., "On Engineering Students, Creativity and Academia", 1987 ASEE Conference Proceedings, Reno, June 1987.

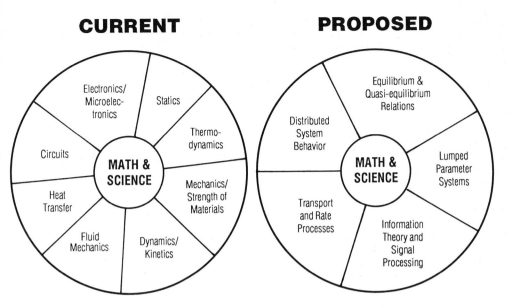

Figure 1 — The engineering science core — existing and proposed

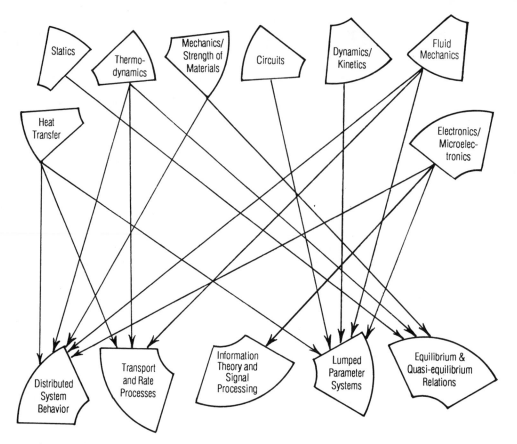

Figure 2 - The concept of coordinate transformation

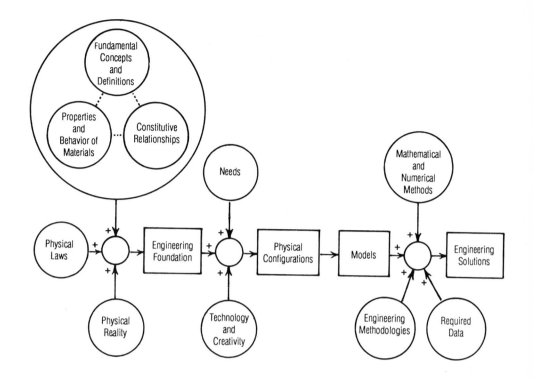

Figure 3 —   Interrelationships among the types of information

C377/007

SEED—a unique formula for grassroots improvement in design

M D KIMBER, CEng, FRAeS, FIED
Design Centre, School of Engineering, Hatfield Polytechnic
D G SMITH, BSc(Eng), DLC, CEng, MIMechE
Institute of Engineering Design, University of Technology, Loughborough

SYNOPSIS SEED - 'Sharing Experience in Engineering Design', an unique association of design lecturers in Polytechnics, Universities and other Institutes of Higher Education, has emerged as a significant force for improvement of design education. This paper outlines its educational and organisational philosophies and also its activities and achievements.

1 INTRODUCTION

The quality of undergraduate design education is crucial to subsequent design performance in industry. The continuous debate and controversy surrounding design and the associated exhortation for improvement in the 1970's[1,2] had not resolved the speculation nor yielded working literature to assist the teacher. It was to rectify this situation that an association of people unde the acronym of SEED was formed and grew from modest beginnings at Hatfield Polytechnic in 1979.[3]

SEED - 'Sharing Experience in Engineering Design', whose primary objective is embraced in its name, was mentioned as a source reference at ICE '81 in Rome.[4] Since then it has attracted ever increasing participation an support throughout the United Kingdom as an independent and significant force for improvement in design education with consequential benefits to design in industry.

This paper deals with the SEED educational and organisational philosophies, achievements and policies which represent a formula that may be of interest to the wider audience represented by an international conference such as ICED '89.

2 AIMS, ORIGINS AND EVOLUTION

The initial philosophy was to ventilate the problems encountered in design teaching through the medium of informal, low-cost seminars with minimal administration. This was achieved by placing emphasis on an environment which encouraged free discussion in small groups and by holding the events each year in different educational establishments. These not only acted as host-organisers but also afforded opportunity for delegates to view design facilities. It was found that the personal contacts made at these seminars were considered by delegates to be extremely valuable and confirmed that there were common problems associated with design teaching which were unresolved and could benefit from shared experience.

It was realised that there was a wealth of expertise and experience residing with engineering design academics, many of whom had practised in industry. They had reviewed their experience and knowledge and attempted to distil this for transmission to aspiring designers and engineers. It was, unfortunately, widely scattered throughout various educational institutions and this militated against its being harnessed for the common good. It became apparent that SEED had the potential to remedy this situation and, within a few years, was moving from a loose-knit association to a body of design educators wishing to express informed opinion through it. To ignore this would not only have been wasteful but would also have constituted a breach of faith with those who had supported SEED in its formative years. Consequently, in 1985, the aims and organisation of the association were revised and extended to respond to the needs being identified, whilst retaining the characteristics that made it attractive to design educators and would continue to contribute to its success.

The aims are:

* to encourage the sharing of experience in engineering design education by those involved,
* to arrange visits to various design teaching departments,
* to provide a forum for the discussion of matters of concern in design education,
* to represent the collective and informed view of members to further a better understanding of design and improvement in the quality of design education.

The management committee was enlarged and reconstructed to effect balanced representation between the Polytechnics and the Universities. Tangible recognition of the potential of SEED to effect improvement in design at the grassroots level were the grants awarded to it. Those from the Department of Trade and Industry were linked to particular scholastic undertakings or specific projects while those from the Smallpeice Trust, a registered charity which promotes design and efficient manufacture, were awarded for the general prosecution of SEED's activities.

The ethos of SEED has undoubtedly been significant in its evolution. It has quite deliberately adopted a neutral and independent position in the United Kingdom national design scene to encourage a meeting of minds and a free unbiased exchange of views and information. This has enhanced co-operative teamwork across all sectors of education and with other national bodies. The essential qualification for participation was a genuine interest in design. Thus, the polarised and defensive attitudes so often present at major conferences, which act as barriers to progress, have not been in evidence. Participants have responded to an encouragement to talk to, and not at, one another and to share their expertise and experience. From the outset, SEED has enjoyed a co-operative working relationship with, and adopted a complementary role towards, established engineering and design organisations. Nevertheless, it holds a neutral and somewhat unique position in Higher

Education, which it intends to maintain so that it may freely reflect the views of its membership and continue to improve the quality of design.

3 EDUCATIONAL PHILOSOPHY

3.1 Foundations

The SEED educational philosophy arose from recognition of the need for a systematic body of design knowledge, allied to the need for promoting an improved understanding of the activity of design. In the post-Feilden[5] era, with the advent of the new Polytechnics, extensive development of design courses took place in these and the Universities. There was thus a need to review the overall design situation and communicate the best information and practice emanating from these developments. Concurrently, there was, in the mid-eighties, a greater awareness and growth in interest in the importance of design and design education and a sense among SEED members that they could and should formulate concensus views on various aspects of design teaching. This came to fruition when one of the first areas identified was the lack of, and hence the need for, a core design curriculum for engineering undergraduate courses. SEED Seminar delegates considered it essential to respond to this challenge and meet this need. Several national reports[1,2,6] had been published emphasising the importance of design and the necessity for improvements in design teaching, but did not, and presumably were not intended to, provide a suitably structured design curriculum.

A SEED working party was, therefore, set up to make recommendations on the content of a curriculum for design in engineering under-graduate courses. To avoid the pitfalls that hamper far too many design dissertations and conferences, the working party established at the outset an agreed definition of the design activity and an acceptable means of representation for purposes of communication.

Design was seen as 'The total activity necessary to provide an artefact to meet a market need, that commenced with the identification of the need and is not complete until the resulting product is in use, providing an acceptable level of performance.'[7] Furthermore, the main characteristics associated with the activity of design were identified as being:

* central to engineering,
* transdisciplinary,
* highly complex,
* highly iterative,
* highly interactive.

A further element underpinning SEED's educational policy was the need to identify the attributes demanded of design practitioners and these were seen to be:

* ability to communicate,
* creative as well as analytical skills,
* ability to integrate,
* ability to make judgements,
* management ability.

A most significant element in the identification of a core curriculum was the choice, from alternatives available, of a suitable model of the design activity. The representation shown in Fig 1 was selected because it:

*   reflected the SEED view of design,
*   identified the main elements of the design activity,
*   brought these elements together and showed the relationships between them in a structured manner,
*   is unambiguous and will not confuse students.

Subsequently, further evidence in support of the model came from academic design teachers who recognised it as representing what they are trying to teach, and practising designers who recognised it as what they are seeking to achieve in practical terms.

As a result of these deliberations, the working party produced two draft reports which were discussed in detail and accepted respectively by the SEED '84 and SEED '85 Design Seminars. The subsequent proposals for a 'Curriculum for Design - Engineering Undergraduate Courses'[7] thus represent the views of a large percentage of those who teach design in the United Kingdom at tertiary level, the majority being chartered engineers, many of whom continue to practise as consultants. They see the proposals as being directly relevant, both to teaching and practice, and not the untried exhortations of non-practitioners. It is believed to be the first document delineating an agreed core curriculum for design at engineering undergraduate level. The proposals were intended to strike a balance between generalisations on the one hand and over prescription on the other. They are believed to represent a sound basis for the development and implementation of design teaching in first degree engineering courses. As such, they provide a framework which lends itself to adaptation[8] to suit particular disciplines and educational establishments.

3.2 Principles

The educational philosophy has been developed by SEED on the basis that the majority of engineering students are likely to have contact with design, either directly or indirectly, during their subsequent careers. It, therefore, seeks to provide all engineering undergraduates with at least an appreciation of design. However, it also seeks to instil into students the challenge and satisfaction that design work can offer in the hopes that those with potential will be encouraged to take up a professional career in design.

In order to satisfy this basic objective and build upon the foundation elements stated earlier SEED incorporates the following main principles in its educational philosophy:

Firstly, design must be shown to be the central and integrating theme in engineering business, and students must be provided with a balanced and comprehensive appreciation of the total design activity, compatible with the current highly competitive product environment. Furthermore, this should demonstrate the fundamental principles of design which are both discipline and product independent. This will place the student in an informed position for project work, whatever the discipline, in his subsequent academic or industrial career. This is in accord with the CNAA report[9] which identifies such a need stating that 'A typical complete design process should be examined from beginning to end, ... and most of the designers interviewed felt that it would be helpful ... to study a generalised model of a design process.'

The second principle is the necessity to teach skills required in carrying out design projects which are not generally included elsewhere in the curriculum of the majority of undergraduate engineering courses. These have been identified[7] as information retrieval, communication, market analysis, specification formulation, ideas generation, evaluation, decision making, quantitative analysis, costing, economic analysis, computing, aesthetics and ergonomics. The level to which these skills are taught will depend upon the particular emphasis given to the course and the interests of the lecturers. However, it has been found essential to strive for an acceptable balance between all these skills in order to give a balanced appreciation of design. It cannot be over emphasised that all techniques should be taught in a design related manner. For example, it is not sufficient to teach communication in isolation, it must be taught so that it is related to specific aspects of design such as project reports, seminar presentations and person-to-person discussions.

The third principle is the necessity to teach students how to handle the interface between the activity of design and specialised subject areas. It would be desirable if all subjects in an engineering course were taught in a design related manner, but for various reasons this is seldom possible. However, students need to be taught how to apply their specialised knowledge in a design context. For example, most students when faced with a realistic design situation find difficulty in applying their engineering science knowledge. Students need to be shown how to handle such situations.

The fourth principle, which possibly has universal acceptance, is the necessity for students to undertake design assignments and projects. These enable the student to bring practical reality to bear upon the philosophical model of the design activity presented to them and enable them to utilise and gain experience in the basic design techniques and skills. Furthermore, they provide a vehicle for teaching the student to manage the complexity of design and to integrate the many topics and factors which must be taken into account in any design project. The projects and assignments should ensure that students have some practice in all phases of the total design activity. These include, market investigation, product design specification, conceptual design, detail design and, ideally, manufacture.

SEED recognises that the above principles must be carried out against a background of the student's developing maturity and increasing knowledge. The student will enter an under-graduate course with A levels or equivalent qualifications, generally directly from sixth forms, without engineering or design experience. He will finish the course well qualified, particularly in analytical subjects. Courses in design appreciation enable students to utilise many areas of their knowledge and put the whole undergraduate course in perspective. As such, it is essential that design teaching be a core activity throughout every engineering undergraduate course. This point is also made in the Moulton report which states that 'Design should be a thread running through all the normal engineering degree courses.'[1]

4 CURRENT PROGRESS

The activities of SEED, managed by an executive committee elected by and responsible to the membership, include the following main elements:

*   direct sharing and discussion effected by annual design seminars or by special seminars in response to matters of current national concern on the design front,

*   provision of written texts to disseminate design knowledge in the form of publications.

4.1 Seminars

Seminars were the raison d'etre of SEED, for it was from one such event that it germinated and subsequently flourished. They have been seen to be the means of ensuring that the widely scattered design expertise was brought together and that those who possessed it met face to face and became known to one another. To guarantee that this occurred, attendance needed to be maximised and the seminar programme to be recognised as relevant and worthwhile. It was believed, and subsequently confirmed, that the former could only be achieved if the cost charged to the delegate was low enough to encourage attendance. The latter was a matter of good intelligence and judgement for what was of concern to design lecturers in their teaching and the suitable structuring of the event itself. SEED ensured low cost charges, thus support for events became related almost entirely to their value as perceived by the participants. Eleven seminars have been held at educational establishments throughout the United Kingdom organised by the host-institution in conjunction with SEED, as shown in Fig 2. Attendance grew rapidly and has for some time exceeded one hundred persons per event which is an indication of the perceived value.

SEED '87[10] saw a new initiative in the broadening of the base across disciplines. Although SEED was, from its inception, intended to be interdisciplinary, it attracted predominantly those in the mechanical engineering field at the outset. The Seminar confirmed that the philosophy adopted by SEED was in fact interdisciplinary and was now being implemented by those in other disciplines.

4.2 Publications

Publications, another major activity, stemmed from the need for material of direct value to the design teacher which could be drawn from more experienced or expert colleagues. To this end, SEED set up working parties having across academia representation with the aim of producing suitable publications to be available at comparatively low selling prices. They would reflect tested and proven approaches rather than be esoteric or untested hypotheses. Thus, SEED has developed a policy of providing an integrated and coherent system of publications for design teaching as illustrated in Fig 3.

4.2.1 Design Curriculum

The first and key report mentioned earlier and to which the other publications relate, is entitled 'Curriculum for Design - Engineering Undergraduate

Courses'.[7] The report identifies the main areas of the design activity
model, such as the core phases, product design specification, information,
techniques and management and defines the terms used in the proposed
curriculum. It includes a teaching strategy which is shown in Fig 4 based on
a progressive build up of knowledge, techniques, skills and experience which
is reflected in the proposed sequence for presentation of the various topics
to the students. The material can conveniently be presented in three stages,
thus providing the opportunity for a 'design thread' throughout the course
which, for example, could be ensured by teaching one stage each year. It
should be noted in Fig 4 that the topic titles are given in stages in order of
their introduction, whilst the arrows indicate that they are further utilised
in subsequent stages.

It is emphasised that the topics listed in the areas of techniques and
information must not be taught in isolation from design requirements, but in
context. They must, therefore, be presented in such a way as to provide a
bridge between traditional subject material - such as thermodynamics,
hydraulics, control theory, materials technology, electronics and structures -
and its application in design. The curriculum provides for the progressive
development of necessary attributes and the progressive integration of topics
leading to total integration in the final stage.

4.2.2 Preparation Material for Design Teaching

In the key report the topics relevant to the total design activity are
identified and to assist the teacher, whether experienced or inexperienced,
in presenting them to the students, a series of guides entitled 'Preparation
Material for Design Teaching'[11] are being published. Each booklet deals
with one topic only. That on 'Specification Phase' deals with the
considerations that arise in determining a comprehensive product design
specification, which is critical to product success. Another on 'Information
Retrieval' recognises that information is the 'raw material' of design and not
only provides an appreciation of the many and varied information sources
available, but also gives practical advice on exploiting some of the most
useful sources. The third booklet that has been published at the time of
writing is on 'Communication' and puts this in the context of design,
providing guidelines for tutors. It recognises that communication skills are
as important to the professional design engineer as any others associated with
engineering courses. SEED has a planned programme to produce further booklets
of 'Preparation Material' relevant to its core curriculum, including
Specification Formulation, Conceptual Design, Detail Design, Manufacturing,
Costing, Computing and Quality and Reliability. This is an ongoing co-
operative activity drawing on recognised expertise.

4.2.3 Compendium of Engineering Design Projects

As stated earlier, the need for students to undertake practical design work is
considered essential in any design course. The nature of such projects ranges
from those required to reinforce a specific topic to industrially based
activities, it being essential in any course to keep a balance between them.
The generation and development of projects can be an exceedingly time
consuming task and for this reason SEED has published the first comprehensive
'Compendium of Engineering Design Projects'[12], for use by teachers of

design. It covers both assignments which relate to a single or limited integration of topics through to projects requiring a time span of two or three terms and involving the integration of many topics. In setting up projects as vehicles for learning, the importance is emphasised of striking a balance between those aspects which develop understanding of the more formal aspects of the course and those which lead to design in an industrial context. This is significant in maximising motivation and giving a good preparation for a professional career. Important considerations are explored in the SEED Compendium, such as the nature of the learning experience using projects, from initial brief, through design specification, concept generation, evaluation, detailed engineering and presentation. Assessment is identified as a critical component of project learning due to its formative and summative functions and the need to determine the student's grasp of the real objectives and how effectively these have been met. These projects, many with industrial origins, come from the experience of SEED members and have already been proved suitable as vehicles for learning in design practicals.

4.2.4 Design Procedural Guides

In carrying out design projects information gathering can be very time consuming and within the time constraints of courses the exercise of itself may be of only limited educational value to the student. Furthermore, most information is not in a form of direct use for design purposes but requires manipulation and reference to several sources when dealing with any specific topic. In order to assist in presenting information of direct use for design purposes SEED is publishing a series of 'Engineering Design Procedural Guides'[13] covering the following areas:

* The selection of system concepts.
* The selection of bought-out devices and components.
* The design of devices and components.
* The selection of materials and manufacturing methods.

The information in the guides accords with current practice but is produced with engineering undergraduates in mind and for this reason is in some cases simplified when it is in the student's interest or for convenience of presentation. The guides should not be regarded as a substitute for fundamental teaching but as an aid to the student in managing design coursework and in making acceptable decisions. There are also indications that they, with other SEED publications, are of potential value to industry.

5 FUTURE POLICY

SEED has achieved something which is unlikely to be realised in any other way; not only is it unique in providing a forum for discussion on design and design education, but there resides in its membership a 'centre of excellence'. Most of its members both teach and practise design and are qualified professional engineers. They represent the majority of Higher Educational Institutions in the United Kingdom. As a result SEED appears to have been instrumental in moving design across traditional barriers between lecturers in Polytechnics, Universities and Colleges and across various academic disciplines.

For the future:

*   It is intended to maintain and develop the present activities such as curriculum development, project learning, preparation and presentation of valid design data and to continue to encourage the use of modern computer aids to appropriate aspects of design. The interdisciplinary nature of the SEED approach, already underway, will continue to be developed.

*   It is considered essential to have a comprehensive and consistent design educational policy which spans the educational system from schools through to industry. SEED already encourages the development of contacts between individual members and local industry but it is hoped to extend such co-operative initiatives. Measures have also been taken to extend the educational philosophy to schools.

*   Contacts have already been formed with some educational institutions overseas and are expected to increase. Furthermore, SEED believes that the adoption of its formula could be attractive to some other countries and would be interested in such developments.

In its understanding of the task facing design educators, in having a network of members in most United Kingdom academic institutions and in having the ability to co-ordinate a diverse educational resource, SEED believes it has an unique formula for grassroots improvement of design through education.

REFERENCES

(1)   MOULTON, A.E., et al. Engineering Design Education. A report by a Design Council Committee, 1976 (The Design Council).

(2)   CORFIELD, K.G. Product Design. A report carried out for the National Economic Development Council, 1979 (NEDC).

(3)   KIMBER, M.D. SEED - A Hardy Perennial. The Engineering Designer, 1985, p 22.

(4)   EDER, W.E. Skills and Abilities: A Role for Design in Engineering Education. Proceedings WDK6 ICED '81, Rome, 1981, p 19-34.

(5)   FEILDEN, G.B.R., et al. Engineering Design. Report of a Committee Appointed by the Council for Scientific and Industrial Research, 1963 (HMSO).

(6)   LICKLEY, R.L., et al. Report of the Engineering Design Working Party. Report to the Engineering Board of the Science and Engineering Research Council, 1983 (SERC).

(7)   Curriculum for Design - Engineering Undergraduate Courses. Proceedings of Working Party, 1985 (SEED).

(8)   McCRAKEN, W. and ION, W.J. Total Design Education at the University of
        Strathclyde. Proceedings of the SEFI Annual Conference, Engineering
        Education in Europe 1988, Leuven, 1988, p 199-206.

(9)   Managing Design - An Initiative in Management Education. Report of a
        project sponsored jointly by CNAA, Department of Trade and Industry and
        The Design Council, 1984 (CNAA).

(10)   Design Across the Disciplines. Proceedings of SEED 87 Design Seminar,
        University of Strathclyde, 1987 (SEED).

(11)   Curriculum for Design - Preparation Material for Design Teaching, 1987-
        1988 (SEED).

(12)   Compendium of Engineering Design Projects,1988 (SEED).

(13)   Engineering Design Procedural Guides,1988 (SEED).

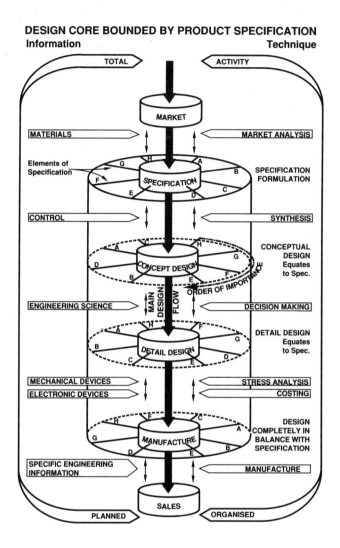

Fig 1   Design activity model

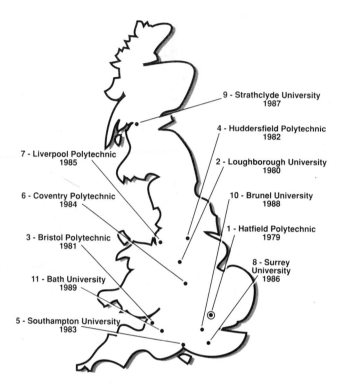

9 - Strathclyde University
1987

4 - Huddersfield Polytechnic
1982

7 - Liverpool Polytechnic
1985

2 - Loughborough University
1980

6 - Coventry Polytechnic
1984

10 - Brunel University
1988

3 - Bristol Polytechnic
1981

1 - Hatfield Polytechnic
1979

11 - Bath University
1989

8 - Surrey
University
1986

5 - Southampton University
1983

Fig 2 Locations and dates of SEED design seminars

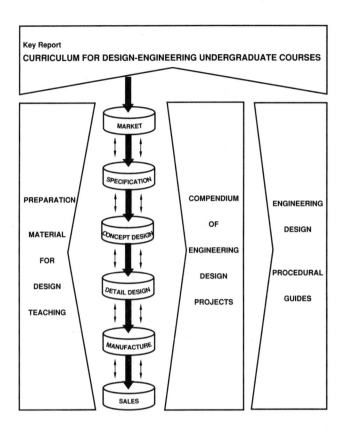

Fig 3   Integrated system of SEED publications

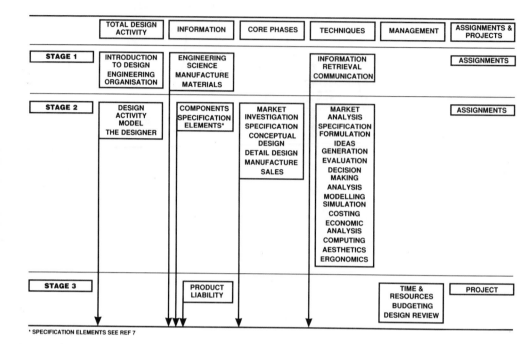

	TOTAL DESIGN ACTIVITY	INFORMATION	CORE PHASES	TECHNIQUES	MANAGEMENT	ASSIGNMENTS & PROJECTS
STAGE 1	INTRODUCTION TO DESIGN ENGINEERING ORGANISATION	ENGINEERING SCIENCE MANUFACTURE MATERIALS		INFORMATION RETRIEVAL COMMUNICATION		ASSIGNMENTS
STAGE 2	DESIGN ACTIVITY MODEL THE DESIGNER	COMPONENTS SPECIFICATION ELEMENTS*	MARKET INVESTIGATION SPECIFICATION CONCEPTUAL DESIGN DETAIL DESIGN MANUFACTURE SALES	MARKET ANALYSIS SPECIFICATION FORMULATION IDEAS GENERATION EVALUATION DECISION MAKING ANALYSIS MODELLING SIMULATION COSTING ECONOMIC ANALYSIS COMPUTING AESTHETICS ERGONOMICS		ASSIGNMENTS
STAGE 3		PRODUCT LIABILITY			TIME & RESOURCES BUDGETING DESIGN REVIEW	PROJECT

* SPECIFICATION ELEMENTS SEE REF 7

Fig 4   Design teaching strategy

C377/272

The use of a three-dimensional modeller in design education at an academy of art

W JONAS, H-J KRISTAHN, A HUNSTOCK and K SCHNEIDER
Hochschule der Kunste, Berlin

The paper describes an 18-months experimental project
in cooperation between the HdK Berlin and IBM Germany
which started in summer 1988. The cooperation was initiated to promote the step towards the superior level
of graphics systems, beyond the field of PCs.
   The aims of the project are: The development and
testing of an educational concept regarding the specific demands of an academy of art, and the profound
application in various disciplines of design (1).

1   INTRODUCTION

The term Computer Aided Design (CAD) is very much occupied with ideas of:
- The reduction of costs and time in the product development process,
- the realization of closed process chains from early concept stages to manufacturing (CIM), and
- the reduction of the importance of the human part and the dependency on human beings in the process.
These are, without exception, unsuitable objectives for us. Therefore we shall
avoid the term in connection with our work.

1.1   Present situation

Computer aided techniques are far less spread in the creative fields than in
the engineering fields. Exceptions are TV, video, computer animation, advertizing, etc. where the picture is the final product and not a medium of communication in intermediate stages of the design process.
   Regarding the situation at our academy in particular, there are two
clearly separated groups:
- On the one hand a few "high-tech freaks", who complain about the backward-
  ness and want as many and as expensive new equipment as possible to be able
  to keep pace with the so-called "development". But most of them do not realize that the development that they are chasing after is nothing else than
  the accelerated progress of technical standards given by the vendors and the
  commercial users.
- On the other hand there are the professed enemies who regard and fear the
  computer as a "devil's instrument" killing creativity and dominating not
  only the design process but also its results.
In their narrow-mindedness both groups confirm each other's point of view and
thus create frustration and paralysation. There is hardly anything in between

except indifference and ignorance. The important demand, especially for an academy of art, of facing the existing world in an offensive and partcipative way and taking the part of the avant-garde can hardly be satisfied in this manner. That is why we are looking for a third way in order to promote this situation in a productive and innovative manner. This way will be a permanent discussion and dispute with the myths obstinately conserved on both sides: Of the computer as a stupid and uncultured machine, as a menace of human sensuality, and of the computer as a creativity machine, able to exceed the limits of the human mind.

Compared with commercial users there is a considerable delay without any doubt. But we should take it positive and accept it as a chance to examine the positive and negative experience, to avoid the mistakes of the others and to find own and really innovative approaches.

Three hypotheses:

- Avoid closed processes structured by the machine. In industry, due to the existing conditions, this goal is not called in question.
  Instead: Allow imperfect interfaces, different "materials" (material and immaterial materials) in different stages of the process. E.g. conserving colour also as touchable and smellable materials and not only as percentages of red, green and blue.
- Develop ways of cooperation between "conventional" and computer techniques. In industry this is tolerated only as a transitional stage on the way to complete computerization.
  Instead: Use the computer beside traditional materials as paper, wood, clay, brush, etc.
- Prevent the increasing separation of man from the actual process.
  Instead: Include as many human senses and parts of the human body as possible. Avoid the sensual and emotional estrangement by "ineffective" working techniques.

Our aim is to define the place of the computer as one further means in the design process and thus to enrich the process, and not to create boredom and monotony by dominating the process by the computer. We know that the second alternative is by far the simpler one.

Otherwise the often propagated (and as a chance maybe existing) overcoming of the limits of space and time through the computer may turn out to be nothing more than the elimination of space and time from human experience. We are well aware that these are merely hypotheses at the moment which have to be verified in the course of our work or which turn out to be unrealistic and wrong.

1.2 Cooperation HdK - IBM

Due to limited financial resources we decided to cooperate with vendors. This strategy is not at all undisputed and certain principles are to be respected as far as possible:

- Sufficient duration to avoid pressure of time and to allow more experimental and basic and not only executing work, and
- no restrictions by the industrial partners concerning the contents of the work.

The 18-months IBM-cooperation "IBM 6150 / CAEDS in education at an academy of art" meets these requirements.

1.3 Choice of system

The question might arise why we use an engineering-oriented 3D-modelling system for artistic and creative applications while it is always claimed that these systems were too inflexible and almost unusable for non-engineers. One possible answer could be: "Because IBM gave us no other system or even could not give us another system." This statement is true, but we do not take it as a shortcoming. We even see it as a logical way into the right direction as will be explained briefly in the following.

Usually commercial users in the creative area have different systems:
2D-graphics-layout systems
- for video input of photos, sketches, real- or computer-generated objects,
- for image-manipulation with a variety of paint-techniques, and
- for the completion with text and other graphics and for the preparation of final print processing.
3D-visualization systems
- for the generation of complex geometric shapes based on simple data-structures (facet- or polyhedron-representation), and
- for the photorealistic representation with light-sources, textures, etc. as still picture or as animation.
To start out from the fact that every representation of real things deals with spatial objects it is reasonable that we use a 3D-system. It generates a complete geometric model, thus offering an almost unlimited potential of representations of the object, while a 2D-system shows only one image of an object at a time. Without any doubt 2D-techniques are necessary for the final rendering of 3D-objects (e.g. the use of 2D-textures in 3D-systems or the postprocessing of 3D-images in paint-systems).

Why do we use a modelling- and not a visualization system? Often both approaches are not distinguished as clearly as necessary:
- Modelling systems as propagated by vendors and as used in industry influence and structure the entire design process. They claim to support the process, but often even hinder it.
- Visualization systems focus on the support of one step in the process, namely the representation of the object in intermediate stages, as final result, for the generation of animation sequences, etc. The use of visualization systems requires certain skills in digitizing and rendering objects whose geometric definition is already completed.
In principle, i.e. with cutbacks in response time and image quality, the most important visualization techniques as ray-tracing, texture-mapping, etc. are available in 100DM-programs on 1000DM-computers today. Basic techniques as the generation of shaded images are standard in all modellers. High-end visualization systems however are always and will always be separate systems due to their specific demands on hardware and software. Efforts are made to standardize the interface between modelling- and visualization systems, i.e. the geometry definition of objects to be rendered, or, at least, to use existing standards from the engineering field as e.g. IGES.

In comparison with the tremendous progress in visualization in the last 4-5 years (complete computer-animated movies) the progress of modelling systems appears rather insignificant at first glance. This is due to the fact that visualization techniques proceed mainly parallel to the development of hardware-technology while the underlying software- and data-structures are uncomplicated. Compared with that, modelling techniques mainly depend on the development of algorithms and data-structures which are able to describe a process which is widely unknown and will probably stay unexplored for a considerable time to come.

These are the essential open questions and the interesting future research fields. And this is the answer to the initial question: We use a modeller, because these systems claim to support the entire design process.

2 THE PROJECT

On the academy's side the project is a cooperation between the faculty for arts
education and science (Kristahn) and the central planning department (Jonas).
This means a combination of long artistic experience in visual communication,
especially poster design, with experience in commercial CAD-applications, mainly in the german automotive industry. An unusual, but maybe just for that reason a profitable cooperation.

There are two main goals for the project:
- The development, testing and evaluation of a concept for the use of a modern
  3D-modeller in design education, and
- the exemplary application in special fields such as graphics design, industrial design, sculpture, etc.

2.1 Universal educational concept

We are developping a course which will be put to the test for the first time
in the winter term 1988/89. One aim is to give an introduction to system fundamentals, working techniques and design potentials of modern computer aided
design systems. An even more important objective is to find answers to the
question: How to teach computer aided modelling at an academy of art (and elsewhere)?

The following conditions shall be respected:
- Didactic orientation to non-engineers, i.e. avoiding mathematical terminolo-
  gies to the greatest possible extent. No assumption of technical knowledge,
  but spatial imagination, phantasy and open-mindedness. Immediate feedback
  after every lesson in form of short protocols with thoughts, problems, suggestions, etc.
- No dissemination of industrially defined criteria of perfection and efficiency. Here the question arises whether this is possible while dealing with
  machines which embody exactly these ideas. On the one hand the computer is a
  product of western culture and technology since more than 2000 years, on the
  other hand its use promotes just this way of linear and algorithmic thinking.
  Looking at it from this point of view the computer is an absolutely conservative instrument.
- Application-independent teaching of geometric modelling, i.e. applicability
  for all disciplines that work with spatial models.
- As far as possible system-independent teaching, using the specific CAEDS
  system as an example for a class of modern modellers and aiming at the standards of future hybrid systems.

The concept uses the system's ability to produce an editable and executable
protocol of the interactive process, the syntax of which is identical with the
menu-options on the screen.

The teaching material consists of:
- A general overview of the essential terms and fundamental principles of geometric modelling, including a brief specific introduction "getting started",
  and of
- 12 lessons as "guided tours" in form of CAEDS-program-files plus a short
  illustrated explanation for each lesson.

The 12 lessons:
1 Basic shapes,
2 Visualization,
3 2D-auxiliary techniques,
4 Sweeping 1,

```
 5  Sweeping 2,
 6  Sculptured surfaces 1,
 7  Sculptured surfaces 2,
 8  Sculptured surfaces 3,
 9  Boolean operations,
10  Fillets and chamfers,
11  Shaping 1, and
12  Shaping 2
```
are partly based on each other concerning the educational contents, but are
completely independent units of software apart from that. This means e.g. that
objects generated in one lesson are not used in the next one in order to avoid
name conflicts, passing on mistakes and wasting time with confused accumula-
tions of objects.

There are no theoretical courses in advance. Students go to the computer
immediately. Learning is to be achieved by looking at, repeating and doing a
lesson and a guided tour. Based on this students are stimulated to work and to
experiment and to explore the system in a trial&error way. The concept as shown
in fig. 1 means a cycle of teaching and learning, a continuous change between
interactive doing and automatized receptive phases. The student can become a
teacher in the next cycle to pass on his own design example as a new lesson for
someone else.

2.2 Example of a "guided tour"

The example illustrates a small part out of lesson 9: Boolean operations. It
deals with the generation of a cube and a sphere and the subtraction of the
sphere from the cube. The result is stored as a new object. The automatic protocol is not readable straight away by unpracticed users:

```
K : "C"
K : "B"
K : 1 1 1
K : red
K : "STO"
K : wuerfel
K : "C"
K : "SP"
K : .65
K :
K :
K : blue
K : "STO"
K : kugel
K : "B"
K : "CU"
K : kugel
K : wuerfel
K : "F"
K : "STO"
K : wuerfelminkugel
```

In contrast to that the structured and commented version of the protocol as
shown below is readable by man as well as by the computer:
- It is executable , automatic or step by step,
- it serves as a detailed guide for the interactive repetition of a lesson,

- and its explanations are usable as an introduction for learning and working
  with the system on one's own, apart from the special example.

```
C : Create a red cube of edge length 1 and store it as "WUERFEL".
K :      CREATE BLOCK   1 1 1   RED
K :      STORE WUERFEL
C : Create a blue sphere of radius 0.65 and store it as "KUGEL".
C : Approximate the exact geometry by 16 facets in both main directions.
K :      CREATE SPHERE   0.65   16   16   BLUE
K :      STORE KUGEL
C : Subtract the sphere from the cube / cut the cube with the sphere.
C : Store the resulting object as "WUERFELMINKUGEL".
K :      BOOLEAN CUT_OBJECT
K :      KUGEL
K :      WUERFEL
K :      FIRST_CUT_SECOND
K :      STORE WUERFELMINKUGEL
```

3 APPLICATIONS

In view of the early stage of the project there are only few well-founded experience and results of the educational experiment at the present time. The
participants of the courses come from various disciplines and the main application areas are graphics design, industrial design, and sculpture.

Graphics design. Dealing with the creation of logos and typographic applications. We realized the extreme difficulty to overcome the "solid" appearance
of computer-generated objects, known from advertizing and TV. Fig. 2a shows the
wireframe model of a complex object built from a savings bank's logo (consisting of a capital S with a circle symbolizing a coin on top of it). Figs. 2b
and 2c show a shaded image of the object and a first variation of perspective
to explore new spatial situations and visual effects.

One of the first system extensions we made was the programming of a macro
to generate a typeface (FUTURA), see fig. 3a. Fig. 3b illustrates the incorporation of the third dimension. Experiments with perspective, space and lighting as shown in fig. 3c can be regarded as small but important steps towards
new modes of graphic expression , overcoming the massive solidness of objects.

Industrial design. There are no problems at all concerning the acceptance
of the computer as a new means in the design process. Sometimes missing modelling capabilities lead to modifications of the initial idea about the object.
Smaller well-prepared projects can be realized in rather short time. E.g. the
input and visualization of a student's telephone design was completed in one
day (with assistence of a tutor). The design won a price in a competition about
"the telephone of the nineties" offered by a german telecommunication company.

Sculpture. The computer is regarded as a well-suited spatial simulation
tool for complex, time-consuming and expensive installation projects. Computer
images turned out to be a very effective aid for the visualization and presentation of concepts, e.g. to potential private or public sponsors. On the other
hand, the fact that a computer is available for use does not at all mean a stimulation to use it as a new artistic medium of expression.

Finally some general experience after the first 3 months of interdisciplinary courses:
- Designers and artists can learn together, as far as geometric modelling is
  concerned. Students from different disciplines can stimulate each other in

exploring the system and finding new spatial situations. E.g. walking inside a torus can be an exciting experience.
- The 3D-modeller is not usable in early conceptional shape-defining stages of the design process, but it is well-suited for the externalization and visualization of a well-defined mental model of an object.
- In the applications mentioned above comprehensive and user-oriented geometric modelling techniques turned out to be more important than most sophisticated photo-realistic visualization techniques.
- The rather restricted model-size of a 3D-modeller, due to its storage-consuming data structures and the missing of additional 2D-paint techniques for the final rendering of 3D-images must be regarded as real shortcomings.
- The missing of closed process chains from input to output is not a shortcoming, as expected in advance. The output process is based on color slides taken from the screen. They can be printed on a laser copier up to A3-format, offering the possibility of final manual postprocessing. In order to preserve sensual involvement in the process the combination of computer- and conventional techniques should be further explored and developped.

The report about the experience gained during the courses, including a lot of images from the three main application fields, will be an essential part of the oral presentation.

4 INDUSTRIAL RELEVANCE

The educational concept integrates learning and teaching in a circular process. It is valid for a variety of disciplines and a class of modellers. Its focus is on geometric modelling concepts and not on system handling skills.

What do we expect beyond an educational concept for an academy of art and application examples in singular fields?
- New insights about the integration of computers into well-established design processes. Avoiding the reduction of human sensual involvement. Exploring possibilities of keeping the freedom to use or not to use the computer.
- Ways beyond the commercially dominated definitions of efficiency, which are mainly shaped due to the requirements of production planning and manufacturing and not due to the requirements of the design phase.
- To gain experience to be able to formulate requirements concerning the future development of systems.
- New impulses concerning the industrial application of computer aided techniques, which is to be outlined in the following:

Since the very first ICED-conference again and again the questions occur: What means creativity in design, and how to stimulate creativity? On the one hand it is more and more evident that creativity is the essential prerequisite for facing and solving today's problems, on the other hand it is obvious that hardly anyone really knows what it is and how to influence it (2). Those who pretend to know appear too exotic to find a broad acceptance for their ideas (3), and maybe their methods are a bit too expensive for a general solution?

It is the time of summarizing and evaluating the results achieved up to now (4,5). Design is not a flow-chart-like process. This is just wishful thinking of people who would like to map the process to computer algorithms and thus "solve" the problem once and for all. The link between design theory and CAD has been much too close in the past to be helpful any more. Computers per se have nothing to do with creativity in design, especially they do not promote the creative process. Almost nothing is known about the mental process of generating and formulating a conception of a shape, and thus existing design

methods might even hinder the designer. There is a great deal of scepticism and even some resignation about the question whether design will ever become a science; as if it were a fault to deal with a non- (or not yet-) scientific subject. Keywords for future research efforts are :
- Empiricism, to examine learning- and working processes in design with and without computer-assistence, and
- interdisciplinarity, to put design activities into their historical, social and individual (psychological) context.

The essence of my own ICED 88-contribution (6) was:
- That it is high time to supercede the historically conditioned, but more and more hindering and unnecessary separation of design and engineering.
- That efficiency is more and more a matter of creativity and holistic, inter-disciplinary ways of working and thinking. That growth should be redefined in terms of quality instead of quantity, and that function (as used in "form follows function") has to be defined in a much broader sense than usually done today. See also (7).
- That engineering design education should be more oriented to that of desig-ners. See also (8). The spatial, holistic way of thinking has to be suppor-ted. Old knowledge and skills have to be rediscovered, such as the meaning of colours or the feeling for shapes and the aesthetic appearance of objects.
- That the importance of the standardized drawing has to be reduced according to existing and foreseeable changes in the product development process. The code "technical drawing", which was necessary as a communication medium be-tween design and manufacturing for a long time also means a barrier in the externalization process from the mental model to the visual representation of shapes. Through the medium of the computer model the code "technical drawing" is rapidly losing significance. Ideally the computer model contains the drawing information in full.

At this point we should do the step towards new ways and new codes of ex-ternalization . Modern 3D-modellers can be the tools suitable to promote a re-integration of design and engineering. They might have the chance to become a new, generally accepted sign system in design. The CAD-model establishes one unique numerical description of an object, and, at the same time, allows a great variety of individual visual representations according to the individual mental models of the users.

And why not proceed one step further and learn from the "creatives"? Respected design scientists speak of the "too one-sided postulate of rationa-lity" (4) or of the "neglected right hemisphere of the brain" (9). Creativity, whatever this may be, is certainly far less blocked and restricted as with en-gineers. A well-known and successful example of such an approach were the ideas of the german "Bauhaus", aiming at an integration of arts, handicraft, industry and science (10), leading to completely new concepts of life- and product-quality. Regarding the problems of the present time we need radically new approaches of this quality.

(The North Sea will not be saved by engineers!)

5 REFERENCES

(1) Kristahn, H.-J. / Jonas, W. / Hunstock, A. / Schneider, K. Einsatz eines 3D-Modellierers für die Herstellung von Bildern in der ästhetischen Praxis - IBM 6150/CAEDS. Wissenschaftliches Forum 88, Informationsverarbeitung in Lehre und Forschung, München, 26.-28. Okt. 1988 (IBM Germany).

(2) Leyer, A. Betrachtungen über das Thema Konstruktionsförderung. International Conference on Engineering Design ICED 83, Kopenhagen, 1983, Vol.1, 35-42.

(3) Matchett, E. Trusting in the Unknown: A Key Element in Design Education. International Conference on Engineering Design ICED 88, Budapest, 1988, Vol.1, 473-479.

(4) Müller, J. Gegenwärtiger Stand konstruktionsmethodischer Forschung - offene Probleme. International Conference on Engineering Design ICED 88, Budapest, 1988, Vol.1, 201-207.

(5) Schregenberger, J.W. Eine neue Systematik konstruktionswissenschaftlicher Aussagen. International Conference on Engineering Design ICED 88, Budapest, 1988.

(6) Jonas, W. Computer Aided Industrial Design - The Re-Integration of Industrial Design and Engineering Design. International Conference on Engineering Design ICED 88, Budapest, 1988, Vol.3, 46-54.

(7) Schürer, A. Koordination von Konstruktion und Gestaltung - Tendenzen zukünftiger Maschinengestaltung. International Conference on Engineering Design ICED 88, Budapest, 1988, Vol.3, 158-165.

(8) Schierbeek, B.B. Mechanical Engineering within an Industrial Design Faculty. International Conference on Engineering Design ICED 88, Budapest, 1988, Vol.1, 506-511.

(9) Jorden, W. Zur Ausbildung des Menschen im Studiengang Konstruktionstechnik. International Conference on Engineering Design ICED 88, Budapest, 1988, Vol.1, 459-465.

(10) Zec, P. Kunst und Technik heute. Umbruch, Zeitschrift für Kultur, 1988, 3, 53-57.

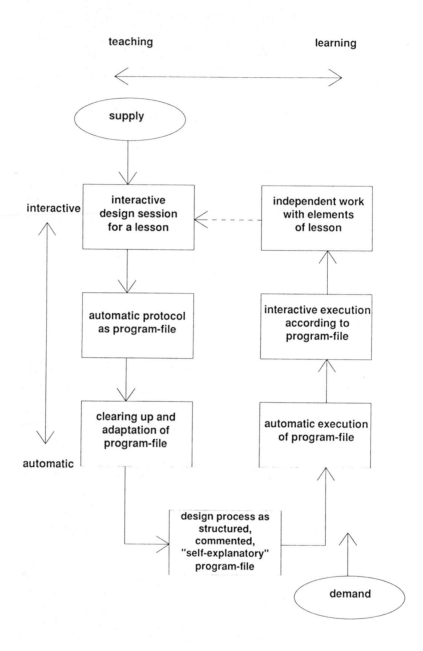

Fig 1   Educational concept

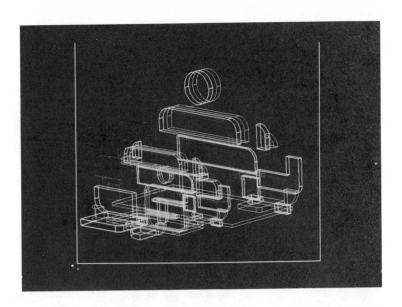

Fig 2a   First example from graphics design (1)

Fig 2b   First example from graphics design (2)

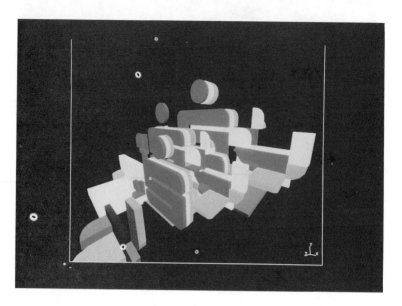

Fig 2c   First example from graphics design (3)

Fig 3a   Second example from graphics design (1)

Fig 3b   Second example from graphics design (2)

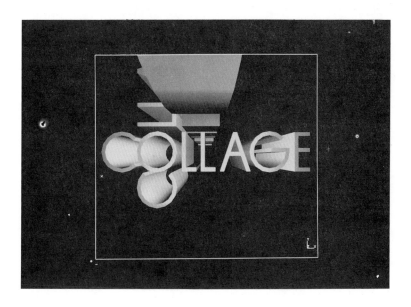

Fig 3c   Second example from graphics design (3)

Revitalization of machine elements in machine design education

K HONGO
Mechanical Engineering Department, Kogakuin University, Tokyo, Japan

SYNOPSIS For the Mechanical Engineering Design Educa-
tion(MEDE) in Japanese Universities the traditional
machine-element-oriented subjects have been reserved
and still continue. Nevertheless Engineering has entered a new cycle of development and definition. The
traditional pedagogical methods of the introductory
lectures and design exercises of machine elements are
now too obsolescent and frustrating. New attractive
pedagogical methods for machine-element-oriented
subjects in MEDE are needed. In today's technical
society we need a revitalization of machine-element-
oriented subjects in the MEDE. The following approach
is proposed : (1)The undergraduate students should
have an introductory lecture on Machine Elements during their first two years. (2)Advanced undergraduate
students as well as graduate students should then be
introduced to case studies, CAD, applied engineering
science and the methodological design of machine
elements. It is hoped that the teaching staff of mechanical engineering design around the world will
consider this approach and support the revitalization
of machine-element-oriented subjects in MEDE.

1 INTRODUCTION

Machine Elements and Design has been a part of Mechanical Engineering Design
Education(MEDE) since Reuleaux(1861) and Unwin(1877). Machine-element-oriented subjects based on the analytical description of geometric and material
parameters such as screw-thread fastenings, power transmission shafts , gears
and bearings were common in the pre-World War 2 curriculums of MEDE. The introduction of mathematics, physics, and engineering science caused a drop in
the traditional design subjects of undergraduate curriculum. This was done at
the expence of machine-element-oriented design subjects. The traditional design of machine elements continued to a lesser or greater degree depending on
the Universities and Colleges in the World, however, the less flamboyant machine-element-oriented subjects tend to become so obsolescent that teaching
and learning of the subjects are now a painful obligation.
The utilization of better machine elements is now required in order to achieve
superior performing machines in this highly technical society. It is clear

that a machine designer with poor knowledge of machine elements will not be successful, so the respectability of Machine Elements in MEDE should be restored. The revitalization of Machine Elements in this new age is important.
The author discusses these problems and proposes pedagogical methods for the revitalization of machine-element-oriented subjects in MEDE from his experience in Japan.

2 SITUATION IN JAPAN

The Design Education Committee of The Japan Society of Design and Drafting (JSDD) sent a questionaire to the Mechanical Engineering Departments of 113 Universities in Japan in 1981.   They were asked about the Design Education System in their own Departments.   134 Departments from 84 Universities responded.[1],[2].   The present situation, in 1988, has not been drastically changed from the result of the survey in 1981, except for the increased introduction of CAD subjects.

2.1 Machine Elements

According to the results of the survey, the lecture on Machine Elements is given at almost all undergraduate courses of Mechanical Engineering Departments of Universities in Japan.   This lecture follows the traditional presentation of tracing machine element dating to the early history of teaching engineering. The topics covered in this lecture in Japan are shown in table 1.   They are based on Japanese text-books which are equivalent to Shigley[13].   The average lecture time is about 60 hours.   The student level is the 2nd and 3rd year in most Universities.

Table.1 Topics covered by "Machine Elements"

Standardization of Machine Elements, Stress concentration, Factor of Safety, Tribology, Tolerance and Fits, Welded Joint, Threaded Joints, Key, Spline, Serration, Variable Speed Drives, Belting, Gearing, Chain Drives, Shafts, Couplings, Clutches, Brakes, Bearings, Springs, Frames, Pipings, Pressure Vessels

The results of this survey indicated that, almost undergraduate Mechanical Engineering curriculum in Japanese Universities have the curriculum with title "Machine Design and Drawing".   The student levels range from the 1st to the 4th year depending on the structure of whole curriculum.   One example is shown in table 2.   Under this title, fundamental drawing office practice is taught at a lower student level, and Machine Design Exercise or Design Projects are assigned to students at higher levels.   Example of Machine Design and Drawing topics are listed in table 3.   The students submit assembly and datail drawing along with a report including calculations for their Design Exercises.   Machine elements are taught through the course of Machine Design and Drawing.

2.2 Evaluation and Criticism

In Japanese Mechanical Engineering Departments, Mechanical Engineering Design Education(MEDE) with machine-element-oriented subjects has retained a high

Table.2 Machine Design and Drawing
(Mech. Eng. Dept. Uni. of Tokyo,1981,1985)

Title of Course	S.L[1]	Semes.	hours/week×week
Machine Design and Drawing 1	2.	4.	3×15
Machine Design and Drawing 2	3.	5,6.	3×30
Machine Design and Drawing 3	4.	8.	3×10

1) S.L.:Student Level

Table.3 Topics of Machine Design and Drawing(1981)

S.L[1]	1st, 2nd year		3rd, 4th year	
Rank	Title	C.F[2]	Title	C.F[2]
1	Bolt,Nut	52	Internal Combustion Engine	55
2	Gears	49	Centrifugal Pumps	55
3	Shaft and Coupling	44	Reduction Gears	32
4	Winch	41	Boiler	27
5	Ball and Roller Bearing	34	Machine Tools	26
6	Gear Pump	23	Cranes	16
7	Jack	22	Water Turbines	10
8	Handle Wheel	17	Steam Turbines	8
9	Cluch	15	Hydraulic Machine	7
10	Springs	10	Air Compressor	4
11			Heat Exchanger	3

1) S.L.:Student Level,   2) C.F.:Cited Frequency

respectability.[2].   Engineering entered new cycle of development and definition after W.W.2.   More mathematics, physics and engineering science were
introduced into Mechanical Engineering curriculum in Japan.   This change was
not a radical as in the U.S.A. during the post war period.   The traditional
lectures and redesign of the existing machine elements still seem far behind
the progress of modern Engineering.   The respectabiliy of machine elements
is mainly sustained from the old pre war tradition.   In some cases, teaching
the machine elements is a monotonous and tiresome duty assigned to the teaching staff. As a result both the teaching staff and students underrate the importance of machine elements in machine design.   These criticisms are clearly expressed in the survey of the teaching staff.   They need a new pedagogical method for MEDE.

3 AUTHOR'S PROPOSAL

3.1 Significance of Machine Elements

The significance of machine elements are recognized from two points of view.
The first is the responsibility of Mechanical Engineering in today's technical
society, namely Designing and manufacturing of hardware that satisfy human
needs.   The hardware should perform the required function safely, with high
reliability and should be produced with the most cost effective techniques.

Machine elements are decisive for a machine system to perform the required function. The technological advantage and competence of a machine system are increased when superior machine elements are used. They contribute to the reliability or cost reduction of the machine. It is therfore necessary to recruit a brighter, more motivated student body, who are either interested in machine elements, or who appreciate machine elements. The second point of view is pedagogical. The teaching of such a vast sphere like Machine systems, represents two extreme possibilities. One is based on real examples while the other is abstract. The traditional approach is derived from real examples. The machine element is the lowest level of Abstraction and of the lowest level of complexity. If a student is to learn the design process beginning with the awareness of the goal to the completion of the design drawing in a limited time interval, it is necessary to use a machine element. Machine elements have been utilized in the MEDE in the past and should be utilized wisely in the future.

3.2 Courses of Machine-element-oriented Subjects

For the revitalization of Machine-element-oriented subjects in the MEDE, the author presents here examples of courses in undergraduate and graduate levels of education in Japan along with other experiences reported at previous ICED conferences.

3.2.1 Course Goals

The course goals are as follows:

Undergraduate course

The machine-element-oriented subjects are given to the undergraduate students,
-to develop an overview and appreciation of machine elements,
-for compensation for troubling eventually false aspects of Machine Elements as follows.

The subject "Machine Elements" is an easy subject which is unnecessary among the undergraduate curriculum of Mechanical Engineering Department of University. The information or knowledge are easily accessible, whenever one wishes to get at. In the area of Machine Elements there are few attractive problems. It is very difficult for a researcher to get Prizes or Grant-in-Aid for a study of Machine Elements.

Graduate course

The machine-element-oriented subjects are given to the graduate students,
-for their theses of master or doctor research works (dissertation).

3.2.2 Examples of courses

(1) For undergraduates at 1st or 2nd year :

Introductory Lectures and Design Exercises of Machine Elements,

Remark: The Design and Study of Machine Elements have developed over 100 years to attain their current state of art. Machine Elements are particular mechanical systems with a low degree of complexity. The sphere of Machine Elements is rather wide, encompassing a large number of elements which at times differ greatly from one another and require specific time-consuming and indepth studies. From the pedagogical point of view it is a difficult problem to teach this wide sphere of Machine elements within a limited time interval, say less than 60 hrs to at most 200 hrs.   It is impossible for the teaching staff to explain every machine element in-depth.   They must discuss machine elements in which they are most interested in and have expert knowledge.   Evoke the inductive spirit in student's mind, and draw out a general rule from a specific example.

(2) For undergraduates at 3rd and 4th year:

Case Studies

Introduction of interesting case studies for demonstration of the significance of machine elements.

Remark: One of the best methods for demonstrating the significance of a machine element in a machine system is a Case Study.   Some examples are i) tragic disasters due to the negligence of the significance of machine elements, such as break of rear pressure bulk head of a civil aircraft due to the fracture of riveted joint,   ii) the explosion of gases that leak through the gaskets of a flange of pressure pipings or vessels,   iii) the falling of movable lighting system from the ceiling in a Discotheque club due to the break of roller chain , etc.,   iv) Microcomputer-controlled Machine system;   The critical technique may be to focus on the machine element in a microcomputer- controlled system. The author found, the design of a belt system was the most time-consuming task in the design process of a microcomputer-controlled screw selecter.(Fig.1,2)[3].

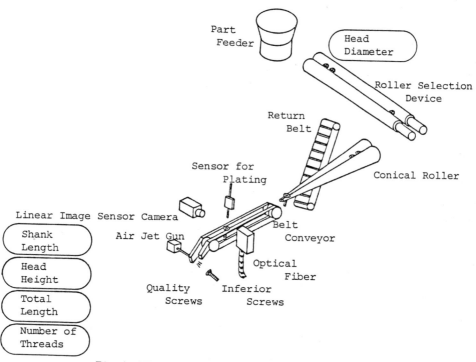

Fig.1   Microcomputer-controlled Fastener Selector
(NHK Spring Co.) [3]

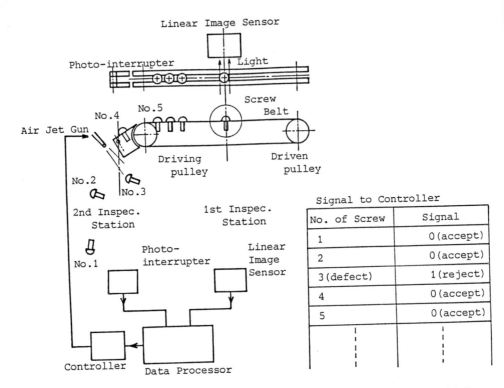

Fig.2 Belt and Two Inspection Stations(NHK Spring Co.)[3]

(3) For undergraduates in their final course and graduates in Master and Doctor courses:

Graduate students may investigate machine element problems for master or doctor thesis. In some cases these studies may be assigned to advanced undergraduate students as graduation research work. Examples are:
-Application of Engineering Science to the analysis of functions of Machine Elements.
-Methodological Design of Machine Elements.
-Effective CAD for designing machine elements.
-Expert system of designing machine elements.
Remark: The application of Engineering Science to the analysis of functions of machine elements is widespread. For reference a number of published articles are cited. Recent examples from the author's Laboratory are reported in [5],[6]. Examples of the research work of the Methodological Design of Machine Elements are reported in [9],[10],[11]. An Example of research work on an effective CAD system for designing machine elements is reported in [7]. An example from the author's Laboratory is shown in Fig.3 [8]. Examples of research work on Expert systems for designing machine elements are given in [20],[21]. The proposed curriculum will vitalize the machine-element-oriented subjects in the MEDE in Universities and Colleges.

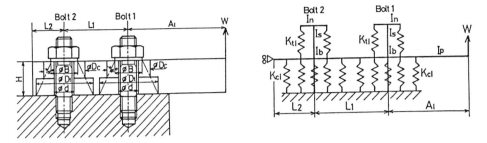

Fig.3 Spring-beam model for multi-bolted joint [8]

4 INTERNATIONAL COOPERATION

Many reports on the CAD of machine elements, the application of engineering science to the analysis of function of machine elements, and the methodological design of machine elements have been presented in the past ICEDs. These reports are very good references for development of the Mechanical Engineering Education in Japan. The reports of revitalization, expansion, and in some cases, reintroduction of machine-element-oriented design subjects in Mechanical Engineering Design Education in Universities and Colleges in the USA are good news for Japan[12]. The older editions of Shigley's text, 'Mechanical Engineering Design'(1963,'73,'77) had a good influence on the machine design instruction in Japanese Universities and Colleges. The revised, metric edition of Shigley's text [13] which appeared in 1986 after a period of silence is a pleasant surprise for the Japanese teaching staff. The new edition of Niemann's text books 'Maschinenelemente'[14] and other new text book of special machine elements [15]-[19] are stimulating and encouraging our Japanese teaching staffs in spite of the fact that in this post war period German is not as popular as English in Japan. The author hopes that this discussion will enable more cooperation among teachers of Mechanical Engineering Design in the world and assist in the revitilization of machine-element-oriented subjects in the Mechanical Engineering Design Education.

5 CONCLUSIONS

(1) Mechanical Engineering Design Education(MEDE) in Japanese Universities and Colleges is based on the traditional machine-element-oriented subjects which have been preserved and still continues.
(2) Since Engineering has entered a new era of development and definition, the introductory lectures and design exercise based on mechanical drawing of machine elements are obsolescent and frustrating. New pedagogical methods for machine-element-oriented subjects are now required.
(3) The author recognizes the significance of machine elements in machine systems design in the highly technical society. He proposes the revitalization of machine-element-oriented subjects in the MEDE. The author's proposals are as follows: After the introductory lecture of machine elements for lower student levels, the students will be exposed to interesting case studies. These include utilization of CAD, application of engineering science, methodological design and expert system for design of machine elements.

(4) These developments are evident in the many reports on the design of machine elements presented in the past ICEDs and the publication of new text books of machine elements in USA and West Germany. The author looks forward to future cooperation among the teaching staff of the MEDE around the world for the revitalization of machine-element-oriented subjects.

REFERENCES

[1] The Japan Soc. of Design and Drafting: The present situation of Design Education in Mechanical Engineering Departments of Universities in Japan, 1982 (in Japanese).
[2] K.Hongo: A Survey of the Mechanical Engineering Design Education in Universities in Japan in comparison with Examples from Europe and North America. Proc. of ICED 87, Boston 898, 1987.
[3] K.Hongo: Reality and Abstraction, Proc. of ICED 88, Budapest, 451, 1988.
[4] K.Hongo: Studies on the Popping of Sprags of overrunning Clutch, Int. Symp. on Grearing & Power Transmission, Vol.2, 431, Tokyo, 1981.
[5] K.Hongo, T.Nemoto, H.Otaki: The Deformation and Distribution of Load in Screw Threads. J. of Faculty of Eng., Uni. of Tokyo. Vol.27(B) No.1.49, 1983.
[6] S.Yazawa, K.Hongo: Distribution of Load in Screw Thread of a Bolt-Nut Connection subjected to Tangential Forces and Bending Moments, JSME International J., Series 1, Vol.31, No 2, 174, 1988.
[7] K.Hongo, N.Nakajima: Designer and CAD, A Report from our Laboratory, Proc. of ICED 83, Copenhagen, 631, 1983.
[8] M.Tanaka, K.Hongo: The Stress Analysis of Bolted Joint with Model using Spring-beam Elements, J. of JSPE, Vol.53, 664, 1987 (in Japanese).
[9] U.Pighini: Methodological Design of Machine Elements, Proc. of ICED 87, Boston, 254, 1987.
[10] K.Ehrenspiel, J.John: Inventing by Design Methodology, Proc. of ICED 87, Boston, 29, 19-87.
[11] K.H.Roth: Konstruieren mit Konstruktions-katalogen, Springer-Verlag, Berlin, 1982.
[12] R.W.Mann: Engineering Education: US-Retrospective and Contemporary, Trans. ASME, Vol.103, 696, 1981.
[13] J.G.Shigley: Mechanical Engineering Design, McGraw-Hill, New York, 1986.
[14] G.Niemann, H.Winter: Maschinenelemente, Band 1. 1981, Band 2. 1985, Band 3 .1986, Springer Verlag, Berlin.
[15] H.Wiegand, K.H.Kloos, W.Thomala: Schraubenverbindungen, Springer Verlag, Berlin, 1988.
[16] J.Looman: Zahnradgetriebe, Springer Verlag, Berlin, 1988.
[17] F.Schmelz, H.C.Graf von Seherr-Thoss, E.Aucktor: Gelenk und Gelenkwellen, Springer Verlag, Berlin, 1988.
[18] H.Peeken, C.Troeder: Elastische Kupplungen, Springer Verlag, 1986.
[19] S.Winkelmann, H.Hartmuth: Schaltbare Reibkupplungen, Springer Verlag,1985.
[20] N.Nakajima, T.Murakami, K.Oikawa: Basic study on Computer Aided Design Diagnosis, Proc. of ICED 87, Boston, 648, 1987.
[21] K.Ehrlenspiel, K.Fliegel: Application of expert systems in machine design, Konstruktion, 39, H7, 280, 1987.

Design and quality

R S de ANDRADE, PhD
Universidade Federal do Rio de Janeiro, Brazil

A comparison between product life cycle models and the quality loop, as defined by the International Organization for Standardization, ISO, is made in order to establish the similarities of the activity models used by specialists in quality and in design. The interrelationships of the parameters of quality, namely, quality of design, quality of conformance and quality in use are discussed in respect to their consideration in the product life cycle. The influence in that cycle of costs of failure, appraisal and prevention, the quality costs, is also considered. A comment on the effect in the design process of the compromise between the value of quality and the cost of quality is made. To finalize, the achievement of quality through design is discussed.

1. INTRODUCTION

In recent years companies began to put great emphasis on the quality of their products. Quality is regarded as one of the major factors for competitiveness. The traditional approach of controlling the quality of a product mainly at its production phase, in the belief that was there where product failures were originated, has evolved to the quality control in all activities pertinent to the company with the understanding that quality has to be cared for in all steps necessary for a product to exist and be in use. Actually, quality should be a matter of attitude. The exercise of this attitude within the whole collection of activities of a company is known as total quality control, company wide quality control or quality assurance. That requires the existence of a quality system for the implementation of the quality management which is responsible for coordinating the actions towards quality in all departments in order to attain the desired quality in the final product. The collection of the entire activities through which quality is achieved, no matter where these activities are performed, is defined as quality function(1). Design plays a key role in this function. In order to develop a discussion about design and quality it is necessary to state the meaning adopted in this paper for the latter. Definitions of quality are as abundant as definitions of design. Juran defines quality as "fitness for use"(1). Crosby as "conformance to requirements"(2) and Feigenbaun as "the total composite product and service characteristics of marketing, engineering, manufacture, and maintenance through which the product and service in use will meet the expectations of the customer"(3). The definition presented by Taguchi is "the loss a product causes to society after being shipped, other than any losses caused by its intrinsic functions"(4). The American National Standard Institution, ANSI, together with the American Society for Quality Control, ASQC(5), define it as "the totality of features and characteristics of a product or service that bear on its ability to satisfy stated or implied needs". The European Organization for Quality Control, EOQC(6) defines it as "the totality of features and characteristics of a product or service that bear on its ability to satisfy a given need". The latter is the definition used here, with "product and service" refered simply as "product".

2. PRODUCT LIFE CYCLE AND THE QUALITY LOOP

A representation of the interactions between the activities involved in the progress in quality is the quality loop introduced by the International Organization for Standardization, ISO(7), and shown in Fig.1. This loop suggests that the assurance of quality is achieved by the constant action on a set of activities that are interconnected in a closed loop in the sense that new needs of customers evolve from previously satisfied needs. Actually, the quality loop is derived from the spiral of progress in quality(1) which represents the ascending evolution of the quality of a product. The process initiated by a collection of needs and completed by the final deactivation of the product designed and built to satisfy them can be referred as the product life cycle. If the quality loop is opened and rearranged as shown in Fig.2., it can be noticed a great similarity between it and product life cycle models proposed by several authors in the field of engineering design. Just for the sake of comparison, the model presented by de Andrade(8) is pictured in Fig.3. Other models(9, 10, 11, 12, e.g.) could be taken to the same effect, though some are concerned mainly with the steps related to the engineering design activity. Actually, the quality loop is a product life cycle model with which the strategy of ataining quality by spreading an atitude of quality along the whole of the development process is emphasized. An important remark to be made from the comparison between the quality loop and the other models is that all try to convey the necessity to consider the life cycle of a product as a whole, i.e., considering the sequence of all activities and their interrelationships in order to arrive at a successful satisfaction of the needs.

3. QUALITY PARAMETERS

The quality of a product is assessed by means of its quality characteristics, i.e., the set of product attributes provided to satisfy the needs. For example: speed, energy consumption, colour, texture, reliability, durability etc. Those characteristics are commonly grouped under three aspects: quality of design, quality of conformance and quality in use known as the quality parameters(1). The refered author does not use the term "quality in use". However, it seems it summarizes well the characteristics associated to the handling and operation of the final product. The first parameter relates to the adequacy of the design and comprises the product design. The second is about the conformity of the manufactured product in relation to the design prescriptions. The third concerns the delivery and installation; the performance including reliability, availability, capability, operationability and safety; the technical assistance and the disposal. The activities where each of these three aspects are considered in the product life cycle model are indicated in Fig.4.

Although, as part of the quality function , several supporting activities which are not explicitly included in the design parameters such as planning, documentation, communication, and management are executed along the design, manufacture and use of a product, it can be said that the key parameter of quality is that of design since the other two are dependent of it.

Conformance of manufacture to the design specifications are made feasible or facilitated if in the design stage the manufacturing constraints are considered. The achievable performance in use is set by the design and also are the requirements for installation and technical assistance. The delivery and disposal, though not always totally controllable by the design, are very influenced by it.

The interrelationship between design, manufacture and use is a necessary consideration of the design process where the manufacture requirements and user needs are incorporated as part of the specifications for the design.

Refering to the approach of assuring quality mainly by the control at the manufacturing stage it can be understood that, at most, it guarantees that what was planned is done properly. Refering to the interplay between the parameters considered above, it can be said that if good quality is to be achieved, it has to be designed in the first place.

4. QUALITY COSTS

Quality costs are related to making, correcting or avoiding failures to achieve the desired quality level. The elements of these costs are usually grouped into the following categories(13):

Prevention - costs of any action taken to investigate, prevent or reduce defects or failures.
Appraisal - costs of assessing the quality achieved.
Internal failure - costs arising within the organization of the failure to achieve the quality specified (before transfer ownership to the customer).
External failure - costs arising outside the organization of the failure to achieve the quality specified (after transfer ownership to the customer).

Prevention costs are generated by the actions of avoiding failures such as the planning and management of the quality system, process planning and control, design and maintenance of equipment for test and quality control, product design reviews, quality motivating and training programmes. Preventive actions support the development process and are performed along the whole of it, even at the stage of use, although at that moment the associated costs are not always computed by producers since they may be incurred by the customer (training for operation or preventive maintenance, for instance).

The costs of appraisal are associated with the verification of the attained quality and comprise tests, measurements and inspections. These actions are mostly concentrated at the production stage of the development process. Model and prototype testing in the design stage are also appraisal efforts. The action of appraisal is that of checking the planned, represented by the design specifications, against the executed, represented by the resultant manufacture and performance.

Internal failure costs are the result of defects found by appraisal actions before the product is transfered to the customer. Scrap and rework are elements of these costs which, in the development process, occur from the production stage until the possession of the product by the customer. Damages caused during transportation by inadequate packaging or unsuitability of the product for the handling to which it is submitted should also be accounted as internal failures if delivery is the producer's responsability.

External failures occur at the stage of use and the costs associated to them comprise the consequences of products repaired, rejected or returned and warranty replacements due to errors or inadequacies in the design, the manufacture or the installation. They also include the consequences of product liability.

Few of the cost elements considered above can be directly associated to the design activity (product design review, prototype testing and external failures costs due to design errors or inadequacies, for example). This is so because costs are commonly related to the consequences and not to their causes. Then, the effect of decisions taken at the design stage are not always appreciated and cost reductions are attempted by other means such as tighter controls or more intensive inspections.

Gryna(14) states that "there is increasing evidence
many fitness-for-use problems can be traced to the design or ~..
product" and reports some of that evidence. He also suggests
that 40 percent of the fitness-for-use problems are caused by
errors during the design stage.

5. VALUE AND COST OF QUALITY

A useful concept for design consideration is that of value and
cost of quality. In this situation the cost of quality refers to
the resources spent in order to provide a product with a certain
level of quality. That comprises not only the quality costs,
which are associated with the failures, as described in the
previous section, but also costs of workmanship, materials,
components, processes, facilities, administration etc.
Differences in the level of quality can be observed by comparing
similar products like a luxury car and a popular model. Each
car could present exactly the same characteristics but treated
in such way that in the luxury version the quality would be
higher. Better fittings, finishes and performance are examples
of differences in the quality level. The value of quality is
the price the customers are prepared to pay for the product and
is related to how they appreciate its quality characteristics.
Some authors represent the relationship between value and cost
of quality by means of a model similar to that in Fig.5.(15).
It shows that increases in the level of quality are followed by
increases in cost and value, though the latter, after a certain
point increases at a decreasing rate. The aim of the producers
is to maximize the difference between value and cost. The task
of the designers is to interpret the customers evaluation of
product quality and devise an adequate solution at an
economically atractive cost.

6. DESIGN AND QUALITY

The considerations in the foregoing were intended to present a
rough picture of the general aspects of product quality. There
is a great amount of literature on the subject and for those
interested in more information a good start is the Juran's
Quality Control Handbook(1).
    From the definitions presented in the introductory section
it can be said that a product with assured quality is developed
with the aim of "attaining the most satisfactory results for
the fulfilment of a set of needs ". Not surprisingly, that
would be also a suitable aim for design. Quality and design
have identical objectives.
    Perhaps it is necessary to make a comment on the meaning of
needs in order to facilitate the appreciation of the importance
of design in the product life cycle. The term is used as
requirements but not restricted to the necessities of the
customer. It should also comprise those requirements related to

every stage of the product life cycle; those related to all organizations and people affected by the product and its cycle; those related to factors external to the life cycle such as society and environment but which could be influenced by the product and its cycle. The broadness of this interpretation of the needs may seem exagerated, but we are daily confronted with bad consequences of design decisions that, maybe, could be avoided if a more adequate consideration of the needs was done. Take, for instance, the depletion of the Earth's ozone layer blamed on the gases released from aerosols and refrigerating systems or the acid rain provoked by industrial processes. Surely, these consequences are not evidences of good quality.

Designing for quality requires, first of all, a comprehensive understanding and evaluation of the needs from which the specifications for the product design are formulated. Quality is one of the elements of the specifications. This has been pointed out by Pugh(16) who stresses in his work the importance of the specifications for the design activity. In a further work he discusses the specific aspect of quality in the design process(17).

In the design stage, the consideration of quality aspects is initially done with the inclusion of the quality characteristics into the specifications. The incorporation of quality characteristics into product design specifications is part of the quality function deployment used by some companies as support to company wide quality control. Kogure and Akao discuss some of the features of the quality function deployment (18). One of the obstacles that may be encountered is the difficulty to quantify some of those characteristics. Development in fields like reliability, tolerancing and process capability makes possible to define, by means of measurable specifications, several characteristics that could not be quantified before.

Quality is also considered in the design phase by the assessment of the design decisions. Techniques like failure mode and effect analysis and fault tree analysis provide designers with means to predict consequences of faults or failures in the conceptual stage. Simulation, another tool for prediction, can be applied throughout the whole design process(8).

In recent years a great deal of attention has been drawn by the Taguchi methods(4) which apply statistical methods to evaluate the combined effect of design parameters in order to minimize the variation of performance characteristics around the target values since that variation is considered to be directly proportional to the quality losses caused by the product. The main features of his methods are the philosophy of off line (off production) control and the use of statistics for planning experiments and analysing data. Taguchi methods are applied after the conceptual stage in the design process.

Other means of quality assurance during the design phase are, for instance, tests with models and prototypes in-house or in the field.

Apart from some techniques that have been developed or introduced by motivation of the search for a better quality, the appreciation of quality in the design process is the same as has been always happening with any of the other specification elements. The current importance given to quality leads to the impression that in order to achieve quality, design has to be done in a different way. That is not so. If design is done properly in the first place, good quality will be a consequence.

7. CONCLUSIONS

The main objectives of this paper were to explicit the importance of the design activity in the effort of product quality assurance and to emphasize the identical aims of quality and design. This identity comes from the fact that both require a total and an integrated view and consideration of the needs and their satisfaction in order to achieve efficiency. A very important aspect of this approach , in this world of increasing availability of information and professional specialization , is the necessary and unavoidable interdisciplinary work and cooperation. The author believes that professionals working with quality and design could play an important role in stimulating this integration. After all, what we are trying to design and produce is a better quality of life.

8. REFERENCES

1...Juran, Joseph M. "The Quality Function". In Juran's Quality Control Handbook, 4th ed., chap.2. McGraw-Hill Book Company, New York, 1988.
2...Crosby, P.B. Quality is free. McGraw-Hill Book Company, New York, 1979.
3...Feigenbaun, Armand V. Total Quality Control, chap.1. McGraw-Hill International Editions, Singapore, 1986.
4... Taguchi, Genichi. Introduction to Quality Engineering, Designing quality into products and processes. Asian Productivity Organization, Tokyo, 1986.
5...ANSI/ASQC A3-1987. Quality Systems Terminology, Milwaukee, 1987.
6...European Organization for Quality Control. Glossary of Terms Used in Quality Control, 5th ed., Berne,1981.
7...ISO 9004 - Quality management and quality system elements - Guidelines. International Organization for Standardization, 1987.
8...de Andrade, Ronaldo S. "The Role of Simulation in Design". Proceedings of the International Conference on Engineering Design, pp 401-407 Budapest, August 1988.
9...Pugh, Stuart. "The Engineering Designer - His Tasks and Information Needs". Proceedings of the Second International Conference on Information Systems for Designers, pp 63-66. Southhampton, March 1977.

10..Ortrofsky, Benjamin. Design, Planning and Development Methodology. Prentice Hall Inc, Englewood Cliffs, 1977.

11..Hales, Crispin. "Analysis of the Engineering Design Process in an Industrial Context". Ph.D. Dissertation. University of Cambridge. Gants Hill Publications, Eastleihg, Hampshire, 1987.

12..Pahl, G. and Beitz, W. Engineering Design. Edited by Ken M. Wallace. The Design Council, London, 1984.

13..BS6143:1981 - Guide to the determination and use of quality related costs. British Standards Institution, London, 1981.

14..Gryna, Frank M. "Product Development". In Juran's Quality Control Handbook, 4th ed., chap.13. McGraw-Hill Book Company, New York, 1988.

15..Kirkpatrick, Elwood G. Quality Control for Managers and Engineers, chap.1. John Wiley and Sons, New York, 1970.

16..Pugh, Stuart. "Engineering Design - Towards a Common Understanding". Proceedins of the Second International Simposium on Information for Designers, pp D4-D6. University of Southampton, 1974.

17..Pugh, Stuart. "Quality Assurance and design: the problem of cost versus quality". Quality Assurance, vol4, no1, pp 3-6, March 1978.

18..Kogure, Masao and Akao, Yoji. " Quality Function Deployment and CWQC in Japan". Quality Progress, pp 25-29, October 1983.

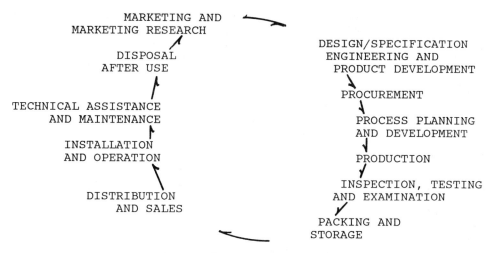

Fig.1. The loop of quality

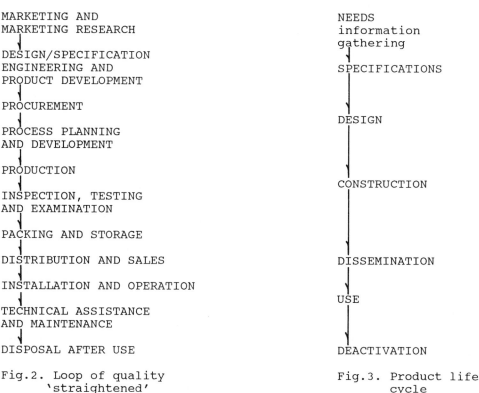

MARKETING AND
MARKETING RESEARCH

↓

DESIGN/SPECIFICATION
ENGINEERING AND
PRODUCT DEVELOPMENT

↓

PROCUREMENT

↓

PROCESS PLANNING
AND DEVELOPMENT

↓

PRODUCTION

↓

INSPECTION, TESTING
AND EXAMINATION

↓

PACKING AND STORAGE

↓

DISTRIBUTION AND SALES

↓

INSTALLATION AND OPERATION

↓

TECHNICAL ASSISTANCE
AND MAINTENANCE

↓

DISPOSAL AFTER USE

Fig.2. Loop of quality
      'straightened'

NEEDS
information
gathering

↓

SPECIFICATIONS

↓

DESIGN

↓

CONSTRUCTION

↓

DISSEMINATION

↓

USE

↓

DEACTIVATION

Fig.3. Product life
       cycle

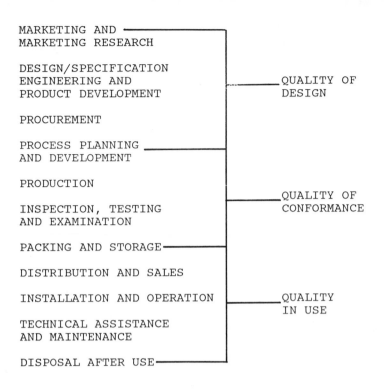

MARKETING AND
MARKETING RESEARCH

DESIGN/SPECIFICATION
ENGINEERING AND   QUALITY OF
PRODUCT DEVELOPMENT   DESIGN

PROCUREMENT

PROCESS PLANNING
AND DEVELOPMENT

PRODUCTION

INSPECTION, TESTING   QUALITY OF
AND EXAMINATION   CONFORMANCE

PACKING AND STORAGE

DISTRIBUTION AND SALES

INSTALLATION AND OPERATION   QUALITY
                                    IN USE
TECHNICAL ASSISTANCE
AND MAINTENANCE

DISPOSAL AFTER USE

Fig.4. The quality parameters in
       the product life cycle

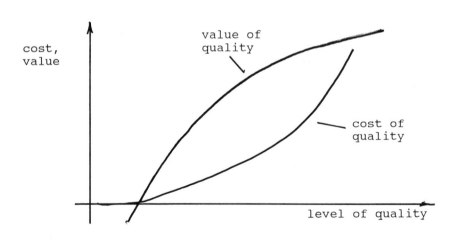

Fig.5. Value and cost of quality

Graduate design training in industry

R BASSIL, MA, C M CARTER, MA and P J W SAUNDERS, MA, CEng, MIMechE
APV Baker PMC Limited, Peterborough
K M WALLACE, BSc, MA, CEng, MIMechE
Department of Engineering, University of Cambridge

In 1986 a design-based Teaching Company Programme was
started between APV Baker Printing Machinery Company
in Peterborough and Cambridge University. The main
reason for undertaking this joint venture was to introduce a new approach for training graduate design
engineers. This paper describes the training approach
used, the project undertaken and the benefits to the
graduates, the company and the university.

1 INTRODUCTION

1.1 The industrial and academic partners

APV Baker Printing Machinery Company (PMC) operates in the highly competitive
heatset web offset printing market and has been successful in gaining a large
share of both the British and North American markets. The company's annual
turnover is around £67 million and most of its 1100 employees work at its
manufacturing centre in Peterborough. The company has recently been acquired
by Rockwell International, which has an annual turnover of around $12 billion
and employs 100 000 worldwide.

APV Baker PMC recognises the importance of high quality engineering
design in capturing and retaining its competitive advantage. It also
recognises the need to train future designers to the highest standards in the
shortest time.

The Engineering Department at Cambridge University has spent a
considerable amount of time developing and implementing new methods for
teaching engineering design (1) and has had experience of running many student
design projects in conjunction with industry.

The opportunity presented itself to undertake a significant design
training project and a Teaching Company Programme was chosen as the
basis for this cooperative venture.

1.2 The Teaching Company Scheme

The Teaching Company Scheme was established in 1975 by the Science and
Engineering Research Council (SERC) and the Department of Trade and Industry
(DTI). The Scheme aims to foster active partnerships between academic

institutions and industry by setting up Teaching Company Programmes. Once a Programme has been approved, able graduates are recruited jointly by the company and the academic institution as Teaching Company Associates. These Associates are employed by the academic institution on two-year fixed contracts, but work almost exclusively at the company on clearly defined projects of real significance to the industrial partner. The formal aims of the Teaching Company Scheme (2) include: training graduates for industrial careers; increasing the relevance of academic teaching; and raising industrial performance.

In October 1986, a design-based Teaching Company Programme was established between APV Baker PMC and the Engineering Department at Cambridge University. The first two Associates recruited were both graduates of the Manufacturing Engineering Tripos at Cambridge University and were aged 22 and 23 when they joined the programme.

The Manufacturing Engineering Tripos is a four-year course. In the first two years, the students study engineering fundamentals and in the second two they specialise in manufacturing engineering. During the course they must undertake a number of projects based in industry during which they are expected to produce significant results. Over the years these projects have demonstrated that the students can accept and respond to a high level of responsibility. This led to the view that design engineers might be trained more rapidly and effectively if they were to be given responsibility for a significant design project as early as possible in their training.

1.3 The training approach

The company's existing graduate design training consisted of first providing the basic experience required to qualify the recent graduates for chartered engineer status followed by a series of short and unrelated design projects of increasing complexity. Under this scheme the trainee design engineers would not be given full design responsibility for at least the first 15 months of their training.

As part of the Teaching Company Programme, the following main ideas were to be tested:

- The Associates were to be given full responsibility for managing their own project.

- The project, which had to be of commercial significance to the company, was to be presented to the Associates on their first day of training and taken through from confirming the initial market need to commissioning the resulting product.

- A systematic design procedure was to be used to provide a structure for the project and to ease communication between the Associates and the company's engineers who had to provide much of the technical support.

The aim was to maximise the ability and enthusiasm of young graduates while at the same time giving them the confidence and support to undertake a significant project. The traditional training approach attempts to build up the necessary technical experience, through working on a number of unrelated small tasks and through special training, before giving managerial responsibility for a project. With this approach many young graduates become extremely frustrated, lose their motivation to become designers and move into other areas where they can achieve responsibility and promotion more rapidly.

It was believed that young graduates could manage a project effectively while they acquired the necessary technical experience. The technical support for their project had to be carefully planned and monitored, and responsibility for key technical decisions had to remain with a senior engineer in the company. Effective communication between the Associates and the staff at the company and at the University was therefore essential. To provide a structure for the project and to ease communication, the systematic design approach recommended by Pahl and Beitz (3) was used. The University staff have considerable experience of applying this systematic approach (4). Because of its comprehensive nature and the condensed experience contained in the methods and guidelines described, it provided a clear and consistent procedure to which everyone associated with the Programme could relate.

By giving the Associates 'ownership' of a significant project, it was believed that they would be much more committed to its successful completion. It is difficult to teach the importance of detail design without providing a high level of motivation. Knowing that one is going to see the results of what one has created is the best motivator, particularly when confronting the many detailed design tasks necessary to see a project through to a successful conclusion. Once the desire to succeed has been created, supplying the necessary technical expertise and specialist training becomes relatively easy.

Based on these ideas, the Associates were presented with their project on the day they joined the Programme. This was the design of an On-Run Lap Adjuster for a paper folding machine. They were given managerial responsibility for this project, which was to result in an important new feature for a high-speed folder. Within 20 months the project was completed and the first folder to incorporate the new feature was operating in a customer's premises. Since then another four folders have been built to the new specification, and the feature has been adopted as standard on all APV Baker PMC C2 folders to be sold outside America.

1.4 Managing the Teaching Company Programme

Although the Associates were to be given responsibility for managing their project from beginning to end, a formal structure was also required to provide the necessary managerial and technical guidance for the Teaching Company Programme.

Technical assistance was provided by formally establishing a multidisciplinary support team. Members of this team, who were drawn from all the key areas within APV Baker PMC, took a specific interest in the project and could be consulted at any time by the Associates. Immediate technical support and guidance was also provided by appointing a senior engineer from APV Baker's staff to act as coordinator for the Programme. This was a key role, providing a rapid source of advice for the Associates and handling the considerable administration and communication generated by the Teaching Company Programme itself. To meet the aims of the Programme, a

careful balance had to be maintained between the training and the technical
objectives.

Two important committees were set up: a Management Committee and a
Coordinating Committee. The Management Committee, chaired by the Managing
Director of APV Baker PMC, met regularly once a quarter. The company, the
university and the SERC were represented on this committee which laid down the
overall policy for the Programme. The Coordinating Committee, chaired by the
Technical Manager of APV Baker PMC, met regularly once a month and was
responsible for all technical matters. The Associates made formal
presentations on the progress of their project at every committee meeting.

2 THE PROJECT

2.1 Design of an on-run lap adjuster

A major component of a printing press is the paper folder. Printed paper
enters the folder at speeds of up to 10 m/s and is cut and folded into
signatures. The first transverse fold made in the paper is known as the
tabloid fold, and typically this is not made centrally. Instead it is offset
so that the leading edge of the paper protrudes beyond the trailing edge as
shown in Fig 1. This protrusion is known as lap. It is desirable that it can
be adjusted whilst the machine is running, but this was not possible on the
company's existing C2 folder. The project was therefore to produce a design
which would make this facility available on the C2 folder. Following the
systematic approach, the project was split into four main phases:
Clarification of the Task, Conceptual Design, Embodiment Design and Detail
Design.

2.2 Clarification of the task

This was essentially a marketing exercise, involving discussions with APV
Baker personnel, printers and bindery equipment manufacturers. From these it
was possible to generate a list of reasons why on-run lap adjustment was
required. An important reason was that the sales force were facing a
situation where all the competitors' folders had on-run adjustment on the
tabloid fold and potential customers were using this as a bargaining point in
negotiations.

Having determined the nature of the task, a design specification
incorporating the requirements was produced. In this document the
requirements were listed and classified as either a Demand, which had to be
satisfied, or a Wish, which was desirable but not strictly essential. The
areas covered in the design specification were based on a checklist (3) which
was modified to be more appropriate to the task in hand. The requirements
were specified in quantified terms wherever possible in order to provide well
defined limits for the design. The originator was recorded alongside each
requirement so that he or she could be consulted should a change be necessary.
This design specification was updated twice as design work progressed.

2.3 Conceptual design

At the beginning of this phase it was necessary to determine the functions
which had to be performed by the design and to link them together in a logical

fashion. This was achieved using a function structure, shown in simplified form in Fig 2, which details the flow of energy, materials and information through the system. The function 'Vary Pin to Tuck Distance' was seen as critical to the design. Since the pins locate the leading edge of the paper and the tucker blade determines the fold position, varying the pin-to-tuck separation will change the length between the leading edge and the fold, and hence alter the lap.

Over 50 solutions for this critical function were generated by a literature search, a brainstorming session and a study of existing designs. A formal evaluation procedure (3) was used to select the preferred solution idea, which is shown in Fig 3.

As new features are designed to satisfy confirmed customer orders within short lead times, it is essential that the minimum of development is required after manufacturing is completed. Hence the potential risk involved in any proposed solution carries a high weighting in the evaluation procedure. As a result, the preferred solution idea was based on an existing working principle. The cylinder is split into two segments, one carrying the pins and the other the tucker blades. These two segments interlock and by moving one relative to the other the pin-to-tuck distance can be altered. The layout of the final conceptual design is shown in Fig 4.

The adjuster handle turns a shaft supported between the folder frames. Being mounted on a screw thread this shaft is displaced axially by the rotation. Two opposite hand helical gears are mounted on the end of the shaft, one meshing with a gear attached to the pin segment of the cylinder, the other with one attached to the tucker segment. When the adjuster shaft is moved axially, the angular relation of the cylinder gears changes, thus moving the pin segment relative to the tucker segment.

2.4 Embodiment and detail design

During the embodiment or concept development phase, the initial step was to assign each of the functions in the function structure to particular components. The design specification was also examined in order to identify those requirements and constraints relevant at this stage of the design. The assemblies and components were then designed to fulfil the relevant functions within the necessary constraints, paying attention to the embodiment design guidelines (3).

Towards the end of the embodiment design phase, a series of layouts was produced showing major dimensions, tolerances and materials for all the components. These layouts formed the basis for the detailing phase, when each component and all the major assemblies were drawn. Production documentation and instructions for assembly, inspection and testing were also produced. A number of detailed drawings were produced on APV Baker's CAD system so that design data could be used directly for the programming of machine tools.

2.5 Manufacture, testing and commissioning

As manufacture progressed, a number of problems occurred. Some required redesign effort, while others resulted from incorrect manufacture. In all cases the Associates bore direct responsibility for the remedial action, though guidance and support were always available from within APV Baker PMC if required.

The most significant problem occurred during the assembly of the outer cylinder segment which carries the tucker blades. Although the cylinder had been correctly stressed for its operating conditions, it was not stiff enough to withstand the heavy forces experienced during manufacture. As a result it was impossible to manufacture the cylinder within the necessary tolerances. The immediate problem was to salvage the existing cylinder which was urgently required to fill a contract, and involved several castings with long lead times. This was achieved by designing a set of internal braces which were fitted temporarily during the critical machining operations. As a result, the cylinder was saved and successfully installed in the first modified C2 folder. Equally urgent was the need to redesign the outer cylinder segment since another folder fitted with on-run lap adjustment had been ordered. Cross-sections were thickened and location faces increased. In addition, an alternative manufacturing procedure was adopted. Since these measures were taken, four further cylinders have been produced and no stiffness problems have occurred. A retrospective investigation was undertaken to determine whether a finite element analysis could have predicted the stiffness problem. It was concluded that a very complex FE model was needed and that the effort to create this model would have been out of all proportion to the size and timescale of the overall project.

Testing was carried out in April 1988 and was supervised by the Associates. The tests were highly successful and the design proved to be fully operational, meeting 97 per cent of the design specification requirements.

The first C2 folder fitted with on-run lap adjustment began production in May 1988. The Associates visited the press house several times during commissioning to ensure satisfactory operation under site conditions.

3 RESULTS OF THE DESIGN TRAINING

3.1 The Associates

The systematic design approach gave the Associates a management structure and a means of communication. It also increased the general level of confidence in the project by enabling them, despite their inexperience, to identify key areas where the knowledge of experienced designers was required. This helped to minimise the amount of day-to-day supervision required throughout the project.

It was possible for the Associates to attempt a more challenging design task than would typically be given to new APV Baker PMC graduates. This in turn accelerated their design experience and has enabled them to become valuable members of APV Baker PMC's design team on completion of the Teaching Company Programme.

3.2 The Company

By adopting this approach APV Baker PMC was able to train two design engineers rapidly and gained a new, commercially significant product. The company was introduced to the systematic design approach of Pahl and Beitz and was able to judge its value in its own design environment. This may lead to the adoption of some techniques into the company's normal working practice. The project provided the opportunity to review the company's design process; including project team structures, project responsibilities and interdepartmental links. Finally, a number of practical lessons about split cylinder design and manufacture were learnt. It is worth noting that this approach to design training required a considerable time commitment from both the management at APV Baker PMC and the staff at Cambridge University.

3.3 The University

The project provided Cambridge University Engineering Department with further validation of its training approach. It also allowed the application of systematic design methods in an industrial situation to be observed through the full spectrum of design activities and detailed data were recorded throughout the project. These data have now been fed into a computer database and the results are being analysed. The project also acted as a case history offering a number of useful examples for future teaching.

4 CONCLUSIONS

The approach to design training adopted in this Programme was highly effective. The main advantages were:

- Commercially significant work was gained from inexperienced designers very early in their careers.

- An accelerated rate of training was achieved with a low level of day-to-day supervision.

- Communication, both internally and externally, was eased through the use of a systematic design approach.

APV Baker PMC is continuing with this approach to design training and changing some of its design procedures. Two further pairs of graduates are currently being trained and the first pair are now undertaking further design projects within the company.

5 REFERENCES

(1) WALLACE, K. M. Developments in design teaching in the Engineering Department at Cambridge University. *The International Journal of Applied Engineering Education*, 1988, 4, (3), 207-210.

(2) BRADLEY, J. A. The Teaching Company Scheme – the first ten years. *Proceedings of the Institution of Mechanical Engineers*, 1986, 200, (B4), 277-284.

(3) PAHL, G, and BEITZ, W. (Editor WALLACE, K. M.). *Engineering design*, 1984 (The Design Council, London).

(4) WALLACE, K. M. and HALES, C. Some applications of a systematic design approach in Britain. *Konstruktion*, 1987, 39, (7), 275-279.

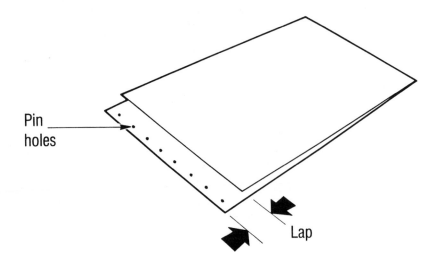

Fig 1   Lap on the tabloid fold

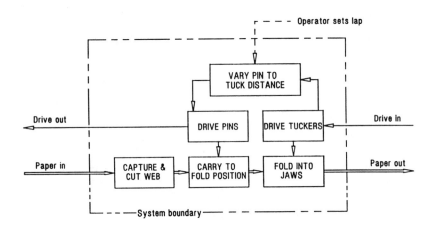

Fig 2   Function structure for the
on-run lap adjuster

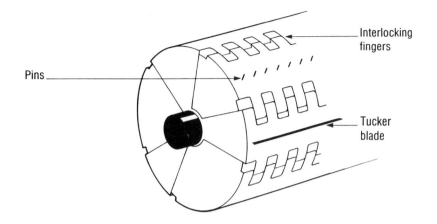

Fig 3   Solution idea using interlocking
        fingers

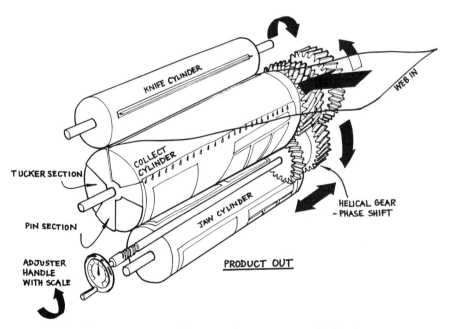

Fig 4   Final conceptual design

Educational programmes for design engineers

J HAVEL, Ing
Skoda, Concern Enterprise, Plzen, Czechoslovakia

The educational system in ŠKODA Plzeň with its
various forms available to the employees in
the in-house courses or at the State Schools
is described. Special attention is paid to the
courses for the beginning design engineers
the curriculum of which is given in greater
detail.

1 INTRODUCTION

The role that the design engineers have played in the creation
of products has always been a very important one. But the intensive progress of science and technology, and the ever shortening cycles of product innovations, brought about by unrelent-
ing competition, make this role more important still. A quick
variation of assortment of technical products which, if they are
to stand up to competition in the world markets, must possess
newness, reliability, and economy in operation. All this brings
about a process leading to a deeper specialization both in products and in the designers profession on the one hand, and an
indispensable synthesis both of specialized products (greater
universality) and of cognition and experience of designers (mul-
tiprofession), on the other hand.

Moreover, in contrast to the growing demands placed upon
the knowledge of engineering designers and the demands for new
and more complex products, the statistics show that the numbers
of engineering designers do not increase but actually are decreasing constantly. One of the reasons for this decreasing trend
can be seen in the low social status of the designers. First
steps have already been made to better this situation but still
much work lies ahead. Another way of relieving this difficulty
is to enhance constantly the quality of designers work while
making good use of their knowledge, experience, and available
means. All this calls for well-educated design engineers, oriented towards the needs of industry, who will keep abreast of the
constantly developing state of the art in the particular fields
of their activities. Also the initiation of the design engineers new in the profession must be done with an eye on the demands placed on the industrial concern on the one hand, and on the

engineering designers on the other.

2 THE EDUCATIONAL SYSTEM IN ŠKODA

Technical education in ŠKODA, Concern Enterprise, Plzeň has a long tradition. It is organized as an in-house educational system in the ŠKODA Schooling Centre, but it makes ample use of the possibilities offered by close cooperation with the State schools. This cooperation is supported by a long-term agreement concluded between the School Department of the District National Committee and ŠKODA Plzeň. The cooperation with the Technical Collge and some of the Faculties of the Universities seated in Plzeň is also based on this agreement.

The cooperation with the schools is manifold. In the field of technical activities, which form the major part, it is directed especially towards science, technology, and engineering and is, therefore, implemented with the Technical University Plzeň. This cooperation takes on various forms ranging from experimental forms of tuition to joint research and development programmes. In the educational area, the aim is to educate young specialists whose specialization will correspond as closely as possible with the needs of ŠKODA Plzeň, one of the major industrial concerns in Czechoslovakia.

The ever growing demands of industrial concerns, including ŠKODA, on technical preparation of young design engineers require novel approaches to tuition. In addition to traditional courses, new forms of tuition have been introduced lately. They include:

- group preparation meeting ŠKODA requirements in the field of technology. Courses are organized at the Department of technology, machining, forming, and casting. They include preproduction practice. The students spend the last year of their study in the ŠKODA Works where they do their semester project and their dissertation. Students in this group obtain a grant from ŠKODA (starting from the third year of study) with the proviso that they will take employment with ŠKODA for a term of 5 years. They get to know the environment they will work in and can prepare accordingly. Their dissertation theme is chosen such as to meet the needs of the particular production plant they will be employed in.

- in the framework of the programme AIP 2000 (Automation of Engineering Work) a schooling workplace has been organized for regular students as well as for practicing engineers, meant as a re-qualification course. The students are taught subjects ranging from drafting to statics, dynamics, kinematics, and mechanics of machine tools. They also do their dissertation there. Again the dissertation themes are prepared in cooperation with ŠKODA.

- laboratory practice in the subject "industrial machines". This is a joint workplace that has been organized in the Machine Tools Plant where ŠKODA experts take part in the tuition on the shop floor so to say. The students have an opportunity of taking non-standard measurements under practical conditions. This form of tuition has been found very useful because theoretical knowledge can be complemented appropriately with everyday production practice.

- another means of putting production practice to use in tuition is the so-called "Students' design office", affiliated to the Heavy Engineering Plant. The working themes are taken from the production programme of gear boxes for a major Czechoslovak producer of pumps. A class of 20 students in average is divided into two groups. The students in each group are all assigned the same problem to solve. The best designs are selected and further elaborated to the stage of a complete manufacturing documentation which is paid for.This system represents a suitable form of incentive for the students.

The ever growing demands of industry on technical preparation of young engineering designers are also reflected in the growing demands on specialized knowledge of the lecturers in the field of technology and associated subjects. Thus a need arises for specialists with an industrial or scientific background to participate directly in the teaching process, opening another field for cooperation of ŠKODA experts with the Technical College and the Technical University Plzeň. This gives opportunity to about a hundred specialists from industry and research to teach there as external teachers or professors. But this kind of cooperation is not new altogether. There has always been a certain number of ŠKODA experts who lectured at the Technical University and eventually devoted themselves to pedagogical work only. At present there is a number of professors on the University staff who are former ŠKODA employees.

The trend to bring school and industry closer together shows in the fact that, for instance, in the school-year 1987/1988 some 239 dissertation and college-leaving examination themes were prepared in cooperation with ŠKODA and about 250 ŠKODA experts acted as supervisors, opponents, or examiners.

The question of enhanced or supplemented specialization to meet challenging tasks in the industry also applies to the practicing engineers. One of the means leading to the enhancement of their knowledge or, in some cases, re-qualification are 3-6 semester postgraduate courses. The Technical University Plzeň organizes such courses keeping an eye on the existing and future needs of ŠKODA. These courses are quite popular. In 1987, a total of 179 practicing engineers finished various postgraduate courses.

Lately there have been instituted new courses, two of them suitable for design engineers, namely "Modern design methods" and "The principles of programming and computer graphics".

Although it has no direct bearing upon the education of design engineers in ŠKODA, perhaps it is worth noticing to what limits the cooperation of ŠKODA with the schools goes. The effort to solve problems in a complex way resulted in a special type of cooperation - joint workplaces pursuing the following fields of activity:

- electrotechnology
- electric driving mechanisms
- reliability and technical diagnostics
- machining and manufacturing processes
- adaptive control systems
- computer systems
- cybernetics and automatic control
- machine tools
- electronics.

A joint workplace for nuclear engineering has been prepared and a joint working team has been set up in the field of cryogenic engineering. The joint working teams work on long-term projects which are evaluated annually and refined as need arises.

The best results are expected from an integrated joint workplace associating ŠKODA Plzeň, Technical University Plzeň, and Czechoslovak Academy of Sciences which has been called the "Institute of Technology and Reliability of Machine Parts". This institute will work on projects included in the State Plan of Research and Development as follows:

- computer-aided optimization of machine systems (Engineering Faculty)

- automated systems of control of technological processes (Department of Technology, Electrotechnical Faculty)

- research into new materials and technologies (Czech.Acad.Sci.)

Of course, the requirements on technical education of young engineers meeting specialized needs of ŠKODA are best met by the in-house educational system. The costs in operating such an extensive system are considerable but the philosophy underlying the educational programmes is that both the employees and the concern benefit from the enhanced knowledge of the employees. The training and the education of the employees is considered as an investment of the concern for the future.

The ŠKODA Schooling Centre organizes many different courses in a variety of professions. Here also belong courses teaching subjects such as product planning and development (i.e. engineering preparatory operations) for project engineers, design engineers, and technologists, courses in calculus methods and higher mathematics, computers and computer systems, computer graphics, programming, and electronics. Neither teaching of languages is omitted, for proficiency in languages is appreciated more and more. There are organized basic 4-year language courses terminated by final examinations. These courses are supplemented with extended specialization courses.

The "Institute of Education" forms an independent part of this educational system. A wide variety of 6-7 semester courses at university level are available to employees meeting special conditions (40 years of age, more than 15 years of practice in the relevant field). Here also the specialization "Engineering designer in mechanical engineering" is taught to selected employees.

An important element in the ŠKODA educational system is the in-service course "Training for the beginning engineering designers". This course is a part of the so-called "adaptation process" in which all newly employed designers take part in the first year of their employment and which is intended to help them to find their feet in their new environment.

The young designers in this course get acquainted with the organization chart of the concern and the incorporation of the various departments and their functions. The course aims at giving them an overview and "cross-sectional" knowledge of engineering design which they will complete with specialized knowledge in their respective production plants. They are acquainted with the system of engineering preparatory operations concerning the planning and development of new products, its various stages and their rationalization. Great emphasis is laid on imparting to the young designers the notion that hand in hand with professional knowledge it is highly important and necessary to acquire a method of work in the designing process, that it is necessary to know how to work with various sources of information. In this connection they are acquainted with the system of technical, scientific and economic information and are persuaded of the importance of systematic work with them. Also the importance of standardization is stressed.

Compared with the past years of existence of this course an endeavour has been made to bring out the importance of computers in the work of the design engineer, especially in computer-aided design and, therefore, the curriculum of the course has been modified accordingly, bringing in more material from the field of computers, computer-aided design, and computer graphics. Young designers are acquainted with a variety of topics in computer hardware but especially in computer software with respect to computer-aided design. Among the lectures in this field one is of major importance dealing with the philosophy of an information system about CAD software as an informational kernel of a developed system of rules for an effective creation and utilization of CAD software in a large industrial corporation. The considerable efficiency of this system has been proved over a wide range of objects, objectives, solution levels and extents of CAD problems in the engineering design of machine tools, rolling mills, electric locomotives, electric generators, etc. This lecture is given personally by the author of a paper presented at this conference under the title "An information system about CAD software for a large industrial corporation".

The attention of young design engineers is focused on the importance of system approach to solving design problems. In general, any system in its solution can be decomposed into the following parts:

- introductory part
- functional part
- processing part
- technological part
- organizational part
- economic part
- ecological part

Applied to the manufacture of a product the solution of the problem is decomposed into:

- 3D design (model) of the product with its functions specified
- selection of material
- technological processes
- means of manufacture (tools)
- production costs
- ecological questions (the degree of possible pollution of environment).

In order to bring the problems of CAD and computer applications to a practical level for the young designers, lectures giving examples of the application of computers in mechanical engineering, electrical engineering, and electronics have been incorporated.

The effort to connect the traditional methods of work with the computer-aided methods in young designers' minds resulted in the introduction of new subject material grouped into sections as follows:

- the section <u>organization, management and planning</u> gives information on

  - the organization of the concern and the cooperation of its various sections

  - the importance of designer's work and his main duties

  - the system of engineering preparatory operations and its inclusion into the automated control system (product planning and development)

  - the activities and course of engineering documentation in design departments and departments of technology (i.e. information flow and material flow)

- the section <u>information and material provision</u> informs on

  - the system of scientific, technical, and economic information

  - patent literature and its utilization

  - automated information systems (e.g. INIS, COMPENDEX, INSPEC, WPI, etc.)

- the system of standardization and its importance for designer's work
- assortment of normalized materials and standardized elements the designer uses in his work

- the section <u>methodological provision</u> gives a survey of methods of creative work such as:
  - intuitive methods (brainstorming - A.F.Osborn, synectics-W.Gordon)
  - systematic methods (F.Hansen, W.Rodenacker, morphological analysis, etc.)
  - decision processes (operational research methods)
  - heuristics
  - the principles of value analysis and value engineering (used by professional analysts in the various factories)
  - technological feasibility of constructions
  - industrial design

The preceding sections containing traditional information for design engineers are rounded out by a short section containing an introduction to industrial property, patents and rationalization, and innovation proposals.

From the viewpoint of updating the course, the most important section in the syllabus is the section <u>rationalization and automation of design work (computer-aided design)</u>. In addition to the usual trends in rationalizing designer's work with the help of various aids, this section contains topics oriented towards the utilization of computers and computer methods. These topics include:

- rationalization of computations
- rationalization of work with texts
- computers in design offices (hardware)
- computer-aided design (software)
- computer-aided drafting (computer graphics)
- computer-aided computations
- computer-aided work with textual documentation (work with manufacturing documentation in automated systems)
- methodology of effective creation and utilization of software in design offices. This is the most important part of this section for the beginning designers' orientation in the applicability of computers in designer's work
- examples of application of computers in mechanical and electrical engineering, and electronics
- handling of engineering documentation and the importance of reprography in this field.

In the advanced courses it is planned to include lectures on psychological aspects of designer's creative work and the psychological barriers hindering his work. Therefore, the section containing a survey of creative methods of work also contains a brief outline of psychological aspects of creative work and psychological barriers such as fear of criticism, conformism, etc.

It seems inevitable that the stormy development of science and technology in the years to come, the rapid outdating of scientific and technological knowledge, and the replacement of the present-day know-how by new discoveries and technologies will force design engineers more and more towards a continuous study, and technical preparation of design engineers will become an life-long business. Therefore, it will be necessary to ensure that the conditions necessary for study will improve. To this end it will be necessary to organize training and courses in a close cooperation of the state and in-house educational institutions in order to meet the challenge of future technologies.

C377/153

Electronic computer aided design—the training crisis

S MEDHAT, MSc, PhD and **A POTTON**, BSc
Dorset Institute, Poole, Dorset

Abstract:

Methods used to design and manufacture electronic circuits and systems are changing rapidly. This paper explains the importance of Electronic Computer Aided Design (ECAD) in this change. There is evidence that the United Kingdom is falling behind its main competitors in taking up the new techniques. Training is identified as a major factor in rectifying this situation.

The objectives of a novel and intensive 14 week ECAD training course, organised by the Dorset Institute are described. The paper outlines the structure, teaching programme and practical work which accommodates trainees with a wide range of qualifications and experience. Results of trainees' achievements are described together with recommendations for future training provision.

Introduction:

In the electronics industry, design, development and manufacturing methods are changing rapidly. In many companies, design and development activities are moving from the workshop and laboratory to the screens of computer workstations. The falling cost of engineering computing workstations and CAD software has now brought this new technology within the reach of small and medium size companies for the first time. A key area of technology is the design and implementation of Application Specific Integrated Circuits (ASICs).

Recent statistics indicate that the number of engineers in Britain, capable of designing ASICs, is no more than a few hundred which is proportionally a good deal less than the number in our major overseas competitors.(1)

While the demand for engineers with new electronic design skills is increasing, other engineers are being made redundant, such engineers would be ideal candidates for retraining in modern electronic design and manufacturing methods to allow them to re-enter the technical field or alternatively to be more effective in a technical management role. Other suitable candidates might come from persons with a scientific or a non electronic engineering background who wish to reorientate their career path.(2)

This paper describes novel approaches to training engineers and technicians in the use of ECAD techniques. A programme of training is outlined which takes trainees from design concept to realisation on silicon in less than 14 weeks. The practical results of students' design work is fabricated on silicon during the duration of the programme allowing for the full evaluation, design, simulation, prototyping and test cycle to be completed within the 14 week period.

How can D.A. tools reward Industry?

The classical sequence from 'order' to 'product' is a linear sequence of events typified by Figure 1. The interesting fact is that the time scale for this chain of events varies from a minimum of two years for the 'white goods' industry to as much as ten years for sophisticated military projects. Considering these time scales with those of the technological developments (typically a doubling of complexity on a chip each year) then clearly a mismatch is an understatement. Testing and evaluating your first system three years after design and finding fundamental problems of not meeting the specification, is disastrous.

The increasing complexity of many electronic systems and circuits suggests that we must have the tools available which allow validation of specification, design and testability at a much earlier stage in the product generation. The only way this can be addressed is by utilising modelling techniques in the CAD environment.

This provides an interactive capability to evaluate the system with the designer fully utilising his experience, intuition and imagination to experiment, thus gaining knowledge and understanding which are vital to his/her creative activity. However, this is not easily achieved and indeed some people will argue not desirable (3) in these times of high unemployment. However for the progress of society they are inevitable. It should be recognised as a goal and framework at a very early stage of planning a D.A.

The past few years have conditioned CAD users to think of the machine as a processor delivering reams of paper for them to pour over. Once again, this is a product of our emphasis on the niceties of analysis with the neglect of synthesis and the abandonment of the original objective to ASSIST THE DESIGNER.(4)

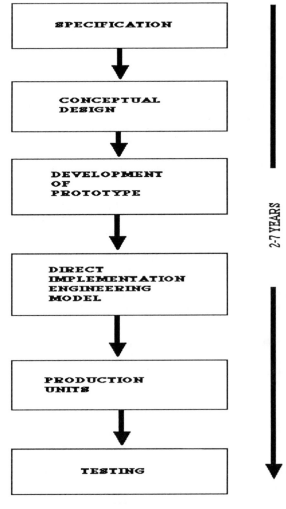

FIGURE 1

In an EEC survey of the use of CAD in the electronics industry worldwide (some ninety companies and institutions were covered) (5), a number of reasons were given for using CAD systems. They are listed in order of priority in the table below.

REASONS FOR USING CAD IN ORDER OF IMPORTANCE

1. Improvement of design quality
2. Increase of complexity
3. Saving in 'design' manpower and time
4. Savings in 'test' manpower and time
5. Provide a common database for all to use
6. Documentation
7. Research and Development tool
8. Ability to change specification
9. Savings in 'production' manpower and time
10. Increase of workload
12. Shortage of skilled manpower

When one delved more deeply it was found that only the Japanese respondents could actually quantify some of the 'savings' factors.
In other words, despite the automation there were no attempts to monitor performance, the excuse given is that it is rather difficult.

Perhaps a better reason can be found in this list together with the answers to the question on what products do you use CAD tools.

A large percentage of the returns indicated that they referred only to LSI designs even though within the same company the volume of work was on PCBs. In many cases, there was far more functional complexity on the PCB than there was on a single chip.

This is a failure to recognise the potential payback on the existing technology that CAD could have. It also re-emphasises that CAD tools are seen too much in association with R and D activities. By asking at the inception of CAD 'what is the payback?' and ensuring throughout its growth in the company activities that this question is asked and results quantified, then we would have much more integrated CAD environments than we have today.

Another aspect of which management must be aware when structuring CAD systems is, that the cost of software is far greater than that of hardware. In consequence, it is important to ensure that software systems are transportable and efficient, otherwise the advances in hardware will have to be foregone.

In essence, it is the responsibility of senior management advised by technical and middle management staff to endeavour to plan a design automation environment to ensure that cognizance is taken of effects from the customer to field trials and maintenance engineers.

Those organisations that have put their manpower effort into an integrated design automation system have been well rewarded with reductions in lead time of over 50%, higher quality design, easier multi-site working, better production control and lower modification costs (6). To summarise, here are some of the advantages of using CAD :

1.	Design alterations can be made quickly and efficiently by the engineer, when and where he needs refinement or is required to correct, find and fix errors.

2.	Only one design cycle is needed to get a design right first time. Therefore, all redesign loops involving outside groups and operations are eliminated.

3.	Better documentation means that it is easier to start production sooner.

Software Tools for ASICs in Education and Retraining

The majority of new designs of application specific integrated circuits (ASICs) exploit semicustom techniques of gate arrays and macrocells to enable engineers who are experienced in the use of standard parts to make a smooth progression to the benefits of higher levels of integration, and to take a concept for a new product all the way to a single chip solution. Conversions of existing designs to achieve higher performance more cost effectively are also straightforward. In parallel with recent developments in software tools, several wafer fabrication plants in the UK and worldwide can now offer access to higher levels of integration and lower unit costs, even at small volume. Thus the marketing of CAD to the user has become a highly competitive field in respect of both hardware and software.

Hence the value-added earning potential of powerful workstation CAD systems more than justifies the initial outlay and on-going maintenance. The fabrication houses are looking toward an upturn in the volume of designs rather than long production runs to fill their current excess capacity. The retooling required for the economic manufacture and test at low volume has already been completed. The vital ingredient missing is the predicted large demand from OEM customers.(7)

To retrieve the situation, by equipping our present and future electronic engineers with the skills and awareness needed for competitiveness in world markets, widespread education and retraining is urgently required.

The current position is comparable with that existing to the start of the microprocessor education programme at the beginning of the decade. Only by education and retraining, organised on a national scale, can the number of ASIC design specialists be raised to match a shortfall in UK industry which is believed currently to run into several thousand. In the present economic and political climate it seems that the policy most likely to find favour is for partnership between industry and academia to bring about a new microelectronics revolution. This must surpass in scale the earlier government-backed initiative on microprocessors if the indigenous electronic equipment manufacturing industry is to survive into the 1990s (8).

A Design Automation Training Course

As a response to the need for engineers and technicians with experience in the use of design automation tools, discussions took place in the early part of 1988 between the Training Commission and the Dorset Institute to formulate plans for a course in this field for unemployed persons. Local and regional officials of the Commission were quick to grasp the importance of the proposed training course and the employment opportunities which it offered participants.

The technical content of the proposed course was clearly centred on electronic circuit/system design tools, a subject area with the most clearly identified prospects of future employment. The aims were ambitious in that unemployed persons with no specific background in the use of design automation tools were required to contribute actively in a realistic design exercise resulting in a physical product (an ASIC) within a period of 14 weeks. Agreement on sponsorship was reached with the Training Commission and the course started in March 1988.

Marketing the course and Recruitment of Trainees

The target market was seen as unemployed persons with sufficient technical background and potential to benefit from and survive the course. Previous experience in offering other more general courses for unemployed persons has indicated that there are specific sub groups of individuals who find the opportunity of technical training appealing and helpful. One such sub group consists of middle or junior managers whose technical skills and knowledge base has become out of date and/or very company specific. Reorganisation and closure of companies leaves such individuals at a severe disadvantage in the employment market if they become redundant. Perhaps a message here for all managers is the need to keep abreast of all technical developments rather than just those of interest to the current employer.

A second substantial sub group of potential clients for technical retraining courses is composed of technicians with ability but whose career has become stuck at a fairly low level. The normal pattern of qualifications and education for senior technicians consists of an Ordinary National Certificate or Diploma followed by a Higher National Certificate or Diploma obtained by full time or part time day release study. We have found a good deal of evidence that many employers are reluctant to recruit junior technicians and offer the opportunity of gaining higher qualifications by day release study when the person applying for a post is over an age limit which may be quite low. Persons seeking employment in this situation are less likely to be successful and more likely to be restricted to routine work than those qualified to Higher Certificate level or above.

Having identified the main target markets, suitably designed advertisements were placed in appropriate newspapers to recruit locally. For advertising the course further afield, advantage was taken of the on line training information network which can be accessed from Job Centres. The marketing strategy proved effective with over 50 candidates of which 14 were selected as suitable.

All applicants were interviewed and assessed using the following criteria:

(i)   High motivation
(ii)   A technical background related to electronic design
(iii)   A recognised technical qualification at technician level or above

Of these criteria, high motivation was seen as the most important for what would be a highly intensive and demanding course of study.

Course Objectives and Structure

The single most important aim of the course was to enhance the trainees ability to gain employment. Although primarily of a technical content it was recognised that potential employers would be more impressed if trainees could relate the technical content to the business, management and commercial implications.

To cement the link between technical and non technical studies, the course was structured around the concept of product realisation, the product being a PCB, a programmed logic device or a custom designed ASIC. Figure 2 shows the relationship between technical and non technical content. The technical skills element consists of a number of interrelated subjects that comprise the knowledge base of the practising ASIC designer. The disciplines of digital, analogue and software design are taught during the initial stages of the course.

The student is then equipped to implement a design using appropriate ECAD tools. The business, management and commercial elements of the course consist of a series of lectures designed to increase the students ability to manage both the technical and commercial aspects of projects and to liaise with the customer and Silicon Vendor to ensure that projects are successfully completed.

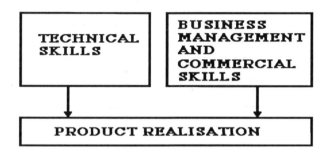

FIGURE 2

Since the theme of product realisation permeates the course it is important that the trainee is able to see a physical product before the end. This ensures that all trainees complete the design - implement - test cycle for the product. The 14 week duration obviously imposed limitations on the depth of study in some aspects of the cycle.

Figure 3 shows the relationship between the three defined product areas and the technical studies which provide the underpinning. An understanding of the basic techniques used to design digital circuits is vital prior to the start of design implementation, for this reason two weeks of the course are devoted specifically to this discipline. The techniques of designing for testability are covered in detail as are the available technologies with emphasis on CMOS logic. The use of ECAD tools to design PCB's is included and students fabricate a PCB with which to test their ASIC design after it has been fabricated.

Students further participate in the design of a programmable logic device using ECAD tools based on standard computer hardware. A reasonable ability to use computer systems and an appreciation of operating system utilities is clearly desirable. A working knowledge of a High Level language allows students to use the simulation modelling facilities available on many ECAD systems within a short time.

The reader will no doubt appreciate that the choice of products is clearly influenced by the commonality of technical studies required to carry through the design process.

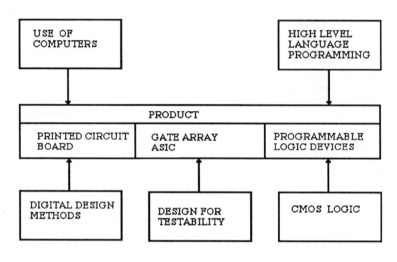

FIGURE 3

In each of the defined product areas, trainees were able to complete and test a prototype within the course time scale. The ASIC design is obviously the most ambitious. In this case, designs were carefully partitioned with different functional areas of a chip being allocated to different groups of trainees. All students completed their part of the design work in time for the complete design to be sent to the silicon vendor (MicroCircuit Engineering) for fabrication and returned for testing by the trainees. Trainees saw this as the pinnacle of their achievements and were justifiably proud of their contributions. As part of this exercise, trainees had to learn how to negotiate with silicon vendors to obtain the best price/performance ratio for the final product.

The emphasis on completion of a serious design exercise added coherence and credibility to the course for students and also for potential employers. This was particularly the case with the ASIC design. Another feature of the course content which added credibility was the use of industry standard design automation hardware and software design tools, identical to that used by many companies. Trainees gained experience with a variety of software packages including AMAZE for programmable logic design, REDBOARD and REDLOG for PCB design, and BX for ASIC design. Hardware platforms consisted of Apollo graphics workstation and PC 'AT' type microcomputers.

The Course Structure

In addition to the technical content, the course has also dealt with personal skills, business skills and management implications of the new technology. This was a minor part of the course which was intended to further enhance the employment prospects of trainees. At various points in the course, strategic *Key Seminars* allowed trainees to discuss technical and other issues with engineers and managers from industry. This added further coherence and relevance to the rest of the course content and also provided a valuable boost to the motivation of trainees. Part of the end assessment consisted of a *viva voce* which was video recorded. Apart from the contribution to the final assessment, it was intended that the procedure should provide a highly realistic simulation of a technical job interview.

Course Topics

Subjects covered during the 14 week course included the following:

Use of Computers.
Review of Digital Design Techniques.
Introduction to ECAD.
Designing a Gate Array.
CMOS Logic.
Printed Circuit Board Design.
Programmable Logic Devices In System Design.
Use of ECAD Development System.
Design for Testability.
Selecting an Appropriate Silicon Vendor.

Business and entrepreneurial awareness content

To introduce the student to modern business methods, and to develop personal skills which would make the student more employable. To help students plan their careers and their self development. The business content was integrated within the technical subjects and included topics such as Managing your time, Communication Skills, Managing People, Career Development (& applying for jobs), Formulating A Business Plan, How Businesses are Managed, Industrial Marketing, Managing Finance, Managing in a Dynamic Market Place, Running a Small Electronics Company, Review of Business Function.Selling Industrial Products and Services.

Level of Achievement

The course demonstrated convincingly that staff at technician level can perform effectively in using complex ASIC design tools. This has an important implication for industry, in that it allows senior engineers to concentrate on top level and conceptual design considerations.

It was very evident that trainees developed confidence and stature as the course progressed. The personal skills and business/commercial content provided a basis for further career development in industry or small businesses. Trainees were required to pass an assessment for at least six of the ten units of study of the course. All trainees taking the assessment were successful in this and 75% of students passed the assessments for all units of the course.

This is regarded as a highly satisfactory result reflecting credit on the ability and motivation of trainees.

Conclusions- Lessons for the future

It was also concluded that given adequate training, technician level staff can contribute effectively to ASIC design projects and relieve more senior staff of the routine aspects of design tool operation.

References:

1. Michael Shortland and Associates, Butler Cox & Partners, Custom Integrated Circuits, The User Company Perspective, <u>A Study of The UK Custom Integrated Circuit Market</u>, This Study is Commissioned by The Department of Trade and Industry, Published by The Controller of Her Majesty's Stationery Office.

2. Grice P, Bowsfield R, Oliver K, The Conversion of Non-Technical Graduates into Engineering, <u>International Conference on Training for Change- the Revolution in Commerce and Industry, IEE Publication No 283, December 1987</u>.

3. Reger JD, The Development Cycle of Typical Semicustom IC, <u>Journal of Semicustom ICs, Volume 4, No. 1, September 1986</u>.

4. Sheldon F, Retraining of Industrial Personnel in CAD/CAM - User Experiences, CADCAM Training and Education Through the 80s, <u>Proceedings of the CADED'84 Conference</u>.

5. Musgrave G, Ed. Computer-Aided Design of Digital Electronic Circuits and Systems, <u>North-Holland 1979</u>.

6. Cooley MJE, Some Social Implications of CAD, <u>MICAD80, Paris, September 1980 SIOB</u>.

7. Jones PL,and Pritchard TI, The United Kingdom Integrated Design Education Programme, <u>ECAD DTI/UGC Initiative London, 1986</u>.

8. Jones PL, Software Tools In Education and Retraining, <u>Colloquium on 'IC DESIGN : SOFTWARE TOOLS' , IEE Digest No: 1986/68, May 1986</u>.

C377/243

The influence of first year's use of CAD work station on the attitudes of employees towards computer aided design

E SIIVOLA, MSc
T-Consult Limited, Tampere, Finland

SYNOPSIS   The actual purpose of the study was to find out what grounds one could choose suitable persons to do planning work on CAD equipment and how the working conditions should be arranged.   The study was carried out in three different industrial plants at the same time by using forms for interviews.   Similar equipment and programmes were used in each industrial plant.

In the study the attitudes of the staff towards CAD (computer aided design) before using CAD equipment, and the changes in attitudes after an experience of about one year's use were cleared up.   The backgrounds, i.e. age, sex, education and linquistic abilities of those to be interviewed were specified.   The CAD questions handled e.g. CAD schooling, planning in general, different part areas in planning, working conditions, working hours and willingness to work with CAD equipment.   Also the influence of CAD on some illnesses was questioned.

INHALTSANGABE   Die Absicht dieser Untersuchung war zu erfahren, mit welchen Gründen man geeignete Personen auswählen könnte, Planungsarbeit mit CAD–Geräten auszuführen und wie die Arbeitsverhältnisse arrangiert werden sollten.   Die Untersuchung wurde gleichzeitig in drei verschiedenen Industrieanlagen mit Befragungsformularen durchgeführt.   In jeder Industrieanlage standen gleichartige Geräte und Programme zur Verfügung.

In der Untersuchung wurden die Einstellungen des Personals zu CAD (Computer Aided Design) klargemacht, ehe Inbetriebnahme der CAD–Geräte und die nach Betriebserfahrungen von etwa einem Jahr stattgefundenen Änderungen in Einstellungen.   Die Hintergrundsangaben der Befragten, wie Alter, Geschlecht, Schulung und Sprachkenntnisse, wurden aufgeklärt.   Die CAD–Fragen betrafen u.a. CAD–Schulung, Planung im allgemeinen, verschiedene Teile der Planung, Planungsqualität, Arbeitsräume, Arbeitszeit und willigkeit mit CAD–Geräten zu arbeiten. Ebenfalls wurden Einflüsse von CAD auf einige Krankheiten aufgeklärt.

1 BACKGROUND OF THOSE INTERVIEWED

In the first interviews 151 employees in designing offices were addressed and in the second round 69 employees. Those to be interviewed were divided into five different age groups (Table 1). Table 2 shows the share of women and men in percentage. The results were studied also based on the education of the interviewed (Table 2). The respondents estimated their own knowledge of English (Table 2).

1st interview		2nd interview	
Age group	%	Age group	%
under 26 years	6	under 27 years	14
26 - 30 "	21	27 - 31 "	18
31 - 35 "	24	32 - 36 "	25
36 - 40 "	16	37 - 41 "	19
over 40 "	33	over 41 "	24
total	100	total	100

Table 1. Age structure of the interviewed

Sex	a (%)	b (%)
women	14	10
men	86	90
total	100	100
Education	a (%)	b (%)
apprentice or corresponding	19	7
technical school or corresponding	25	40
technical collece or corresp.	36	38
institute of technology or corr.	20	15
total	100	100
Knowledge of English	a (%)	b (%)
bad	28	26
fair	38	43
good	34	31
total	100	100

Table 2. The sexes, education and knowledge of
English of those interviewed
a) in the first interview round
b) in the second interview round

2 EXPECTATIONS OF CAD

There was a great interest in CAD before use. 86 per cent of those interviewed wanted to participate in CAD schooling. The eagerness was greatest
amongst the age group 26-30 years and decreased in older people. Basic education didn't have a very big influence on the willingness for schooling. In
women the eagerness was less than in men.

30 per cent of those interviewed wanted to work with CAD equipment continuously and 45 per cent occasionally. Education didn't have much influence
on willingness to use the equipment. 55 per cent of those under 26 years of
age wanted to work with CAD continuously. As age increased the willingness
decreased, in the age group of over 40 years old only 27 per cent were willing.
Women were more eager to work with CAD than men.

CAD argument	agrees		disagrees	
	fully	partly	partly	fully
changes my job	17	41	28	14
increases efficiency of the company	36	49	14	1
makes working speed stricter	5	35	43	17
makes working easier	16	43	30	11
makes working more interesting	20	46	23	11
improves planning quality	38	46	13	3
causes arrangements in working hours	30	46	17	7
makes planning work faster	36	49	20	4
makes it possible to study several alternatives	71	26	3	0
causes backache	10	27	37	26
causes headache	11	32	37	20
causes eye tiredness	29	43	17	11
causes risk of cancer	3	12	31	54

Table 3. Preliminary expectations of CAD

Planning zone	agrees		disagrees	
	fully	partly	partly	fully
offer planning	67	27	6	0
device planning	29	53	18	0
pipeline planning	66	32	2	2
lay-out planning	67	27	5	1
building planning	45	45	9	1
scheme planning	76	22	2	0
strength calculation	55	31	11	3

Table 4. Applicability of CAD in different
planning zones (%)

CAD was believed to cause eye strain no, other medical damage. Only a few were afraid of getting cancer.

Table 3 shows some preliminary expectations of CAD.   Table 4 shows some CAD expectations of suitability for planning zones.   It can be seen that the biggest expectations were directed towards scheme planning. Next came offer, pipeline and lay-out planning.   The smallest expectations were directed to-wards device planning.

3 CHANGES IN OPINION CAUSED BY EXPERIENCES

The second interview round was made when the CAD equipment had been used about one year.

3.1 CAD in general

Willingness for CAD schooling was still strong.   As many as 81 per cent said they wanted to get CAD schooling. Most eagerness was found among the age group of 27-31 years.

In women the willingness for schooling had increased considerably, being now about the same rate as in men.   It was noted that none of the women abso-lutely objected to eventual CAD schooling. People with different basic educa-tion had no big differencies in their willingness for schooling in the second round.

The willingness to work with CAD had clearly increased.   Now 47 per cent preferred continuous working. The increase was as much as 17 percentage units. The share of those who wanted to work with CAD only occasionally decreased by 12 percentage units, being now 23 per cent.   People in age groups of 27-31 were most eager to work with CAD.   The willingness of women to work with CAD had decreased a little, but it was in percentage, anyway, greater than in men. The graduates from an institute of technology were clearly less willing to work with CAD than people with other education backgrounds. Most positive at-titudes were from those who had studied at technical colleces.

CAD argument	Experiences of CAD use after an operation of one year (%)				Changes in percentage between the 1st and 2nd interview rounds			
	agrees		disagrees		agrees		disagrees	
	fully	partly	partly	fully	fully	partly	partly	fully
changes my job	24	28	33	15	+ 7	− 13	+ 5	+ 1
increases efficiency of the company	38	55	7	0	+ 2	+ 6	− 7	− 1
makes working speed stricter	11	29	45	15	+ 6	− 6	+ 2	− 2
makes working easier	31	43	23	3	+ 15	0	− 7	− 8
makes working more interesting	18	39	33	10	− 2	− 7	+ 10	− 1
improves planning quality	53	38	6	3	+ 15	− 8	− 7	0
causes arrangements in working hours	42	46	12	0	+ 12	0	− 5	− 7
makes it possible to study several alternat.	69	28	3	0	− 2	+ 2	0	0
makes plann.work faster	43	46	10	1	+ 7	+ 6	− 10	− 3
causes backache	7	25	36	32	− 3	− 2	− 1	+ 6
causes headache	5	41	29	25	− 6	+ 9	− 8	+ 5
causes eye tiredness	16	51	19	14	− 13	+ 8	+ 2	+ 3
causes risk of cancer	5	7	34	54	+ 2	− 5	+ 3	0

Table 5.   Experiences of CAD use after an operation of one year and changes in percentage of opinions relating to CAD between the first and second interview rounds

Belief in CAD's influence on the efficiency   of   the   company   had   become stronger.   38 per cent thought it to have a greater importance   and more than half, i.e. 55 per cent was of the opinion that it had some significance.   The earlier percentage figures   were 36 and 49   which means that there was an in-increase of 8 percentage units.

The opinions about a stricter working speed had not changed noteworthily. Instead   it   was thought   that   the CAD equipment   had made   the planning work easier than expected.   To work was thought to be as interesting as thought in the beginning.   CAD improved the quality of planning more than believed, an increase of 15 percentage units. This was due to the faster planning work and the possibility to study several alternatives.   Working hours had to be arranged more often than expected.

The results are shown in Table 5 as well as the changes of the same subjects in percentage.

3.2 CAD's influence on different illnesses

The study showed that there appeared less backache and headache than feared, as well as eye strain which, true, about half still were afraid of.   The views about risk of getting cancer were not changed. It was not feared.   The results from the second interview round as well as the changes in percentage between the first and second rounds are shown in Table 5.

3.3 Applicability of CAD in different planning zones

The study cleared also how the opinions of applicapility of different planning zones changed.   CAD could be used clearly better than expected in lay-out planning and offer planning but as for strength calculation CAD caused disappointment.   The results are shown in Table 6.

The experiences got from use showed that CAD best suited lay-out planning and offer planning.   It could be used almost as well for scheme and pipeline planning.   In device planning CAD was not useful.   The results are shown in Table 6.

Planning zone	Experiences of CAD use after an operation of one year (%)				Changes in percentage between the 1st and 2nd interview rounds			
	agrees		disagrees		agrees		disagrees	
	fully	partly	partly	fully	fully	partly	partly	fully
offer planning	75	19	6	0	+ 8	− 8	0	0
device planning	33	50	17	0	+ 4	− 3	− 1	0
pipeline planning	71	24	5	0	+ 5	− 10	+ 3	0
lay-out planning	79	21	0	0	+ 12	− 6	− 5	− 1
building planning	51	40	9	0	+ 6	− 5	0	− 1
scheme planning	71	20	8	1	− 5	− 2	+ 6	+ 1
strength calculation	42	35	14	9	− 13	+ 4	+ 3	+ 6

Table 6. Applicability of CAD in different planning zones after using experiences of one year (%) and changes in percentage of opinions between the first and second interview rounds

3.4 Working hours with CAD

Table 7 shows the estimates achieved from different interview rounds for an appropriate number of working hours with CAD. In the first interview round no clear optimum was found, for the working hours changed from two to four.   Experiences from one year's use favour, however, clearly a continuous working of four hours with CAD. A change to work longer with CAD was indisputable.

Working hours (h)	a (%)	b (%)	c
max 1	4	2	− 2
max 2	27	21	− 6
max 3	33	18	− 15
max 4	30	45	+ 15
over 4	6	14	+ 8

Table 7.   Appropriate number of working hours with CAD.
a = the first interview round
b = the second interview round
c = changes in percentage between the first
and second round

4   OPINIONS BY THE CAD–USERS AND NON–CAD–USERS AFTER AN OPERATION OF ONE YEAR

The study also searched answers on how the opinions of the constructors who use CAD differed from the opinions   of the other constructors.   There were 29 CAD–constructors and 40 non–CAD–constructors in the second interview round.

4.1 Planning in general

It can be stated that the constructors using CAD have more positive opinions of CAD than others have. All CAD constructors believed that CAD increases the efficiency in the company whereas in the comparison group the percentage figure was 85.   73 per cent of CAD constructors were of the opinion that the quality of the planning (correctness, cleanness) become essentially better and 24 per cent believed CAD to have some influence on the quality of planning. The percentage figures in the comparison group were 38 and 47.   The comparison groups had the same kind of opinions about the working hour arrangements.   The CAD constructors believed more than others that planning become faster. It is significant that the CAD constructors hoped that the CAD working stations would be decentralized in the planning offices. The CAD users were, more than others, in favour of extended CAD schooling also for non–CAD–users.   They'd like more than the comparison group to have CAD schooling for themselves.   The CAD constructors resisted establishing a special CAD department.   The CAD–users thought more often that their working conditions were good.

The CAD constructors felt their position to be secured more often than others.   They also felt themselves to be better valued than non–CAD–users. Their job and responsibility areas were clearly specified.

CAD argument	CAD-users				non-CAD-users			
	agrees		disagrees		agrees		disagrees	
	fully	partly	partly	fully	fully	partly	partly	fully
my task and responsibility areas are clearly specified	14	41	31	14	18	28	36	18
the continuity of my tasks is secured	28	41	21	10	18	31	36	15
my working conditions are good	38	42	17	3	8	36	43	13
increases efficiency of the company	52	48	0	0	28	57	12	3
makes my work easier	55	38	7	0	12	40	43	5
makes my work more interesting	38	49	10	3	3	30	50	17
improves planning quality	73	24	0	3	38	47	12	3
causes arrangements in working hours	42	48	10	0	43	45	12	0
makes work faster	48	45	7	0	40	45	12	3
decentralized CAD	49	31	17	3	25	30	33	12
separate CAD-section	3	14	17	66	12	12	33	43
I want CAD schooling	87	7	3	3	41	26	26	7

Table 8. Opinions by the CAD-users and non-CAD-users
after an operation of one year (%)

87 per cent of the CAD constructors were of the opinion that their work
had become more interesting after CAD had been taken into use. CAD equipment
had made the planning work easier also for those not using CAD. The opinions
of the CAD users and the comparison group are to be seen in Table 8.

4.2 Illnesses

Regarding illnesses it has to be noted that the CAD users had less backache
and less headache than the comparison group expected. Eye strain was milder

CAD working causes	CAD-users				non-CAD-users			
	agrees		disagrees		agrees		disagrees	
	fully	partly	partly	fully	fully	partly	partly	fully
backache	0	24	31	45	10	31	44	15
headache	0	41	28	31	8	41	33	18
eye tiredness	10	66	10	14	18	43	26	13
risk of cancer	7	10	35	48	3	15	41	45

Table 9. Fears regarding illnesses (%)

but more frequent than other peopla expected. There were no differencies concerning the risk of getting cancer. The results are shown in Table 9.

4.3 Planning zones

When comparing experiences of CAD constructors of CAD's applicability to different planning zones with answers by other constructors, it was found that the biggest difference lied in scheme planning. 93 per cent of CAD users thought that CAD suited very well in scheme planning whereas the percentage figure for the comparison group was only 56. Another significant difference concerned the applicability for device planning. The corresponding percentage figures were 45 and 23. All CAD constructors thought that CAD would suit offer planning either well or very well. The results are shown in Table 10.

Planning zone	CAD-users				non-CAD-users			
	agrees		disagrees		agrees		disagrees	
	fully	partly	partly	fully	fully	partly	partly	fully
offer planning	76	24	0	0	72	18	10	0
device planning	45	41	14	0	23	49	25	3
pipeline planning	72	21	7	0	64	25	8	3
lay-out planning	87	10	3	0	72	25	3	0
building planning	52	38	10	0	41	46	13	0
scheme planning	93	7	0	0	56	31	13	0
strength calculation	38	49	10	3	41	23	23	13

Table 10. Applicability of CAD in different planning zones (%)

5   CONCLUSION

The results of the study show that the willingness for CAD-working increased with time by all designers, as well as the desire to participate in CAD-schooling. The use of CAD influenced the working hour arrangements more than expected.

The CAD-equipment facilitated the designers' work more than presumed but had no influence in the interest of work. It helped to make better design quality by speeding up the procedure and making it possible to study more alternatives.

There occured less backaches and eye strain than feared before. No one was afraid of getting cancer.

The study showed that CAD can be used for lay-out and offer planning more than expected. In scheme planning and building planning the results corresponded to the expectations, but in strength calculations the benefits expected were not achieved. One half of the working day was found to be the appropriate working time.

C377/046

Teaching expert systems as a design tool

J N SIDDALL
Department of Mechanical Engineering, McMaster University, Hamilton, Ontario, Canada

The teaching of expert systems to engineering students is made quite difficult by the unnecessary jargon and obscurity that is widely prevalent in the literature. A new approach is proposed that brings out only the essential aspects of the subject. The author's experience in teaching a one semester graduate course on expert systems to engineers is described. It is concluded that engineering designers should be developing expert systems; that expert systems should be a widespread facet of design activity; and that the subject can be taught to engineers in a one semester course. Curriculum implications are discussed.

1 INTRODUCTION

All of engineering and science is hampered, indeed burdened, by unnecessary obscurity in the literature. It is due to what might be called Siddall's Law of Obfuscation.

THE LESS SUBSTANCE THAT A SUBJECT HAS, THE GREATER THE AMOUNT OF JARGON, ABSTRACT NOTATION, AND UNNECESSARY AND ARTIFICIAL OBSCURITY.

The basis for this law would seem to be a recognition, not just in science and engineering, but throughout history in all fields, that perceived knowledge is power. All of us in science and engineering are driven by this law; and it is a curse of the professions, particularly in academia, where it retards the successful teaching of new ideas to students.

It seems to be generally accepted in the literature that their has been a disappointing amount of substance in the field of artificial intelligence over its 30 odd years of development; which accounts for the rather appalling amount of obscurity in the writings of the field. Expert systems, the topic of greatest application in artificial intelligence, has in fact made substantial progress in recent years; but nevertheless the tradition of obscurity continues to blanket it, and make learning discouraging for engineering students.

Expert systems is an important, useful, and original subject; but that does not necessarily mean that it is a particularly complex one. It is not, at least at this stage of its evolution, despite the difficulty in reading any typical paper on the subject in artificial intelligence journals.

It is also very important that engineers develop an independent body of theory as a subset of the subject of expert systems. Engineering expert systems, particularly in design and control applications, are significantly different. I believe that there are important differences in how they are developed, in the treatment of uncertainty, and the in the use of Boolean algebra methods.

A new approach is required in presenting expert system theory to engineering design students.

2 BASIC CHARACTERISTICS OF AN EXPERT SYSTEM

An expert system is primarily and basically a complex logic system; and has a great deal in common with electronic logic switching circuits, reliability fault trees, and zero-one optimization problems.

The algorithmic IF/THEN/ELSE structure is an important feature of procedural computer languages, such as in this example.

```
IF (X₁=a OR X₂>b) THEN
      A₁ = .true.
   ELSEIF (E₃=.true. AND E₄=.false.) THEN
      A₂ = .true.
   ELSEIF (E₅=.true. AND 97<X₆<220) THEN
      A₃ = .true.
ENDIF
```

We can convert this to a set of simple logic statements by first converting all IF arguments or premises to logic events.

$$E_1 = .true. \text{ if } X_1=a \text{ is true}$$
$$E_2 = .true. \text{ if } X_2>b \text{ is true}$$
$$E_6 = .true. \text{ if } 97<X_6<220 \text{ is true}$$

Now the logic statements become

$$\text{IF } (E_1 \text{ OR } E_2) \text{ THEN } A_1$$
$$\text{IF } (E_3 \text{ AND } E_4) \text{ THEN } A_2$$
$$\text{IF } (E_5 \text{ AND } E_6) \text{ THEN } A_3$$

These IF/THEN statements, or "rules", are the basic tools of expert systems. In Boolean notation, commonly used in switching theory, they look like this

$$E_1 + E_2 = A_1$$
$$E_3 \, E_4 = A_2$$
$$E_5 \, E_6 = A_3$$

We do not create rules by conversion from IF blocks, but the equivalent to nested IF/THEN/ELSE blocks is hierarchical rules, in which the output from one rule may be an input to another. Logic networks are a valuable way to describe systems of rules, using logic "gates"; and show how rules may share inputs with other rules, so that there can be many interconnections and levels.

If we used IF/THEN/ELSE blocks as in conventional programming, an expert system would become a very large and complex nested system of blocks, which are extremely difficult to work with. Whereas in the simple rule form, new rules can be easily added without necessarily disturbing the existing rules. It makes the Boolean representation possible, with all of the scope available in the methods of Boolean algebra. It makes methods incorporating uncertainty easier to handle. And it makes possible the use of what I call data struc-

<u>tured code</u>; in which almost all of the expert system is incorporated into arrays in a data management system which uses pointers and stacks. By this means rules can be added, removed or changed without writing any new code. And the code for executing an application of the system becomes very simple. The rules are not explicitly stated in the code at all. In fact the same code and data structures can be used for other unrelated systems. Only the data itself need be different. We can even use commercial data management software to some extent. It can now be seen that this approach is what makes the famous <u>shells</u> possible. For those of you unfamiliar with the term shells, these are the device by which the keepers of the secret knowledge make available to the rest of us the means of writing our own expert systems, without understanding what we are doing.

Uncertainty is a key factor in engineering design systems. Everything in engineering is uncertain; and we should be able to incorporate uncertainty in our logic systems. Maybe we cannot predict for sure that a rule will "fire" (a nice bit of jargon), even though its premises are true. Or maybe we cannot be certain if a premise is true or not.

The concept of uncertainty in engineering design is basically different than in other fields; and it is essential that we <u>rigorously</u> incorporate uncertainty in design expert systems, just as we should in more familiar design activities. We do not yet widely use probability theory in design, although the theory is available (1, 2, 3); but when we do it is done rigorously.

A rather remarkable thing has happened in the evolution of the treatment of uncertainty in expert systems. Scientists have completely abandoned a rigorous probabilistic approach. They use things like belief functions, fuzzy sets, relative entropy inference, and a non-rigorous version of Bayes' Theorem. I believe that there are several reasons for this.

(a) <u>The general non-acceptance of the concept of subjective probabilities.</u> The commonly accepted definition of probability is that it is essentially a frequency ratio, based on observed samples. Workers in the field of expert systems tend to reject the idea that probabilities can be subjectively estimated. And since frequency data is rarely available, all kinds of obscure techniques are used to disguise the unavoidable fact that uncertainty measures must, in the end, be based on judgment.

(b) <u>The argument that many experts, and most users, are unfamiliar with probability concepts, and are therefore unable to make uncertainty judgments in terms of probabilities.</u> This may be true, but it is no excuse for lack of rigour in engineering expert systems.

(c) <u>"Knowledge experts" are unable to accept the low output probabilities that tend to result when probabilistic rigour is used</u>. Even smallish systems can have a very rapid degradation of probabilities that are propagated through a system, unless the input probabilities are quite high. It follows that not very much uncertainty is acceptable in inputs, if a conclusion is to have a sufficiently high probability of being true to be of practical use. Engineers must accept these hard facts.

In a rigorous treatment, the rules become a set of probability expressions, with the general form

$$P(A) = P(A|E)\ P(E) + P(A|\overline{E})\ P(\overline{E})$$

where

$P(A)$ = the probability that a conclusion or output from a rule is true

P(E) = the probability that the combined input events or premises of a rule are true (either as AND or OR rules)

$P(\overline{E})$ = the probability that the combined input events of a rule are false

P(A|E) = probability that a rule will fire given that the combined input events are true

$P(A|\overline{E})$ = probability that a rule will fire given that the combined input events are false

   The control system, or inference engine, is the algorithm that incorporates the logic which decides the order in which rules are checked to see if they will fire. There are many variations on these algorithms, but unless one is really pressed for computer time, or there are other special problems in the control, a simple iterative sequential algorithm will always work, even when the rules are in any order. In the case of a probabilistic system, it is repeated until all rules are evaluated, and the one or few conclusions with the highest probability of being true are presented to the user.

   I have so far said nothing about expertise, which is generally considered the essence of expert systems. It is, but not in the narrow intuitive sense; where it is an implementation in computer code of an expert's intuitive decision making process in a rather narrow field. This decision making is based on specialized knowledge and experience, and the decisions essentially come to the expert from his or her unconscious mind. However the expertise can equally well be based on observable facts, or experimental results, or engineering modeling analysis; or it can be any combination of these.

   The term "knowledge engineer" is commonly used to describe those who obtain the knowledge from experts and translate it into rules; and the knowledge engineer is generally said to require special skills that enable him or her to extract from one or more experts the subjective generalized decision procedures that constitute the expertise, and create from them the rules and control system that make the expertise available to anyone. I believe that this is basically a creative design process, which is the essential core of engineering practice. For this and other reasons, engineers should be innately well suited to create expert systems. The system builder and the expert would both be engineers, and think and talk alike. The engineer system builder would also commonly have considerable expertise in, or at least familiarity with, the system domain, and would therefore have good insight into its structure and scope. And finally, the engineer has an almost unique training and experience in intuitive decision making, which is the whole basis of "pure" expert systems. I am therefore optimistic that the building of engineering expert systems, by engineers, will turn out to be not nearly as difficult and time consuming as we might expect from reports in the literature about experience with the pioneer expert systems.

   This brief review of expert systems from an engineering viewpoint leads us to the question - should expert systems theory be taught as part of the curriculum in engineering design?

3   THE MYTHS OF EXPERT SYSTEMS

The many myths that occur in expert system writings presumably have their genesis in the Law of Obfuscation; and they must be debunked before the theory of expert systems can properly take its place in an engineering curriculum.

   You must be a specialist in order to build expert systems. My experience suggests that you do not have to be a specialist in artificial intelligence in order to design and develop an expert system. Indeed it is important that non-specialist engineers develop their own. The ability to do so can be a part of engineering training, either taught or self-acquired, just like any other subject such as mechanics, heat transfer, control theory, and so on.

Special computer languages must be used to create expert systems. The myth that special computer languages, such as LISP and PROLOG, must be used has created a great barrier impeding the widespread use of expert systems. I have never seen any evidence whatsoever that the more familiar procedural languages such as FORTRAN, BASIC, PASCAL or C could not serve equally well or even better. Good working expert systems are being written in procedural languages, and no engineer should be inhibited from tackling his or her own expert system because of the time required to learn a new and very different language.

Shells must be used for large expert systems. I believe that it is a myth that use of a shell is a necessity, or even desirable, for any serious expert system development. Shells do not permit the flexibility and creative design input that is important in building expert systems, and which is a special talent of engineering designers. This is closely related to the treatment of uncertainty. The designer is stuck with the uncertainty procedures built into most shells.

You must be a knowledge engineer to develop an expert system. Many authors have emphasized the importance of being an experienced and expert "knowledge engineer" in order to write an expert system. This is not true for engineering design expert systems. Some familiarity of the generalist engineering designer with the subject of expertise, combined with the innate creative talent and intuitive insight of designers, are all that are required.

Probability theory cannot be used in expert systems. The myth that probability laws and concepts cannot be used to treat uncertainty has been discussed earlier.

Expert systems require an enormous number of man hours to build. Although this was true of the pioneer research systems, increasing experience, familiarity with the subject, and the special qualifications that designers have, should greatly reduce the time required. In addition, many valuable engineering systems are relatively small; and can be quite rapidly created.

The proliferation of jargon that barricades the subject of expert systems creates a frightening aura of secret and special knowledge that must be learned before it is possible to penetrate the mystery and create expert systems, at least without a shell. You must know all about abduction, inference engines, predicate calculus, symbolic inference, knowledge base, control systems, abstraction, conflict resolution, production systems, frames, blackboards, schemas, scripts, situation calculus, ontology, property inheritance, slots, isa, ika, demons, Bayes' theorem, MYCIN uncertainty method, fuzzy sets, relative entropy inference, belief functions, and so on and so on. This is all myth; these terms are not important or really mysterious. They need only be explained to a practicing engineer so that he or she can have some hope of reading the literature, and appearing knowledgeable. They are not really necessary.

4   EXPERT SYSTEMS AS AN ENGINEERING SYSTEM COMPONENT

It can be anticipated that the use of expert systems as control components will be a major application. Many of these will be primarily complex logic systems, without necessarily incorporating the intuitive expertise that is sometimes considered essential for a "real" expert system. In the control of very complex systems, the logic used for responding to changing conditions may be well known and understood, and clearly based on modeling; but just too involved for a person to adequately prepare a response. Or time may be of the essence, so an automated response algorithm is required. In such systems the inputs would come from sensors, and the outputs would be control signals going directly to actuate control mechanisms.

It is quite feasible (4,5) to embody expert systems that are control components into custom logic chips, which incorporate Boolean switching circuits that directly model the logic. No inference engine or any software is required at all within the expert system, and only a simple microprocessor is required which would read inputs either from the user or direct from sensors and feed them to the logic chip; and would read outputs from the logic chip and interpret them to the user, or feed control signals directly to the process. The advantages would be great speed and simplicity.

Boolean methods are particularly applicable to these kinds of expert systems.

5 EXPERT SYSTEMS AS A DESIGN AUTOMATION TOOL

There are many potential areas of application of engineering expert systems in design automation.

(a) Synthesis. It is difficult to visualize expert systems ever replacing the designer by automating synthesis in any direct way, except for quite stereotyped systems, such as gear selection. However an expert system could provide guided access to data banks for assistance with decisions related to material selection, choice of a stereotype configuration, manufacturing considerations, marketing considerations, maintenance considerations, choice of specifications, and the like. Good synthesis requires an extensive familiarity with the field of application; and an expert system could interface with an encyclopaedia of knowledge concerning all aspects of the field of application, greatly aiding the designer's more general knowledge. CDROM devices are already available to provide the technology for such systems.

(b) Engineering Modeling. This provides the verification that the design will work. As well as selecting the significant failure modes and appropriate modeling theory, the designer must also substitute for the physical problem a simplified and idealized abstraction that is still a workable representation of the real world system in all of its essential features. It requires engineering skill, experience, intuition and judgment to set up such a model. An expert system could provide valuable advice, and access to data banks on available modeling theory, and an interface with CDROM's containing complex modeling software; but again it could not conceivably replace the designer.

(c) Taxonomy and Morphology. Another part of the expertise required in design is an intimate knowledge of all of the elements that are used, or might be used, to synthesize design. New designs are primarily new combinations of existing elements. Assistance in the selection of the optimum type of a class of elements such as a bearing or a clutch, could be of enormous assistance to a designer. This kind of decision making requires an expertise that is simply not possible for a designer to have, unless he or she is very specialized, and working with fairly stereotyped configurations.

(d) Optimisation. The designer must be an expert on the criteria and procedures for achieving the best possible design. I do not believe that selection of the criteria could ever be delegated to an expert system, although expert advice would be useful. But there is considerable potential for automating most of the procedure if it is analytical. Considerable skill and experience is necessary to formulate a problem and execute it on the computer. This has been a major barrier to more widespread use of analytical optimization, and there is considerable promise that an expert system interface would go far to solve this problem.

(e) Risk Judgment. This is perhaps the most critical and difficult expertise required in design. I am referring to the risk levels associated with failure of a device to meet one or more performance requirements. In the current state of the art, risk judgments are most often coded in simple factors of safety, but the computer has made practical a codification in terms of probabilities, based on the explicit use of probability distributions and probabilistic analysis. An expert system could have an advisory role, giving the designer advice on past similar applications, but in the final analysis these decisions must be judgemental. The designer is again indispensable.

(f) Aesthetics. The sense or judgment of the aesthics of an engineering design would seem to be the basis for judging the overall optimization of a device or system, and not just the external appearance or style. The designer uses this sense of aesthetics to incorporate optimimization criteria that are not amenable to analytical procedures. These include human values such as pleasure from comfort, convenience, playing games, excitement, and so on. I believe that this judgment is based on innate intuitive skills, similar to creativity, and could not conceivably be automated in an expert system.

It is a concern of many engineers and educators that there is a dangerous trend towards the use by engineers of computer packages as black boxes, with prescribed inputs, and from which they uncritically accept outputs, accurate to eight or more significant figures. "Handbook engineer" used to be a derogatory term applied to an engineer who, when confronted by a technical problem, went to a handbook and searched for a likely looking formula, which he or she then applied uncritically by simply plugging in numbers. The concern is that the computer has become a super handbook, and the problem may become critical if design automation becomes dominated by expert systems. However I believe that the opposite will be the case; expert systems have the potential for saving us from the super handbook. The expert system will act as a consultant, interactively advising the user on all aspects of the expertise required for the design problem, as we have noted earlier. Our particular concern now is with complex computerised engineering modelling. The advisory roles would include the following.

(a) Understanding the nature of the failure modes.
(b) Understanding the assumptions upon which the modelling is based.
(c) Understanding the details of formula derivations.
(d) Understanding details of the numerical techniques used.
(e) Interpretation of results.

The design engineer could now gain much better insight into the modelling being used. The expert system could hopefully also help dispel some of the artificial mystery in the modelling due to the Law of Obfuscation.

6   WHAT IS DIFFERENT ABOUT DESIGN AUTOMATION EXPERT SYSTEMS?

Many expert systems in the non-design area have been developed, and they are basically similar in type to non-engineering systems. They include systems for maintenance, fault diagnosis, data interpretation, system operation, and classification. But some kinds of design systems are different in a fundamental way, and must be given special consideration. This is because they are concerned with a hypothetical physical system, rather than a real one. Consider an expert system for selection of bearing types. The user enters a set of inputs that are mostly design specifications, which are performance criteria that the design must meet. The output is a recommendation for one or, more usually, several types of bearings, such as a gas lubricated hydrodynamic journal bearing. The main result of the hypothetical nature of the system is that it is quite possible for the user to specify a set of inputs that result in no conclusion; all outputs would then be false. This would not normally happen with the other types of systems referred to earlier. The inputs are observed real world events, and unless the user has made an error in observation, the system is inherently consistent, and will always have a conclusion, and perhaps most commonly just one. I am considering now deterministic systems. So the design system should have provision for helping the user redefine inputs, in the event that the first trial fails.

There is also the opportunity for optimization in such a design system. Since it will usually give several optional selections that will work, i.e. satisfy the specifications, we can now apply some optimisation criterion, such as cost, to choose between them.

7 TEACHING EXPERT SYSTEMS TO ENGINEERS

I believe that my experience in teaching a one semester graduate course on expert systems for the past three years, to about 45 students, supports my opposition to these myths, and confirms the relative simplicity of the subject. The group has included students from mechanical engineering, electrical engineering, chemical engineering, engineering physics, and computer science. To qualify for entrance to the course a student only needed to be familiar with any computer language and simple probability theory.

The topics discussed were as follows.

1. Introduction to Applied Artificial Intelligence
2. Introduction to Expert Systems
3. The General Structure of Expert Systems
4. Phases of Developing an Expert System
5. Knowledge Base
6. Inputs, Intermediate Conclusions, and Outputs
7. Control Structures
8. Data Structures in Expert Systems
9. Expert Systems Incorporating Uncertainty
10. Machine Learning Expert Systems
11. Boolean Algebra
12. Design Systems
13. Development of Small and Large Systems

The essential topics for writing fairly straightforward systems are one to seven; and these are rather short. Topics eight, nine, 12 and 13 are rather important; and 10 and 11 are valuable for more advanced work.

The theory of data structures is an important tool in the design of expert systems, unless the system is rather small and static. I provide the students with some brief notes on the theory of data structures, which experience has indicated is adequate for small to medium systems. However I strongly recommend to the students that they study the subject in more depth if they become seriously involved in expert system development.

The course is very much hands on. The student quickly start writing code for expert systems, using the language of their choice. And they have used every language referred to earlier. Their programs are based on simple expert systems for which the rules are provided; and cover all aspects except machine learning, which is a little too complicated an exercise. Most of the students have no background at all in the theory of data structures, but they seem to have no trouble picking up the basic concepts needed for an assignment that requires data structures, and the building of a simple shell. They also have a final project, in which they write an expert system for a subject of their own choosing, preferably based on their own expertise, and also preferably based on their thesis work, or industrial work if they are part time students. Several of the projects have been based on actual company problems; and are being used in practice. But it can be on any subject, not even in engineering. They must create their own rules, and write their own code. I do not permit the use of shells, but have rather reluctantly accepted the use of PROLOG in one case. Many have used data structures, using code that they have developed in an earlier exercise. They include provision for uncertainty if it is appropriate to the problem. The number of rules in the projects have varied from about 30 to 150. Feedback from the students indicates that they spend from 30 to 100 hours on the projects.   Some are very good indeed, and with more development have the potential for very useful expert systems. But all are quite realistic and workable. The students seem to enjoy the practical nature of the course.

8   CONCLUSIONS

Should designing expert systems be an activity for engineering designers? There may be some feeling that this is too specialized a job for engineers, and it should be left to specialists in artificial intelligence. I have indicated earlier why I think that engineers are particularly well qualified to do it themselves; and I have tried to demonstrate that it is not all that difficult a field for them to learn, and that shells are not really necessary, or even desirable. I believe also that an expert system can be built better and more quickly if the "knowledge engineer" has some expertise in the subject; and this would commonly be the case in engineering expert systems.

If we assume that the creation of expert systems is a legitimate, and even necessary, engineering activity for designers, then curriculum changes become an immediate consideration. My experience suggests that it can be done in a one semester course, although perhaps a little heavy for undergraduates to get a real hands-on feel for the subject. It may be also desirable to strengthen prerequisite study of probability theory, data management theory, Boolean algebra, and software development.

Expert systems is not really a particularly difficult or complex subject; and should be taught as a design tool in a common sense, down to earth, and no frills manner.

REFERENCES

(1) Siddall, J. N. Probabilistic engineering design; principles and appli-
    cations. Marcel Dekker, N.Y., 1983.
(2) Ang, A. H-S. Probability concepts in engineering planning and design,
    Vol. 1 – Basic principles. Wiley, N.Y., 1975.
(3) Benjamin J. R. and C. A. Cornell. Probability, statistics and decision for
    civil engineers. McGraw-Hill, 1970.
(4) Lu, P., J. N. Siddall and J. Verhaeghe. Expert systems built on a chip.
    In press.
(5) Andert, E. P. Jr. and W. C. Frasher. Real-time expert control system using
    propositional logic. Computers in Engineering 1988, Vol. 1, Proceedings
    of the 1988 International Computers in Engineering Conference, July 31-
    August 4, 1988, San Francisco, pp. 159-163.

C377/298

Post delivery design support for reliability monitoring

D J LEECH, BSc, MSc, CEng, FIProdE, FIMechE and **A J WATKINS**, BSc, MSc, PhD, FSS
Department of Management Science and Statistics, University College of Swansea

The delivery and failure of components in fleets
of vehicles (or other designed systems) is a birth
and death process and there is a need to determine
the product support policy early in the life of
the fleet. We need, therefore, to estimate component reliability before the fleet has reached a
stable population and when only limited failure
information is available. Procedures for dealing
with early fleet data are offered together with
methods for gauging the accuracy of reliability
estimates. The procedures are exemplified by a
recent attempt to bring a newly designed vehicle
accessory into service.

1 INTRODUCTION

A newly designed product may be completely new although, more often, it is
evolutionary and an improvement on earlier designs. But in either case when
it is first delivered, it is unlikely to meet the customer's needs. It may
fail because the design specification that has been agreed between the customer and the supplier does not properly describe the need, or the acceptance
tests derived from the design specification may prove to be inappropriate, or
because the system has to be delivered before it has been shown to meet the
requirements of the specification. If the product has previously been shown
to meet agreed acceptance tests, any support provided by the designer must be
paid for by the customer. Support that is required because the product does
not meet the contracted requirements must be paid for by the supplier, who
will be assumed to have allowed for such costs in his original tender.
Support will be required even if the product meets its specification because
use will show that any product is capable of enhancement. Such post acceptance enhancement of a product will usually be the subject of formal change
procedures.

One aspect of a product that will almost certainly require post delivery
support is its reliability. When a customer commits himself to buy a newly
designed product, it is unlikely that the supplier will have done more than
estimate reliability from published data on generally available components, or
from the known performance of earlier, and hopefully similar, designs. Any
tests before the product is delivered will have been unlikely to have proved
the full, advertised life of the product.

This situation is exemplified by a recent purchase of equipment of high technological content. The product was designed specially for the customer; it is fitted to a vehicle and is delivered, item by item, during the life of the vehicle fleet. The supplier of the equipment had calculated its MTBF from what were believed to be the MTBF's of its component parts (mainly electronic), and for which published lives were available (as in, for example, [1]). This calculated system MTBF was offered to the customer as that which could be expected in service. The customer kept defect reports of the product in service and the life of each product at failure was recorded. Even a cursory glance at the defect reports suggested that the lives at failure were much less than the supplier had predicted - for instance, a simple graphical analysis of the lives yielded an estimate of the MTBF that was about a tenth of the predicted figure. It was appreciated that this simple analysis was made early in the life of the fleet, so that this estimate was derived on the basis of products which had failed at atypically short lives, and should be modified to take account of the majority of products which had not yet failed. There is no great difficulty in allowing for effectively censored lives of surviving equipment, but this necessitated a census of products which had not failed. The customer was able to carry out the required census, although considerable effort was involved and could not, lightly, be repeated regularly. The evolving nature of the population also creates difficulties in simple parameter estimation because the censored lives of survivors may be less than the lives of the failed equipment. There are graphical methods of dealing with this type of censored information (see, for example, Nelson [2]), but they beg the question of the confidence with which life parameters can be determined. It was necessary, therefore, to use a method of parameter calculation which tells us how confident we may be of our estimated parameters.

2 THE NATURE OF THE FLEET

When an organisation buys a fleet of equipment to a new specification it may be making a decision which it must live with for thirty years (consider an airline determining which aircraft they will buy; an army deciding which fighting vehicle they will buy; a driving school deciding which car they will buy). The framework is characterised by the following properties.

i)   Vehicles will come into service one by one, or in small batches.

ii) Vehicles will fail independently, one by one, and be removed from service or be subjected to major repair to give them new life.

iii) It will take time for the fleet to build up to its equilibrium number of vehicles.

iv) Vehicles may improve in quality as defects discovered during development and service are identified and corrected.

(Note that what applies to the vehicles is likely to apply also to such sub-assemblies as power plants or gear boxes.)

What we have, in fact, can be viewed as a continuous birth and death process with, it is to be hoped, a population that improves in quality with each new unit that is born.

3 THE SIGNIFICANCE OF RELIABILITY

The owners of the fleet want to know how reliable the vehicles or sub-assemblies are because

i)   they want to know if the supplier is selling them equipment that is as good as advertised,
ii)   they want to estimate how many vehicles will be available for use at any time (an army with 100 tanks, 80% of which are available will be out-gunned by an enemy with 90 tanks, 90% of which are available),
iii) they want to know how many spares they need to hold to maintain an acceptable level of availability, and
iv)   they want to design a maintenance policy, with an acceptable availability, that minimises ownership costs.

4 WEIBULL LIVES

The life of a product is often assumed to follow the Weibull distribution, and in several families of vehicles or major sub-assemblies, the analysis of lives at failure shows this to be a reasonable approximation. This is exemplified by Figure 1, a Weibull plot of gearbox data.

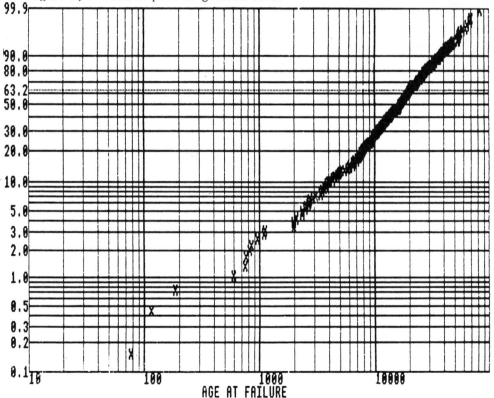

Fig 1   Probability of failure by a given age
(Data from 333 gearbox failures)

The cumulative distribution function of the Weibull distribution is F(t), where F(t) = 1 - R(t), and

$$\text{Reliability} \equiv R(t) = \exp\{-(t/\theta)^{\beta}\},$$

where θ is the characteristic life,
β is the shape parameter,
(and θ and β are to be determined).

A common measure of reliability is the Mean Time Between Failures (MTBF) or the Mean Miles (or Kilometres) Between Failures (MMBF), although this needs qualification.

For the Weibull distribution, the MTBF is

$$\theta \ \Gamma(1 + \beta^{-1})$$

where $\Gamma(.)$ is the usual gamma function; that is, the MTBF is a little less than θ.

The easiest way to determine β and (more obviously important) θ is to plot the lives of the failed products on Weibull paper. Generally, this should allow us to estimate the values of β and the MTBF, while the scatter of the points about the line of best fit will tell us whether our assumption of a Weibull distribution is reasonable. Figure 2 shows a plot of 150 failure lives of a product and it could be argued that

i) the line of best fit is reasonably drawn, although we might question the way that the shorter lives fall away from our line.

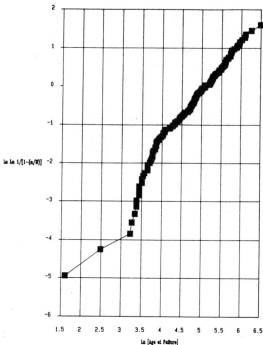

Fig 2   Weibull plot of 150 failures early in fleet life

ii)   β is about 1.3,
iii) θ is about 150, and, hence,
iv)   the MTBF is about 140.

However, these arguments are not acceptable once we observe that there were 300 surviving products, many of which had lives in excess of the lives of many of the failed products. The lives plotted on Figure 2 are, in fact, atypically short because the data had been collected at an early, and transient, stage of the fleet life. Thus, the argument above is akin to estimating life expectancy of a group of people when only the ages of those who died in infancy are known.

5   ALLOWING FOR SURVIVORS

We do not want to wait until the fleet has reached an equilibrium population before we calculate the reliability of the product because we may need to discuss quality with the supplier and design spares holding and maintenance policies early in the delivery programme. We could, of course, argue that the supplier should have designed the product support system in the feasibility or project definition stages of design, but this is probably a council of perfection, and what is required is a programme of supplier and customer cooperation during the early post delivery phase of the project.

The simplest method of allowing for the transient nature of the population is that used to determine the lives of components from experiments (censored tests) which are terminated before all the components have failed. In most censored tests survivors have lives that are longer than those of the failed components, but this is not essential. In a fleet that is viewed as a birth and death process, the lives of some survivors will be short if those survivors are products which only recently came into service. The method used is outlined by Kalbfleisch [3], in which the most likely values (that is, maximum likelihood estimates) of β and θ are calculated. A simple program

Value of BETA

		1.12	1.14	1.16	1.18	1.20	1.22	1.24	1.26	1.28	1.30
	490	.353	.519	.695	.852	.957	.989	.942	.830	.677	.514
	491	.362	.530	.707	.862	.964	.992	.940	.825	.670	.506
	492	.371	.541	.718	.871	.970	.993	.938	.819	.662	.497
Value of	493	.380	.551	.728	.880	.976	.995	.935	.812	.654	.488
THETA	494	.389	.561	.738	.888	.980	.995	.931	.805	.645	.479
	495	.397	.571	.748	.896	.984	.994	.926	.797	.635	.470
	496	.406	.580	.757	.903	.987	.993	.920	.788	.626	.461
	497	.414	.590	.766	.909	.990	.990	.914	.779	.615	.451

Fig 3   Tabulated relative likelihoods
(Data from 150 failures and 300 survivors)

(AMOL) uses numerical methods to estimate these parameters from all the information available. The program is quickly run on a micro-VAX, and is easily used on a desktop PC - using a spreadsheet, for example. The 150 failures shown in Figure 2, together with the 300 survivor lives, when analysed by AMOL estimate that

θ = 500
β = 1.2, and, hence,
the MTBF = 490.

Significantly, the MTBF is about three times as much as failure data alone led us to believe.

6   CONFIDENCE IN THE CALCULATIONS

It is not enough to know the most likely values of β and θ if we intend to argue with suppliers or design a product support system. We have to know how

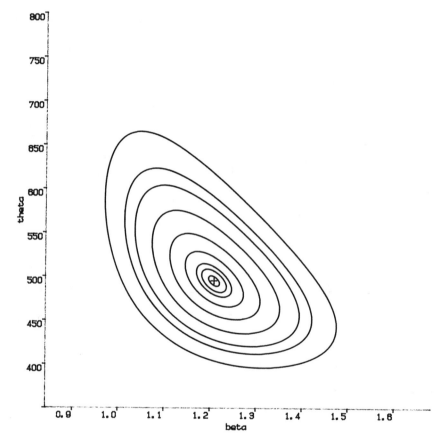

Fig 4   Contours of relative likelihoods
(Data from 150 failures and 300 survivors)

confident we are in our calculations and, generally, the more data we have the
more confidence we will have in our estimates. Usually, too, we derive more
confidence from lives at failure than from the lives of survivors. Once we
know the most likely values of the parameters it is straightforward to construct a grid of pairs of parameter values and determine the relative likelihood of each parameter pair. In principle, one can interpolate relative
likelihood values within this grid to find contours of constant relative
likelihood. Figure 3 is an example of such a grid. In the event, it was
found more convenient to solve the likelihood expression for fixed values of
relative likelihood to obtain the contour values of β and θ. The program to
do this is called AREL and its development is described elsewhere [4]. AREL,
which requires considerably more computational effort than AMOL, is used here
to calculate the 99%, 95%, 90%, 75%, 50%, 25%, 10%, 5%, and 1% relative
likelihood contours for the 450 data available. Figure 4 shows the relative
likelihood contours for this data, while Figure 5 shows how our confidence
reduces when only half the data is available.

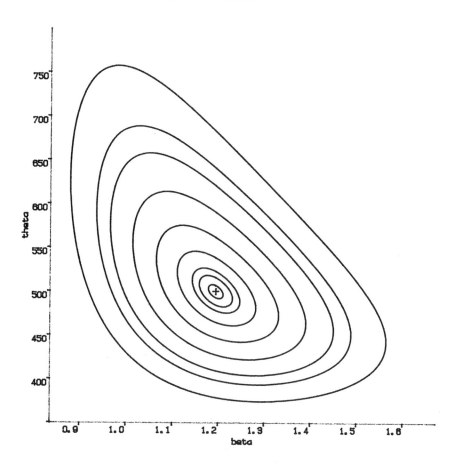

Fig 5   Contours of relative likelihoods
(Data from 75 failures and 150 survivors)

7   COLLECTING DATA AND ESTIMATING PARAMETERS

We need to know the reliability of the product before an equilibrium population has been reached. To calculate the parameters we need information on
the (censored) lives of surviving products as well as the lives of those that
have failed.

We need to know whether the parameters are changing as more products come
into service (are the later units better than the early ones?) and so we need
to determine the parameters periodically while the product is in service.

The sequence of activities may be as follows.

i)   The lives at failure should be recorded on a defect report whenever
      there is a failure. This information is easily stored in a data base
      program which will thus contain a file of all lives at failure since the
      first vehicle came into service.

ii)   Every time (say, every year) it is required to estimate the parameters,
      and hence reliability, it will be necessary to conduct a census of the
      lives of all the surviving vehicles and create a file of "censored"
      lives.

iii) The lives at failure and the censored lives form the input to program
      AMOL which will calculate the most likely values of β and θ.

iv)   If it is desired to determine the confidence with which we accept the
      estimates of β and θ, the two files will also be input to program AREL
      which will calculate the contour values of β and θ for given relative
      likelihoods.

8   CONCLUSIONS

i)   A management procedure should be installed by any operator of fleets of
      equipment to determine the reliability of designed items, and sub-
      assemblies. This procedure must include the storage of data from defect
      reports and a periodic census of equipment which has not failed.

ii)   The supplier of equipment must budget time for the post delivery calc-
      ulation of equipment life, and for the support by senior design staff
      of the customer's efforts to confirm that the equipment meets the
      requirements of the specification.

iii) The Weibull distribution may often be used to describe the lives of
      vehicles and sub-assemblies.

iv)   An important factor in the calculation of reliability is the mean life
      (or miles) to failure of the items, or sub-assemblies.

v)   When the fleet has not reached its stable population, the calculation
      of the mean life to failure of items requires a knowledge of the lives
      of those items that have not failed. Thus, in addition to the failure
      lives which are obtained from defect reports, it is necessary to con-
      duct a census, from time to time, of the life of surviving equipment.

vi)   A simple program (AMOL) is available for calculating the most likely
      mean life of an item or sub-assembly.

vii) A more complex program (AREL) will determine the confidence with which estimates have been calculated.

9   REFERENCES

[1]   US MIL-HDBK-217E

[2]   Nelson, W., Applied Life Analysis, Wiley, New York, 1982.

[3]   Kalbfleisch, J.G., Probability and Statistical Inference II, Springer-Verlag, New York, 1979.

[4]   Watkins, A.J. and Leech, D.J., Towards Automatic Assessment of Reliability for Data from a Weibull Distribution, Reliability Engineering and System Safety, 1989: to appear.

Safety in design—an American experience

T WILLIS, BSc(Eng) and **M P KAPLAN**, BS(Eng)
Willis, Kaplan and Associates Incorporated, Arlington Heights, Illinois, USA
M B KANE, BA, JD
Morris and Stella, Chicago, Illinois, USA

SYNOPSIS This paper discusses the topic of products liability it the U.S. from the engineering viewpoint, and explores the relevance of that experience to the changing European situation in this field.

The first section of the paper outlines the basic structure of American products liability law, explaining the various common legal bases used in lawsuits involving engineering products. The distinction between the legal theories of negligence (now commonly applied in EEC member countries) and strict liability (now being introduced in the EEC) is discussed, particularly as it relates to the design engineer. The U.S. litigation process is briefly described, primarily in terms of personal injury litigation on industrial machinery. Selected similarities and differences between the U.S. and the new EEC laws such as state of the art, the maximum time limitation for bringing suit against the product, post-manufacture alterations by others, subsequent improvements by a manufacturer after production of the subject equipment and compliance with codes are identified and discussed.

The main body of the paper discusses the potential impact on European engineers of introducing strict liability. The past impact on American design is traced, especially in the time period subsequent to the introduction of strict liability in tort (1965). The paper identifies the relevance of the American experience to the EEC at the design engineers' level, in such factors as modifying design methodology to address machine safety. The paper emphasizes the need to design for safety, as traditionally recognized by the engineering community, the larger community of machine users, owners, the legal community, and society, and now in light of the changes in EEC laws pertaining to product safety. No longer is it acceptable to simply show that you did your best and exercised reasonable care: instead the machine safety will also be evaluated on a cause/effect relationship for a particular accident and defect without regard to the practices of the engineer.

1. INTRODUCTION

The need for safe design in equipment has long been apparent to the engineering design profession. However, benefits of safety such as increased productivity, enhanced employee morale and lower insurance and liability costs have often eluded U.S. employers in the manufacturing sector. Historically, the result has been a reticence by some to spend money on safety, both in terms of providing a safe workplace environment (noise free, pollution free hygienic conditions) and in the equipment and safety training aspects.

Fortunately, these attitudes are changing, and some of the factors driving U.S. employers toward even higher safety consciousness include the machine manufacturers insistence on providing safety devices from the outset as part of the equipment, (not as an option) and the increased activities of regulatory agencies such as the Occupational Safety and Health Administration and the Department of Transportation . Other driving factors include the increased awareness of safety equipment and techniques (through safety and standards organizations and in the education of engineers), and the products liability/litigation process.

This paper discusses the topic of products liability in the U.S. from the engineering view point, and explores the relevance of that experience to the changing European situation in this field.

2. AMERICAN PRODUCTS LIABILITY LAW

2.1 Negligence

The concept of negligence as it applies to personal injury litigation embodies the traditional notion that a party who has done something wrong is liable for the damages caused by that transgression. In product liability law suits, prior to the adoption of strict liability, this meant that a manufacturer could only be held liable for the injuries caused to a person if it were shown that the manufacturer did something which a reasonably prudent manufacturer would not have done , or failed to do something which a reasonably prudent manufacturer would have done. In short, it was required that the complaining party prove some fault on the part of the manufacturer.

While negligence is still alleged in nearly all present product liability law suits filed in the U.S., the concept of strict product liability has largely eclipsed negligence principles in litigation involving injuries resulting from the use of products. Typically, the complaining party will file a law suit which sets forth claims based upon both negligence and strict liability in order to obtain information about what went into the design and manufacture of a particular product. By a process known as "Discovery", the parties exchange information, submit
witnesses for interrogation and provide documentation, such as design drawings, to the adversary. It is during this process that the facts supporting the claim are established.

Each party is allowed to submit written questions, referred to as "Interrogatories," upon each other party requesting information concerning the issues in the suit. In addition, by serving on the manufacturer a "Request to Produce Documents" a plaintiff may obtain copies of all relevant design drawings, memoranda and any other writings which were generated in the course of the design and manufacture of a product. The plaintiff's attorney and usually his "expert" are thereby allowed essentially full access to the files of the manufacturer as they relate to a particular product or group of products. Since there are generally allegations of both negligence and product defect, the scope of information which can be obtained during this discovery process is quite broad. If, as a result of the information obtained during discovery, the negligence case appears weak, the complaining party can, and frequently does, proceed to trial only on the strict liability claim.

2.2 Strict Liability

In the U.S., each State may elect whether or not to adopt strict product liability, and each is free to establish the legal basis for that liability. By far, the majority of States have elected to adopt a form of strict liability with respect to personal injury law suits relating to products. In light of this, it might be expected that there would be very little consistency in the product liability law among the various States. This has not proven to be the case however. Those States that have adopted strict product liability have generally based it upon either a breach of an implied warranty of fitness for the product's intended use or upon the concepts set forth in Section 402(A) of the Restatement of Torts

(Second). When based upon warranty, the complaining party must generally show that the product was not suitable for the use for which it was intended or for a reasonably foreseeable use for which the product may be used and that an injury resulted from this lack of suitability.

The second basis, Section 402(A) of the Restatement of Torts, states:

"(1) One who sells any product in a defective condition unreasonably dangerous to the user or consumer or to his property is subject to liability for physical harm thereby caused to the ultimate user or consumer, or to his property, if:

(a) the seller is engaged in the business of selling such a product, and

(b) it is expected to and does reach the user or consumer without substantial change in the condition in which it is sold.

(2) The rule stated in Subsection (1) applies although

(a) *the seller has exercised all possible care in the preparation and sale of his product, and*

(b) the user or consumer has not bought the product from or entered into any contractual relation with the seller." (Emphasis added)

Thus, a manufacturer in the U.S. can employ the highest degree of care in the design and manufacture of its product and still be held liable for a particular accident if it is determined that the product is defective. What is really on trial in a product liability suit, at least one grounded upon strict liability, is not the manufacturer but the product itself.

2.3 Definition of Defect

The concept of defect is key to understanding U.S. product liability law. There are generally considered to be two separate types of defective products for purposes of product liability suits: those containing manufacturing defects and those containing design defects. Manufacturing defects occur when the product is not properly manufactured to its own design, as when a machine contains a badly made weld which allows the part which it secures to come loose , resulting in an injury to a worker. A design defect, on the other hand, implicates some basic aspect of the machine itself. Often, in the context of U.S. product liability suits, design defect cases are based upon a claimed failure to either incorporate proper guards or to properly warn the user of foreseeable dangers. A machine which is properly guarded and functions as intended may, nonetheless, be considered to be defective if it fails to incorporate proper warnings. Far and away the greatest number of product liability suits in the U.S. deal with the issue of design defect.

There is unfortunately no uniform definition of what constitutes a defect in a product since, as noted above, the several States which have adopted strict product liability have been free to develop their own notion of what a defect is. This fundamental definition is generally determined by the courts of each State which have wrestled with the problem. The closest thing to a generic definition would probably be that a product is defective if it is unreasonably dangerous, which of course requires that the term "unreasonably dangerous" be defined. One commonly used definition for an unreasonably dangerous product is one which is unsafe when put to a use or misuse that is reasonably foreseeable considering the nature and function of the machine. Unfortunately,these definitions raise as many questions as they answer.

2.4 Comparison with the EEC Directive

A "Statute of Limitation" is a provision which sets a time limit after an occurrence within which a suit must be filed. The EEC Directive on Liability for Defective Products provides that an action must be commenced within three years from the date of the loss or damage complained of. Each State in the U.S. has a statute of limitation which governs the time within which a personal injury action, including one alleging strict product liability, must be filed. Each State sets its own statute of limitation and they vary from between one and six years. It can be very difficult to defend a suit which has been filed nearly three years after an accident has taken place, which may be years after the product has been provided. For this reason maintaining accurate records of the components provided with a machine, as well as accurate maintenance records, is imperative.

In addition to the three year statute of limitations, the EEC Directive provides that once ten years has passed from the date of supply of the product, the liability of the manufacturer ceases. Such provisions are sometimes referred to as "Statutes of Repose" and they effectively set an outside time limit within which a strict product liability claim may be brought. They have no effect on the plaintiff's ability to file a negligence claim against the manufacturer. Statutes of Repose are not common in the U.S. although some States have adopted such provisions. In those States which have not adopted a Statute of Repose, a manufacturer may be required to defend the merits of a machine which is obsolete when measured against current products performing the same function.

In light of the problems which arise in defending now obsolete products, many States, though not all, allow what is known as the "State of the Art" defense. This allows the manufacturer to argue that his product was designed and built in accordance with the state of the art as of the time of manufacture and that the failure to incorporate a feature which was not available because of the then existing technology does not constitute a defect. For example, if an industrial machine was produced before the existence of photo-electric cells, the machine should not be considered to be defective for failing to incorporate a photo-electrically triggered safety, since such a device was not within the state of the art at the time of manufacture. The EEC Directive has a similar provision . This provides that a product is not defective if the state of the scientific and technical knowledge at the time of manufacture was such that the existence of the claimed defect could not have been discovered by the manufacturer. In those States in the U.S. which do not allow the "State of the Art" defense, the manufacturer is held to the state of the scientific and technical knowledge as of the time of the accident rather than the time of manufacture, which, with an old machine, can make defending a product extremely difficult.

Sometimes, a manufacturer may make changes to his product after the time of an accident. Most States in the U.S. have adopted a rule which prevents a plaintiff introducing evidence of such changes as part of his proof of the existence of a prior defect. The theory behind this is that allowing the introduction of such evidence would discourage the manufacturer from improving the safety of his product. The exceptions to this rule are legion, however, and the rule is honored more in the breach than in the keeping. There does not appear to be a similar provision in the EEC Directive, however since this has been implemented as a rule of evidence in the U.S., this is understandable.

In summary, while the EEC Directive contains some provisions similar to those incorporated in many of the States in the U.S., there exist notable differences It remains to be seen whether the similarities will result in parallel engineering developments.

3. IMPACT OF LIABILITY ON AMERICAN DESIGN

Since the introduction of the strict liability doctrine in 1965, the design engineering profession in general has been called upon to respond to the additional constraints imposed by that change in law. The impact on those companies which were already producing reasonably safe products was, of course, minimal, often amounting to little more than ensuring that their liability insurance premiums were paid up. For the companies that had paid little attention to product safety or who were simply ill-informed on such matters, the ensuing period has been one of developing awareness and skills in the field of safety. It is the experience of this latter group that is of interest here.

Strict Liability laws have impacted the process of engineering design in many ways. Three important effects are:

* The Engineering Industry has developed an increased sensitivity to safety in design

* Documentation of safety aspects has been much improved

* Growth in number of safety specialists, both inhouse and consultants, to handle both design itself and the litigation related matters.

3.1 Increased sensitivity to safety in design

This begins with engineering education. Throughout the United States, many engineering degree programs now include a separate course on Products Liability and Safety. The teaching of safety principles for engineers has always been included in mechanical engineering programs, but these elements are separated and identified much more in today's programs. The American Society of Engineering Education, in conjunction with the National Institute for Occupational Safety and Health (NIOSH) has recently sponsored several case studies on machine safety specifically for use in undergraduate engineering programs.

Additional educational emphasis in safety matters has been made through seminars promulgated by such organizations as the National Safety Council, American Society of Mechanical Engineers and the American Society of Safety Engineers.

The concept of strict liability in tort was introduced in 1965. Machinery sold in the U.S. in that era was not always adequately guarded, even though guarding technology was well developed at that time. Some equipment was produced which did not provide protection, or which did not have adequate warnings. As an example, an Italian machine imported into the U.S. in the mid-sixties, was designed for trimming the bristles in the manufacture of sweeping brooms. This particular equipment had no point of operation guarding, and required the operator to hold the bristles within three inches of the rapidly rotating cutting blades. No safety warnings were incorporated. At that time, there were no regulations in the U.S. which would require that this equipment be made safe.

In 1971, the Federal government enacted the Williams-Steiger Act, and established the Occupational Health and Safety Administration (OSHA). This was similar in form to the Health and Safety at Work Act subsequently introduced in the U.K. in 1974. However, OSHA is applicable principally to employers, and does not address equipment manufacturers. Furthermore, its enforcement history has been less than ideal. Nevertheless, equipment in the 1980's much more frequently comes from the manufacturer equipped with guarding, wherever this is possible. User manuals often go the extra step to spell out specific hazards and safety procedures, even though prior manuals may have contained all the necessary information to determine this. Often these manuals are backed up by warning decals on the equipment, sometimes in a redundant fashion. A major U.S. manufacturer of foot switches who in the early 1970's warned in their literature of hazardous applications of the product has now incorporated an additional and similar decal on the switch itself.

A typical example of a technical issue raised in the context of strict liability involves guarding of high speed cutting blades. One such case involves a mitre saw made in Japan. In this design, the guard can easily be removed, If the machine is then operated, the rotating blade can fly off. An injured plaintiff in an ensuing law suit contended that this can be designed such that when the guard is removed, the saw becomes totally inoperative, so giving protection. Operating this saw without the guard is considered by plaintiff to be reasonably foreseeable, and so suit was filed under strict liability. A U.S. manufacturer of air powered grinders has discontinued a product line in which the machine will run without its guard in place, replacing it with a model which becomes disabled when the guard is removed. Older machines from this manufacturer are still the subject of lawsuits.

Of course, the nature of some equipment is such that the manufacturer cannot routinely provide point of operation guarding, since he is unable to determine the type of product to be machined on that equipment. Mechanical presses and press brakes are examples of this type. In response to this, the last two decades have seen rapid growth of many companies specializing in specific guarding techniques and equipment to assist owners design a safe system for their employees. One major press brake manufacture actually provides a list of such companies in their equipment manual.

European imports to the U.S. in the 1960's were a mixed bag from the safety point of view. A major Swiss manufacturer of paper converting and printing equipment shipped equipment which was securely guarded, using barrier guards, interlocks, safety switches and warnings technology as appropriate. On the other hand a British four spindle drill with a five location indexing work table was imported into the U.S. in the mid-sixties with virtually no guarding present, relying almost exclusively on operators care for providing safety. A mechanic lost several fingers when this equipment was cycled while being adjusted. This accident primarily resulted from mis-communication between mechanic and operator, and also the manufacturer's failure to adequately guard moving parts. Under negligence law, the behaviour of this mechanic and/or the operator would be a factor in determining contribution to this accident. Under U.S. strict liability law, the machine is evaluated as to whether or not it is unreasonably dangerous and so caused the accident.

The trend in equipment protection in the U.S. since the middle sixties has been to redesign the basic equipment for better user protection where possible. If this could not be done, then more complete guarding has been introduced and warnings and instructions provided where necessary. Nevertheless, liability insurance premiums in the U.S. have mushroomed in the same time period and the number of products lawsuits filed against equipment manufacturers seems to continue unabated. Strict liability law has clearly had an impact on equipment design in the U.S.

3.2 Documentation of safety aspects has been improved

Most equipment manuals produced for use in the U.S. today include a specific section on safety. Safety decals appear on almost every type of equipment. One U.S. ladder manufacturer now includes as many as ten safety signs, these giving forty-six different safety warnings and/or instructions regarding use of the product. The American design engineer has become familiar with the requirements for design of appropriate warning decals, including use of key words (Danger, Caution, Warning), approved color schemes and symbols. This is in addition to the need to warn of both the hazard and its consequences, together with necessary action to avoid injury. Basic documentation of warnings is given in the American National Standard ANSI Z-53.1.

Many industrial organizations now provide safety information and advice as part of their services. The National Safety Council, the American National Standards Institute and the American Society of Safety Engineers have long provided such information, but additional groups, e.g. the Conveyor Equipment Manufacturers Association, also produce their own safety literature. The National Machine Tool Builders Association provides films/videos relating to safety on several major classes of equipment (Presses, Press Brakes, Shears etc.).

3.3 Growth in the number of safety specialists

Since the mid sixties, U.S. industry has seen an enormous growth in the safety field. Major manufacturing plants in the U.S. now employ full time safety professionals on their staff. These professionals, often fully trained and certified in this field, oversee in-plant safety in such matters as health and hygiene, compliance with safety regulations and in equipment safety. This is a broad requirement, and it is rare that these people also have engineering qualifications.

Support for these safety professionals in the engineering and technical aspects of the their work is sometimes provided by the in-house engineering staff, but very often outside consultants specializing in safety engineering matters are used. These consultants are typically from independent consulting firms, safety equipment manufacturers or from insurance companies. These people can bring fresh ideas to the problem, are usually unencumbered by inside politics, and have proven to be very effective.

Additionally, the safety aspects of the company's product are often subject to review by outside experts. This is done as design review at various stages of progress, and is performed in many contexts such as foreseeable use/misuse, failure modes and effects analysis and/or safety code compliance. Corporate legal departments in the U.S. are also becoming more and more involved in decisions regarding warnings and manuals, especially relating the text of these documents to potential lawsuits later on.

Perhaps the most significant growth in the U.S. safety engineering sector since the advent of strict liability has been in Forensic Engineering. This is an area of engineering most likely to be unfamiliar to the European engineer. Forensic engineers specialize in the investigation of accidents, analysis of design from the view point of safety and in assisting the legal profession in technical matters relating to their case. The American Society of Mechanical Engineers recognizes the discipline, and frequently sponsors conference sessions on selected technical aspects.

3.4 Forensic engineering analysis

The forensic engineer may be retained by an attorney (plaintiff or defense), and provides technical expertise in preparing the case, evaluation of the opposing party's case, and possibly providing expert testimony at trial to assist the court to understand the complex technical issues. Engineering assignments include review of technical data (blueprints, calculations, test reports etc.), literature searches, physical examination and testing of the equipment involved in the lawsuit and preparation of courtroom exhibits such as videos, slides/posters, computer simulations etc.

The forensic engineer must understand the technical issues involved. He is asked to formulate independent opinions regarding the safety aspects of the equipment and its relationship to the accident in question, and these opinions must be well founded on sound engineering principles. In most cases, the forensic engineer will also be required to give deposition testimony prior to trial. This is a formal discovery procedure wherein each attorney is given opportunity to question the 'opposing' expert about his qualifications, his investigation and his conclusions and opinions. Deposition testimony is given under oath, it is formally recorded as part of the case record and may last from several hours to several days.

Cases involving personal injury accidents are always serious, and may involve damages in the millions of dollars. The attorneys in these matters adopt very strong adversarial positions, and clearly the forensic engineers must pursue their own investigation and prepare their opinions with corresponding rigour and engineering independence. The degree to which an engineering product is scrutinized in this context is thus very intense.

The concept of strict liability, wherein the design of the equipment must 'stand alone', independent of the behavior of the designer, demands a highly concentrated safety effort. The fact that the designer followed industry custom and practice and acted in a reasonable professional manner is not a consideration in this. He must now defend not only his engineering approach but also his product. In the U.S., the widespread use of guarding, interlocking techniques, fail-safe design and warnings technology reflects an evolutionary change over the last two (2) decades, which is subsequent to the introduction of the strict liability doctrine.

4. IMPACT OF STRICT LIABILITY ON EUROPEAN DESIGN

The adoption of the EEC Directive on Liability for Defective Products in July, 1985, which was implemented into U.K. law to take effect March 1st, 1988 established the concept of strict liability in tort. The liabilities introduced under the U.K. Consumer Protection Act May, 1987 (Part I) are in addition to those which already exist, and which deal with the concept of negligence. The strict liability doctrine applies to products irrespective of whether the goods are intended for private or commercial use.

As in its American counterpart, the interpretation of the law involves tests, particularly as to whether the product meets the expectations of the users (reasonableness in its design). The U.K. Consumer Protection Act however allows for a consideration of the intended purpose of the product evaluation of its safety. In the American context, both foreseeable use and foreseeable misuse must be accounted for by the designer/manufacturer. The British manufacturer may be able to show, for example, that since he shipped his product with a guard attached, it is clearly intended to be used with a guard. Operation of the equipment with the guard removed is thus not an intended use or operation. The question of foreseeable misuse must then be resolved. The interpretation of the manufacturer's intent, the effectiveness of his design and his safety warnings and instructions are thus questions to be resolved by the court.

The European design engineer is confronting the situation encountered in the U.S. several decades before. His work is about to be analyzed and dissected after it is beyond his control. It will be evaluated for defects to determine if it provides the safety which a person is entitled to expect, independent of what the designer feels they should expect. Furthermore, this can occur with a product that was supplied as much as ten (10) years previously. There are legal defenses. For instance, it may be that the product has been modified since it left the control of the manufacturer and that the modification is that which makes the machine dangerous. Nevertheless,the effects of a defective design can be very costly, both in personal injury and now in legal consequences.

4.1 The American experience - some suggestions

As was found in the United States, the companies already producing a safe product will experience little impact. A review of current procedures and safety documentation to verify that these meet the standards set out in the new laws may be all that is necessary. Companies with an unstructured and informal approach to product safety however may well reconsider their position. The American experience in this regard would indicate some of the following suggestions.

A first step is to be fully familiar with the responsibilities of the producer. Some U.S. manufacturers have been slow to follow this step. Engineering design, human factors and written communications to the user all play an important role, and are all brought into consideration in the evaluation of the product in a lawsuit after an accident. Clearly, these should be evaluated both for safety purposes and for reduction of exposure to liability before the product is shipped, not after an accident.

Specific safety techniques should be considered. "Fail safe" design is being re-evaluated (BS5304:1988), attention being focussed on the concept of minimizing failure to danger. This is a very important safety tool, and beyond the obvious safety benefits, it is a technique valuable in establishing a legitimate legal defense regarding foreseeable use/misuse of the product. Although sometimes costly, redundancy in design should be considered. For example, the use of a second parallel valve in a pneumatic circuit controlling a press may be necessary to avoid double stroke malfunctions. The redundancy methodology is not often found in European equipment imported into the United States. Single limit switches which can also be improperly used as machine controls can, in some circumstances, lead to serious accidents. This recently occurred in a German made printer-slotter used in the packaging industry, wherein an operator was using an interlocked guard limit switch as an on/off control while clearing a jam. A second switch would have prevented this accident. A Failure Modes and Effects Analysis (FMEA), initially developed for the military, may be applicable to some products, specifically for addressing potential problems which would lead to personal injury and not just operational malfunctions.

The engineering department should consider assigning overall product safety analysis to a single experienced and knowledgeable engineer. This person can then coordinate the work of all others in the safety design, and ensure that proper documentation of the safety effort is conducted. It is often true that engineers have a tendency to document problems and not solutions. Internal memos spelling out safety shortcomings can prove disastrous later in the courtroom, especially if those safety problems were never addressed.

Care should be taken to take safety codes into account in the design process. While compliance with safety codes and standards does not constitute a legal defence in a strict liability suit, such compliance often means that the designer has surveyed the available technology, and is probably designing to a high standard at that time. As a caution however, it should be noted that technical standards only serve to assist in ensuring a reasonably safe product. They don't guarantee this. For various reasons there are occasions when code compliance is not enough. (outdated codes, ambiguously worded safety recommendations , codes which do not address a particular safety aspect etc).

Perhaps the most valuable safety technique that can be used in a design office is that of a peer review. Another engineer, with an independent view point and experience, can most readily spot safety problems in a design. This reviewer can be from inside the engineering department (possibly from another group) or may be brought in as an independent consultant to analyze a product, without the burden of pride of authorship.

As in the United States, the use of independent consultants both for in-house safety, product safety evaluation and for forensic engineering purposes will certainly grow in response to the change in law.

5. CLOSURE

The European engineering product, sold in the U.S.A., is regarded with great respect. Safety standards in this equipment are usually high. However, in the occasional situation where the European manufacturer finds himself in U.S. litigation relating to product liability, he is often surprised when confronting the strict liability doctrine for the first time. His expectation of the safety role of the employer and the machine operator are sometimes higher than this doctrine recognizes. The new EEC regulations will necessitate a change in this expectation. A manufacturer can no longer afford to be caught with his guard down.

Maintenance aspects in design of mechanical systems

A L van der MOOREN, Prof dr ir
Eindhoven University of Technology, Eindhoven, The Netherlands

Summary

The cost of maintaining machines, equipment, vehicles etc. during their
operational life often amounts to a multiple of their cost of replacement,
due, largely, to their design. It is therefore important to solve the
problems of maintenance as early as at the design stage so as to ensure that
both preventive and corrective maintenance tasks are few and can be easily
carried out.
Systematic patterns of thought are proposed to that end and illustrated to
show how the maintenance characteristics, that is reliability,
preventability and, in particular maintainability, can be promoted by
appropriate design. The possibilities open are largely determined by the
design requirements and the design concept. Also the problem, partly
economic, of tracing weak spots in an object and establishing priorities in
eliminating them is dealt with.

1 INTRODUCTION

Yearly maintenance costs of mechanical objects, such as machines, apparatus
and vehicles, vary between 2 and 20% of their purchase costs at a mean of
about 6% [1]. Taken over the operational life the amount spent on
maintenance actions is generally at least of the same order of magnitude as
purchase costs and in that case contributes equally to product costs. Hence
future maintenance costs require as much attention as purchase costs, all
the more because profit losses caused by non-availability also should be
assigned to maintenance. And in choosing equipment the sum of purchase costs
and cumulative maintenance costs, called the cost of ownership, should be
the decisive criterion (Fig.1).

Reducing maintenance costs may be pursued in four ways:
1. Improving the load situation.
   Damage by load is the root of all maintenance. Loads stemming from normal
   conditions of use are inherent to the object's function and purpose, but
   temporary overload, e.g. by contamination of raw materials or by
   operating errors, might be prevented.

2. Improving the **design**.
   Given the load situation the damaging process follows from the object's strength, as determined by its design. Also the possibility for carrying out maintenance actions safely, easily and quickly, is largely dependant on the design.
3. Improving the **maintenance concept**.
   Given the development of the damaging process, the occurrence of unexpected failures and subsequent unplanned, corrective actions may under certain conditions be reduced by planned preventive actions, as part of the objects maintenance concept.
4. Improving the **maintenance system**.
   To what extent the package of preventive and corrective actions can be carried out effectively and efficiently is determined by the quality and quantity of the maintenance resources, such as manpower, equipment, procedures and data.

As each of these four measures costs and saves money, the problem is to find the combination leading up to the largest profit. This optimization has to be carried out on a quantitative basis. As the maintenance behaviour of an object is essentially not a deterministic, but a stochastic process, as far as the number and the duration of actions is concerned, it has to be described by stochastic measures, e.q.:
- the **preventability Rp** as a measure for the number of preventive actions, resulting from the chosen maintenance concept;
- the **reliability Rc** as a measure for the number of corrective actions, resulting from occurring failures;
- the **maintability M** as a measure for the time needed to carry out maintenance actions.

From these measures other useful characteristics may be derived, such as the **availability A**.

2 MAINTENANCE CENTERED DESIGN

The conditions of use and also the maintenance system being generally preconditioned to a great extent, there remain the **design** and the **maintenance concept** to reduce maintenance costs. It seems however rather unsatisfying to develop a sophisticated maintenance concept if maintenance actions are in fact superfluous because of bad design.

Since improving the maintenance behaviour of an existing object by modifying its construction is technically or economically hardly possible, one should therefore pay attention to the maintenance behaviour of an object in the **design stage**. In other words: designing should be **maintenance centered**, briefly **MC**, taking into account future maintenance so that the cost of ownership are minimized. This may be done by (Fig. 2):
1. designing **low maintenance objects** that seldom fail and/or hardly need preventive actions

and as far as this is not attainable by
2. designing **easily maintainable objects**, so that inevitable actions can be done safely, easily and quickly.

Of course in MC-designing the maintenance aspect has to be balanced against other design requirements to find the best compromise. Proper optimization in this way however calls for a method to improve systematically the maintenance behaviour of an object during the whole design process. Fig. 3 shows a simplified working scheme for the proposed approach. The first column states the three main design stages: determining the design specification, choosing a preliminary design and working out the design. The second column lists some of the constructional parameters to be chosen in each design stage, such as the working principle, and the third column refers to corresponding constructional recommendations to be considered from the maintenance point of view.

The recommendations mentioned cannot comprise a compendium of preferred solutions, owing to the diversity and progress of mechanical engineering. The method presented aims at basic schemes of thought covering all fundamental possibilities from wich keywords may be derived which may be developed in universal and specific checklists. In this way the designer is allowed to apply his general know-how about the subject in discovering all the practical possibilities for improving the maintenance behaviour.

It would be very much useful if a set of recommendations could be given, which, applied to any design, straight on led to a MC-object. But on second thoughts this idea doesn't seem feasible. Even if suggestions are not contradictory among themselves, their material realization may lead to imcompatibility; more inspection holes in a vessel e.g. may promote the maintainability, but reduce the reliability. Thus only conditional recommendations can be made and the designer should in each seperate case decide whether and how to apply them, considering technical and economic consequences.

Following the proposed approach the designer is faced with the following questions:
1. How may an MC-solution be included in the design specification?
2. How may an MC-solution be included in the preliminary design?
3. How should the constructional parameters generally be chosen in the design stage to promote a good overall maintenance behaviour?
4. Has the maintenance behaviour of a design to be improved?
5. How can specific improvements regarding the maintenance characteristics R_p, R_c or M be obtained?
6. Which design improvements are attractive, taking into account technical and economic merit?

These six questions will be discussed subsequently as far as main points of the method are concerned. Some simple obvious examples will be given to illustrate the lines of thought. Further details and more examples are given in [2]. As the maintenance behaviour of an object is in fact determined by its detailed design, we will treat the questions 3-6 at first, and use the gathered insights afterwards in discussing questions 2 and 1.

3 GENERAL DESIGN RECOMMENDATIONS

The following "ten commandments" may generally be applied to promote an
object's overall maintenance behaviour.

1. Keep the construction simple
2. Use standardised components
3. Take care of good accessibility
4. Take care of good replaceability
5. Apply modular construction
6. Neutralise human errors
7. Neutralise developing damage
8. Make the condition assessable
9. Aim at "self-help"
10. Provide a maintenance manual.

Mostly their application improves Rp, Rc and/or M without affecting one
of these characteristics. It may be obvious that not all these rules have a
bearing on every design. However, generally a couple of them should be
applicable. Fig. 4 shows a typical example, namely a hydraulic plunger seal.
A cylinder (1) is mounted on the pump housing to guide and seal the plunger
(2). The object is an example of a simple modular construction that can be
replaced easily and quickly, corresponding to the commandments 1, 4 and 5.
The packing (3) is pressed by the bush (4) with a force proportional to the
pressure in the pump housing and automatically adjusted, according to
commandment 9. Pin (5) indicates when the packing should be renewed, by
changing its colour from black to red, as stated in commandment 8. All these
features in some way improve the preventability, reliability and maintainability.

4 THE MAINTENANCE ANALYSIS

A crucial point is to estimate the maintenance behaviour of alternatives in
the design stage. A simple way to get an overall judgement is to compare the
result with a set of recommendations or a checklist as a criterion, e.g. the
"ten commandments". A more effective, but also more elaborate way is to
estimate the objects maintenance characteristics, turning to relevant
components whose maintenance behaviour is already known. Such a maintenance
behaviour analysis is initially a qualitative one and can be split up into
three coherent parts:
- **preventability analysis**: Which components require preventive actions
  according to the maintenance concept?
- **realiability analysis**: Which components are apt to fail and require
  corrective actions?
- **maintainability analysis**: Which maintenance level performs the required
  preventive and corrective actions and with what means?

This qualitative analysis must have a quantitative follow-up if it is
to lead to trade-offs and decisions. With that aim the frequency and
duration of maintenance actions may be estimated roughly. In order to avoid
biased and unbalanced estimations the designer should team up with

specialists from other disciplines, particularly maintenance people. The results of this analysis can be put into a graph, thus providing a "maintenance behaviour profile" that may be used to identify weak spots at a glance (Fig. 5). Vertically the components are listed, horizontally their scores, for the frequency as well as for the duration of preventive and corrective actions.

From the maintenance behaviour analysis in combination with cost factors, a **maintenance cost analysis** can be derived. The results not only supply more significant information about the weak components, but also allow a direct estimation of the future maintenance costs of the object.

5 SPECIFIC DESIGN RECOMMENDATIONS

Now the designer has identified weak spots, he must try to find out all possible improvements. Let's confine ourselves to the maintainability aspect, just to get some idea how recommendations may be based on a model and a scheme of thought. The maintainability of an object can be improved by adapting the design better to the maintenance means at different levels of the maintenance system. We restrict ourselves moreover to the human factor. Apart from physiological action like breathing, the primary functions of a man at work are observing, thinking and acting. From this model all fundamental recommendations for **improving the maintainability** can be derived; they are listed in the next table.

1. Improve working conditions
2. Improve perceptibility
3. Improve conceivability
4. Improve detectability of defects
5. Improve accessibility
6. Improve replaceability
7. Improve workability

We pick out the fifth one, improve accessibility, for further development. A component must be accessible from the outside and if necessary, also from the inside for appropriate action to be taken. Accessibility may refer to a man's hand, his whole body and to personal protective means. Consider also the use of tools and other equipment. The more a component needs maintenance, the better his accessibility should be. Properly functioning components should not have to be demounted in order to get at a defective one. The example in Fig. 6 shows that wearing bushings should not be at the inside, but at the outside to avoid demounting of a linkage.

6 TECHNICAL AND ECONOMIC EVALUATION

When constructive possibilities for improving the maintenance behaviour of an object have been found, the question is how to decide, if these improvements cost extra money, as is often, but certainly not always, the case. Let us at first assume that the benefits can be expressed in yearly financial savings. This assissment could be based on a maintenance cost analysis, as discussed before. Now optimization can take place on the basis of minimum cost of ownership, according to well known economic practice, e.g. applying the DCF-method. Still the producer may try to minimise the sum of his manufacturing costs and guarantee cost, neglecting higher maintenance cost involved. The user should be aware of this discrepancy and inform the producer beforehand about his willingness to accept higher purchase costs as far as lower cost of ownership result.

In many cases however differences in investment cost are fairly well known, but their effect on maintenance behaviour is difficult to express in financial terms. This is certainly the case when considering the aftereffects of failure and maintenance actions, such as loss of production and loss of goodwill. In these cases the best thing to do seems to fall back on a structurised, heuristic approach, e.g. the classical and still up-to-date Kesselring Method [3] which balances technical and economic merits (Fig. 7) The technical merit can in this case refer to the expected maintenance behaviour and may be based on a checklist analysis or on a behaviour analysis, as discussed before. If within the class of equipment where the object belongs to, the cumulative maintenance costs usually equalize the purchase costs, one should prefer a solution near the diagonal in the upper right-hand corner.

7 THE PRELIMINARY DESIGN

It may be clear that the possibilities for realising a detailed MC-design are determined largely in the early design phases. From the stated function of the object the first constructional parameter to be chosen is the objects mode of action, as determined by the actional principle of its components and by the way these components are linked. Compare e.g. the principles of a piston compressor and of a centrifugal compressor, and the linking of the compression stages in series or parallel. In choosing the working principle the designer should e.g. try to foresee whether discontinuous operation is inevitable, sophisticated technologies have to be applied and poor working conditions can be expected. He may especially check whether a large number of moving parts will be necessary, dirt depositions will develop and certain components will be critical, for instance seals. Fig. 8 shows an oscillating sieve and a sieve bend, both fulfilling the same function in dewatering slurry, but it may be expected that the last one has a much better maintenance behaviour because it is stationary and doesn't clog.

Following from the chosen working principle and from the stated capacity the objects mode of construction is selected, as determined by the constructional principle of its main components and the way they are spatially arranged. In choosing the designer should bear in mind a.o. that the orientation of the components and their relative position play a major part in maintainabilty, for instance on the need for demounting horizontally or vertically and on the possibilities of using a mobile crane. The result of these choices may be shown in the preliminary design. One should realise that in the preliminary design the cost of ownership are already fixed for over 80 %. It is therefore very important to take maintenance into account already in this early design stage.

8 THE DESIGN SPECIFICATION

Drawing up the design specifications is the first step in the design process. They are not only the startpoint for generating possible solutions, but also criteria in choosing the best one. As far as maintenance characteristics are concerned, they should sometimes be stated specifically: safety aspects for instance may call for a high degree of reliability. In general, however, explicit specification of these properties seems undesirable at this stage, because in the course of design, they are more or less interchangeable: bad maintainability e.g. may be acceptable as long as reliability is high. So it should be left to the designer to choose the optimal combination in pursuit of minimum cost, including the cost of non-availability. Some general recommendations may be given to reach this objective.

1. Avoid complex and excessive functional requirements. Keep the function as simple as possible. Requiring two or more options from one object, e.g. suitability for processing different raw materials, usually leads to complicated constructions and to compromises, not seldom with reduced reliability and maintainability.
2. Avoid needless restrictions to the design parameters. It can be disadvantageous to specify parts or materials of the objects design too precisely, because this deprives the designer of the possibility of favourable influencing the maintenance behaviour. For a heat exchanger e.g. the capacity should be prescribed, but not its type, unless for clear, functional reasons.
3. Include a description of the maintenance system. For the different levels of the maintenance system data must be given on the available maintenance means, their quality and quantity, for instance in regard to manpower: are contractors or own staff going to do the maintenance work, and are specialists available? Reliability and maintainability can be too low if the designer assumes repair by welding, whereas the future owner lacks the know-how and/or other means.
4. State the optimizing criterion. It is clear that, in the long term, cost of ownership are a better measurement than purchase cost. Nevertheless other considerations, for instance lack of short-term liquidity, may force the customer to prefer a solution that is relatively cheap to buy but expensive to operate, instead of doing nothing.

9 CONCLUSION

This general method for **Maintenance Centered Designing** fits into the normal design process [4]. Also it is in accordance with common technical and economic rules. Anyone who thinks that the method mainly concerns will known elements, is right: reliability, maintainability, cost of ownership are not new ideas. On the other hand we all know too well about maintenance problems of equipment induced by inadequate design. From the presented systematic approach, better results may be expected, especially lower cost of ownership following from savings on maintenance actions and from extra profits due to increased availability. Of course other pros and cons have to be considered as well in applying the method. Initial engineering costs may be higher the first time, because one has to become familiar with the method and has to draw up checklists. The designer needs time to consult the future maintainer and to put forward his wishes to the producer, who may from his part put up the purchase price and extend the delivery time. On the other hand much time and money will be saved during the period of construction and starting up, because less modifications have to be improvised.

The method has been developed at Eindhoven University in close cooperation with industry for over ten years now. Main points proved to be the **maintenance analysis** showing weak spots, the use of **schemes of thought** in discovering all possibilities for eliminating them, and the necessity of anticipating on a MC-solution already in the **design specification** and in the **preliminary design**. The method comprises also detailed quantification techniques not mentioned here, such as the simulation programm "Mainsithe" [5]. Though practical applications were successful, they also showed that full benefits only can be gained under optimal **organizational** conditions, especially allowing maintenance experience to find its way back to the design stage.

REFERENCES

[1] Deutsches Komitee Instandhaltung e.V.: Die volkswirtschaftliche Bedeutung der Instandhaltung in der Bundesrepublik Deutschland. Düsseldorf, 1980.

[2] Mooren, A.L. van der: Equipment and system design for lower maintenance cost. Paper presented at the Conference "Maintenance means money", Petrotech '88, Amsterdam, 1988.

[3] Kesselring, F.: Technische Kompositionslehre. Springer,Berlin, 1954.

[4] Pahl, G., Beitz, W.: Engineering Design, Ed. K. Wallace. Springer, Berlin, 1984.

[5] Both, H.: Maintenance as a parameter in the behaviour of a complex system. Euromaintenance-88, Helsinki, 1988.

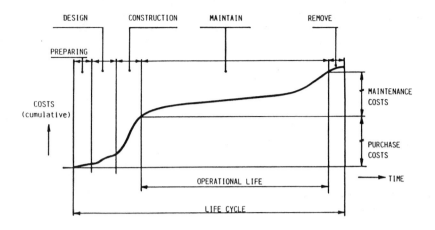

Fig 1   Cost of ownership

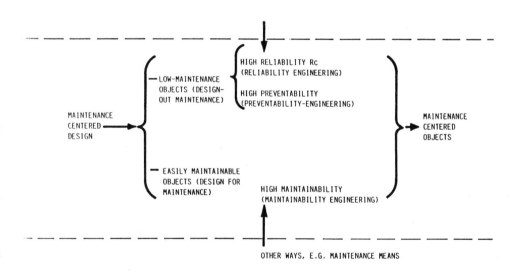

Fig 2   Maintenance centered design

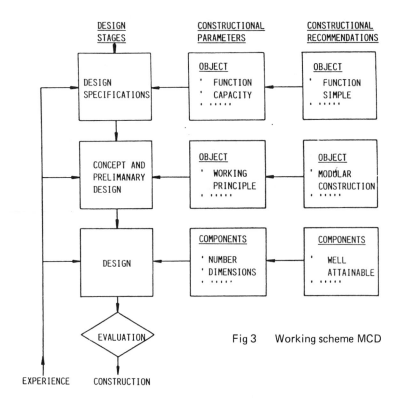

DESIGN STAGES	CONSTRUCTIONAL PARAMETERS	CONSTRUCTIONAL RECOMMENDATIONS

DESIGN SPECIFICATIONS

OBJECT
· FUNCTION
· CAPACITY
· .....

OBJECT
· FUNCTION
  SIMPLE
· .....

CONCEPT AND PRELIMANARY DESIGN

OBJECT
· WORKING
  PRINCIPLE
· .....

OBJECT
· MODULAR
  CONSTRUCTION
· .....

DESIGN

COMPONENTS
· NUMBER
· DIMENSIONS
· .....

COMPONENTS
· WELL
  ATTAINABLE
· .....

EVALUATION

EXPERIENCE   CONSTRUCTION

Fig 3   Working scheme MCD

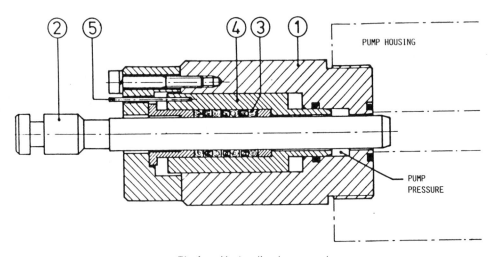

Fig 4   Hydraulic plunger seal

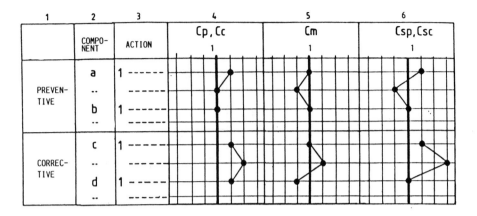

1	2	3	4	5	6
	COMPO-NENT	ACTION	Cp, Cc 1	Cm 1	Csp,Csc 1
PREVEN-TIVE	a ..	1 ------ ------			
	b ..	1 ------ ------			
CORREC-TIVE	c ..	1 ----- -----			
	d ..	1 ----- ------			

Fig 5   Maintenance behaviour profile

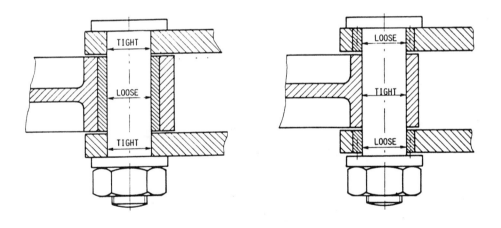

Fig 6   In- and outside placed wearing bushings

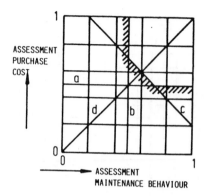

Fig 7   Selection according to Kesselring

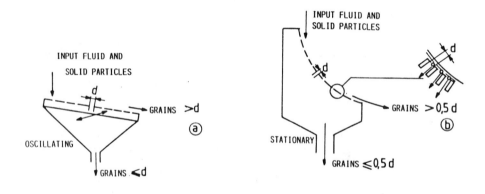

Fig 8   Dewatering slurry

C377/211

The basic concepts supporting a commercial component selection system for designers

G PITTS, BSc(Eng), PhD, CEng, MIMechE, MIProdE and **P A VEDAMUTTU**, BSc
Department of Mechanical Engineering, University of Southampton

SYNOPSIS   This   paper   sets   out   to demonstrate that it is possible to achieve consistency   of   description   for   the characteristics   of   bought-in   components used   by   the   systems   designer.   The operation   of   a   computer-aided   component retrieval system is described, and some of the   considerations   in   using   it   are explained.   The paper concludes by showing the detail to which analogies can be taken between   systems   in   order   to   define   the required design characteristics.

INTRODUCTION

As systems design becomes more automated, through the application of CAD, it is becoming more important that designers have rapid access to bought-in components. At first sight, it would seem that this should be a simple matter of transferring component information onto computer databases. If such a step were to be taken, then it would be a simple matter for the designer to identify suitable component characteristics, enter them as search parameters into the database management system, and hence retrieve the required components. At the moment such a global approach is not feasible. The two main obstacles are: there is no consistency of description across mechanical component characteristics, and, secondly, the characteristics used to describe components are those which the manufacturers of the component have decided to use in order to promote their product. These characteristics are not necessarily those required by the designer. Indeed, some essential characteristics are often omitted from the manufacturer's literature. It therefore follows that before a database retrieval system can be developed, a study must be made of the parameters which are needed by the designer and there must be consistency in their description.

The consistency of description means that components having like functional capabilities must have a common way of describing their characteristics. A good example, which has been investigated and which will be described later, is the rotary motor. The motor can be considered as an input to a rotary mechanical power transmission system and it is therefore a disaster that the different motor types have the same functional output characteristics described in different ways.

The objectives of the work to be described, covered two principal areas. The first was the classification of the characteristics and, the second, the application of these characteristics to the retrieval of component information by the designer. The area selected for the study was the field of rotary mechanical power transmission. This was chosen because it was a topic around which a boundary could be drawn and within which a total system was capable of being designed. The components studied were rotary motors, gearboxes, couplings and the group: clutches, rotary dampers and brakes. The latter was treated as a group because they have the common factor of being rotary energy dissipators.

For purposes of explanation the utilisation of the work by the designer will be described, first to set the work in context, and the fundamentals will be described later.

The retrieval system

The purpose of the system was to develop a means by which the designer can use a PC to provide a quick route to the selection of suitable components, but at the same time ensuring that the components are being selected on the basis of properly defined characteristics. The need to identify the characteristics for functionally similar components has already been mentioned and will be dealt with later.

It was not the intention of the project to become involved in software development. However, having identified the characteristics required to access component information, and defined how they need to be presented, a software specification could be made. A very basic retrieval system was implemented and its development limited to small changes required to make the system user friendly.

The software specification for the database management system was as follows:

i) It must handle alpha-numeric characters and real numbers.

ii) It must cope with inequalities and equalities, i.e. it must be possible to stipulate upper and lower limits to the search parameters

iii) There must be a capability to write a front end to the package to make it user friendly.

iv) The software must operate on most personal computers (PC's).

v) The expected size of a single component database could be up to 50 000 records requiring up to 20 search characteristics to be used in their retrieval. However, some sub-dividing of database into manageable blocks could be accepted.

A search of the software market yielded a database management package which satisfied these criteria. The package was 'Superfile 16' produced by Southdata of London. The retrieval system is driven by a user friendly front end program whose format is common to all components. The selection data presented to the user, within this structure, will change with the component chosen. As the main objective of the project was to develop the characteristics, rather than the software, the front end was written in BASIC for convenience, and later compiled to increase its speed of operation.

Since the intention was to develop a system which could be made available as a commercial service to designers in industry, it was important to optimise between the needs of the designer, in terms of the characteristics used in specifying the product, and the economic viability of entering such data into the system. Technical Indexes Ltd. who were the potential agents for the system, worked very closely with the University and one aspect of this co-operation was identifying the minimum number characteristics which would be acceptable to the designer and the maximum number of characteristics which was commercially acceptable to the company. It is considered that a good compromise was reached in each of the component areas. Normally, a designer would be satisfied with using only a few of the characteristics chosen, in order to identify a suitable product. However, when a wider range of options is open to the component user, selection can become frustrating because of the lack of precision in the search. As an example, one thousand motors of all types had been loaded into a database for the purposes of evaluating the system with designers; this compares with around 50 000 motors, if all those available in the UK were entered into the system. Those testing the system tended to search on the same number of characteristics as when using a small number of catalogues. The result was that many more motors were retrieved than could be readily assessed by the designer. This meant that they found it necessary to increase the number of characteristics used to interrogate the database in order to retrieve a manageable number of motors. With a larger number of motors on-line, all the search characteristics available on the system would be required to achieve the necessary accuracy of retrieval, this is why it is important that a range of characteristics is considered which would be outside those normally used by the designer. A similar argument can be applied to the other items currently on the system, namely: gearboxes and couplings.

The economic benefits of a component retrieval system

Having assigned values to the various component parameters representing the component, the designer is faced with identifying the component which best satisfies the system requirements. A thorough search, using a system such as the Technical Indexes microfilm system, is capable of producing a wide range of component options, but the difficulty is then one of finding the component which most nearly satisfies the specified parameter values. This is complicated by the different ways in which manufacturers represent their data. The situation is also less satisfactory when the designer has to depend on a small library of component catalogues held by the company.

There are three main economic benefits resulting from the retrieval system under consideration. The first comes from the nearness of matching the requirements of the designer and the second from the ability to identify satisfactory competing products. The third is that a computer based system allows these processes to be performed quickly and repeatedly. The user therefore becomes more productive because no time is wasted interrogating catalogues. The final decision can then be made on the basis of manufacturer's quotations for price and delivery times. In the case of matching the requirements, the designer will normally take the next available size above that required. The smaller the range from which the selection is made, the more likely to be the degree of mis-match and, consequently, the greater the margin of over design. Over design not only results in the extra cost of purchasing the component, but normally results in a physically larger component requiring larger supporting structures, bigger enclosures, the cost of a product can readily double.

Using the system

The system is known as ROCCI, standing for the 'Retrieval of Compatible Component Information'. and operates on a PC. The user is taken step by step through a selection procedure. Unlike an expert system the emphasis is on trying to have the designer learn to specify bought-in components in the correct way. The expert system will normally lead the user through a question and answer procedure in order to formulate the enquiry. The ROCCI approach is one of confronting the designer with most of the parameters which will need to be specified. Figure (1) shows the sequence of the principal menus which the user will see during the motor's selection procedure, although shown here in monochrome, the presentation is normally in colour to highlight certain features.

In Figure (1)c it will be seen that the majority of the characteristics listed relate to the performance characteristics of the motor and, particularly, the output characteristics of the motors. Essentially, the motor should be looked upon as an input device to a rotary mechanical power transmission system. This means that the characteristics of interest are the output characteristics, rather than the input characteristics. This is explained more fully in Reference (1). Normally, a designer should not be specifying a motor type, i.e. a.c. or d.c. electric, hydraulic, pneumatic, etc., but specifying what is required of the motor. As an example, most designers consider motors to always be electrical, but if a motor is required for an explosive environment, a fluid power device may be safer and cheaper than an electrical machine with a flameproof enclosure. In many shipbound and heavy plant applications, hydraulic services are readily available and most chemical processes use compressed air for power and control needs.

It would therefore seem that the inclusion of motor type in Figure (1)c is a contradiction of that philosophy. It was decided that to be psychologically acceptable to the designer, the system should include the option of motor type, although this should be low on the list of selection criteria to be used by the designer. Similarly, the name of the manufacturer was included so that the designer could use this as the final filter in arriving at the required motor.

Figure (1)d gives an example of a help screen. The option for help appears when the user has the pointer on the menu targetted on the mounting type. A similar screen is essential when selecting the directions of output shafts on a multi-shaft gearbox. Figure (1)e gives an example of one output record which may be scrolled past the user in order to view the available options.

The second sequence is as follows. Having arrived at the motor characteristics presented in Figure (1)c, the designer enters a value (δ) against each of the characteristics for which a parameter is known. The system then starts a search recording the number of hits achieved. If the number of hits is high then the designer may increase the search precision by either narrowing some of the value bands on the parameters being searched, or by including values from some of the other characteristics. When the number of recorded hits is of a manageable size, then the motor records can be viewed and the selection made.

The search time of a thousand records on a modern PC is around ten seconds.

Finally, the Technical Indexes (TI) reference number on the listing is the TI microfilm frame on which the manufacturer's catalogue is displayed for further study.

The basis for identifying the characteristics

Table 1 shows a listing of equivalent characteristics between different motor types. This list is longer than that displayed on the search screen, but only presents a part of the fully extended list of about fifty characteristics.

The method for identifying the characteristics is to work, by analogy, between the different power systems. Figure (2) shows a matrix of analogies between the different power transmission systems. Some of the rows have been excluded for the sake of clarity. As explained earlier there are two problems in dealing with the identification of characteristics, one is in defining the characteristics of different components, having identical functions, in the same way, and the second, is one of identifying characteristics which are not normally quoted by manufacturers. A good example of the latter is in the study of gearbox characteristics. A gearbox is the equivalent, in a rotary mechanical power transmission system, of the transformer in an electrical power transmission system. Each has power losses, namely: friction and resistance. However, a transformer has an inductive 'loss' and to a very much lesser extent, a capacitive 'loss'. These are not true losses because the energy is not transformed into heat but remains stored within the transformer. In both cases the energy is stored in such a way as to be a liability. At first sight there appears to be no equivalent 'loss' in a gearbox but further study shows that gearbox inertia is analogous to inductance and gearbox twist stiffness is analogous to capacitance. These become important when the gearbox undergoes speed and torque changes respectively.

Discussion of motor analogies

In the case of motors, similar analogies can be investigated. The studies yielded some very interesting analogies between the operating principles of the motors concerned.

In a.c. motors the supply may be single or polyphase (usually three). For selection, knowing the number of phases required from the supply is considered a secondary consideration. In order to keep the research thorough, the analogous characteristics in d.c., hydraulic and pneumatic systems should be established. Potential a.c. motor features can be offered across to the other systems, in an attempt to build up the analogous quantities and, hence, put together a full and justifiable analogy. There are two basic types of a.c. motor, the synchronous motor where the rotor runs at a speed dictated by the supply frequency, and the induction motor where the rotor runs at a speed just under this synchronous speed and will also run within a speed range below this speed dictated by the load on the motor. The action of both of these machines relies on a rotating magnetic field created in the stator windings by the a.c. supply.

It is now possible to develop the analogy between the synchronous motor and a fluid motor. The major problem is that all fluid motors are self-starters, whereas the synchronous motor cannot start by its synchronous action, it requires a starter using another motor principle to accelerate the rotor to synchronous speed (an induction winding may be provided). The postulation is now to analyse the mechanisms of fluid motors to attempt to find some kind of construction that will mimic a synchronous motor.

There are two basic types of fluid motors, the vane type and the piston type. Consider the action of a simple axial piston motor (Figure (3)a), the pistons are supplied in turn by the rotating pintle valve, the pistons force round the swash plate which is keyed to the drive shaft. The drive shaft provides rotary mechanical power at its free end but also rotates the pintle valve within the motor. Thus, the pintle, pistons, swash plate and drive shaft are all synchronised by their mechanical connections.

In the synchronous motor, the rotating magnetic field and the rotor are not physically joined, they are linked by electro-magnetic forces. These forces can be overcome, and once out of phase, they will not produce useful torque and the motor will stall. An experiment can then be performed with the piston motor: each of the mechanical links causing the mechanical synchronisation, is broken and the effect noted.

The mechanical connections that can be broken are the pistons/pintle valve, swash plate/pistons, drive shaft/pintle valve, drive shaft/swash plate arrangements. The swash plate and pistons are not connected because the pistons do not rotate with the swash plate but reciprocate forcing the swash plate round; they must therefore be free to slide over the swash plate face.

1244

If the pistons were not connected to the pintle valve, the fluid power would not be available to the pistons, the motor would not work.

If the drive shaft was to be disconnected from the swash plate, the motor would fail to rotate because although fluid power was available, it could not be ported to the pistons because the drive shaft was not working the pintle valve. If the drive shaft was to be rotated, the swash plate would begin to spin but, of course, no useful power would be transmitted to the drive shaft, so the motor would be useless.

If the drive shaft was disconnected from the pintle (Figure (3)b), the motor would again fail to work because the pintle could not supply the pistons in sequence. However, if the pintle was rotated, the pistons would start to reciprocate. The vital point to now note, is that if the pintle were to be rotated at any significant speed, say 1000 r.p.m, the swash plate would never accelerate to that speed in time to continue to receive the piston pushes to perpetuate its rotation. The piston strokes and the swash plate would be out of phase, and the motor will doubtlessly stall.

Consider the case where the drive shaft was somehow spun up to the pintle speed and then the pintle driven. Although the pistons and the swash plate would be out of phase, the mechanism may well pull up (the swash plate is just lagging the pistons and is able to be accelerated by them to the synchronous speed – the speed at which the pistons drive the swash plate perfectly without leaving its surface), or, drop into synchronisation (the swash plate is retarded by the pistons). A great deal would depend on the amount by which the pintle and the pistons were out of phase. Clearly, if the piston were to come down on the wrong side of the swash plate, it would be akin to putting a stick into a bicycle wheel, or a timing belt on a vehicle engine breaking, a great deal of damage would result. Thus, ideally, the drive shaft/swash plate should be in near synchronisation with the pintle when spun up.

This demonstrates the depths to which the various analogies can be taken to ensure that the characteristics are rigorously defined.

CONCLUSIONS

In establishing any component retrieval system it is essential to define the characteristics which should be used in the specification of the components. The study has shown that those characteristics used by the manufacturers of bought-in components are not precise enough, or comprehensive enough, to satisfy design requirements. It has been demonstrated that the study of analogies can be used to provide proper design definitions for use by the designer.

These definitions have been shown to have practical use in the retrieval of component information by the systems designer.

REFERENCES

(1) PITTS, G. and VEDAMUTTU, P.A. Retrieval of component information for system designers. ICED '87, Boston, U.S.A. 17 August 1987, 707-714.

TABLE 1 .   Analogous Characteristics between Motor Types .

AC Electric	DC Electric	Hydraulic	Pneumatic
Rated Power Rated Torque Rated Speed Starting Torque	Rated Power Rated Torque Rated Speed Starting Torque /	Rated Power Rated Torque Rated Speed Starting Torque /	Rated Power Rated Torque Rated Speed Starting Torque /

Pull Up Torque
Pull Out Torque
Static Friction
  Torque   /
	Pull Up Torque /
	Pull Out Torque /
	Static Friction
	  Torque /
		Pull Up Torque /
		Pull Out Torque /
		Breakaway
		  Pressure /
			Pull Up Torque /
			Pull Out Torque /
			Breakaway
			  Pressure /

Maximum Speed No Load Speed /	Maximum Speed No Load Speed /	Maximum Speed Free Speed	Maximum Speed Free Speed

Electrical
  Efficiency
	Electrical
	  Efficiency /
		Fluid Losses /	Fluid Losses /

Rated Input / Rated Voltage Rotor Current /	Rated Input / Rated Voltage Armature Current	Fluid Power / Rated Pressure / -----------	Fluid Power / Rated Pressure / -----------
Stator Power Rotor Poles	Field Power / " Commutation " /	Fluid Power / -----------	Fluid Power / -----------
Number of Phases Frequency	" Commutation " / --------------	" Porting " / Speed & Porting /	" Porting " / Speed & Porting /
Resistance Start	Resistance Start	Valves /	Valves /

Key :
       / Found by Analogy .   --------- No Analogy Found .

a) The Component Menu .

```
11111111
1 MENU 1
11111111
```

COMPONENTS AVAILABLE :

```
1 - Motors.
2 - Gearboxes.
3 - Couplings.
4 - Others '.
    - Type X to EXIT.
```

ENTER NUMBER >> [_]

b) The Sub-database Menu .

| MOTOR DATABASES. |

Databases available :

```
1 - power range 0 to 1000 kW (WHOLE DATABASE

2 - power range 0 to 5 kW.

3 - power range 5 to 100 kW.

4 - power range 100 to 1000 kW.
```

ENTER SELECTION > [_]

c) The Prompt Menu .

```
Rated Power   (kW) >10/AND/'12.5
Rated Torque  (Nm) /50
Rated Speed   (RFM) >1000/AND/<1800
Starting Torque (Nm) :80
Maximum Speed  (RFM)
Minimum Speed  (RFM)
Length  (mm) <500
Breadth (mm) <350
Height  (mm) <300
Mounting Type _
Motor Type
Company Name
Pull Up Torque  (Nm)
Pull Out Torque (Nm)
Reversibility

MOTOR FRAME CODE
MICROFILM NUMBER
```

INSTRUCTIONS.

Use the '+'and '+' to move the
cursor to the required parameter.
Enter a search type and value
and then press 'RETURN'.
Press 'END' to search the database.

SEARCH TYPES AVAILABLE.

'=n' or '=*' - exact match.
'<n' or '<*' - less than match.
'>n' or '>*' - greater than match.
'?*' - sounds like search.
Use '/AND/' to combine searches.

| MESSAGE WINDOW. |
| >>>>> HELP AVAILABLE <<<<< |
| Press 'ESC' for help. |

d) A Typical Help Screen .

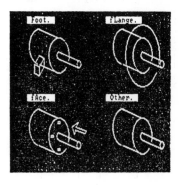

e) The Search Menu .

RECORD MODEL COMPLETE.

SELECT OPTION :

```
1 - Change record.
2 - Search database.
3 - Change database.
  - Type X to exit.
```

ENTER NUMBER > [_]

ENTER CHOICE (F,L,A,O) >> [ ]

f) The Count Menu .

```
1111111111SEARCHING DATABASE1111111111
        Please wait.
```

NUMBER OF HITS = 10

```
To abort the count ;
Please ensure that the Caps Lock button is on
and then press Home.
```

g) The View Menu .

NO MORE RECORDS ''.

NUMBER OF HITS = 4

```
1 - CHANGE Record.
2 - VIEW Records.
3 - Change database.
  - Type X to EXIT.
```

ENTER NUMBER > [_]

h) A Retrieved Record .

```
Rated Power   (kW)      =10.5
Rated Torque  (Nm)      =160
Rated Speed   (RFM)     =450
Starting Torque (Nm)    =
Maximum Speed  (RFM)    =565
Minimum Speed  (RFM)    =
Length  (mm)            =186
Breadth (mm)            =122
Height  (mm)            =95
Mounting Type           =L
Motor Type              =H
Company Name            =Wraxall Fluid Power
Pull Up Torque  (Nm)    =
Pull Out Torque (Nm)    =
Reversibility           =

MOTOR FRAME CODE        =MAF 100
MICROFILM NUMBER        =3850-0732

       Press 'N' for next record.
       Press 'E' for end
```

Fig 1   The ROCCI retrieval procedure

Power Source	COMPONENT TYPE							
	Power Converter	Resistive	Capacitive	Inductive	Transformer	Switch	Other	Power Converter
AC Electrical	Generator	Resistor	Capacitor	Inductor	Transformer	Switch		Motor
Rotary Mechanical	Motor	Brake	Compliant Shaft	Flywheel	Gearbox	Clutch		Pump
Others								
Hydraulic	Pump	Rough Pipe	Accumulator	Fluid Mass	Differential Piston	Valve		Actuator

Fig 2   The analogy matrix

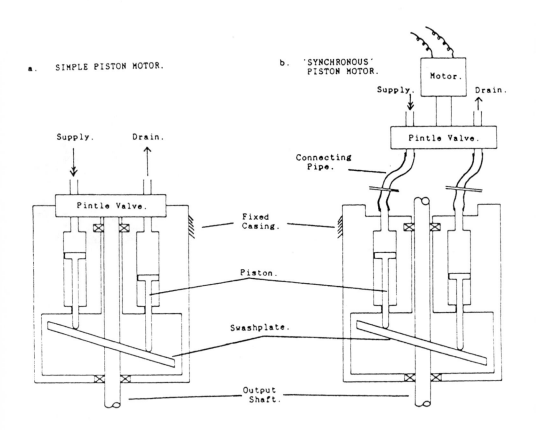

Fig 3   The 'synchronous' piston motor

C377/307

Data and the teaching of engineering design

A A CHASTON, MBIM, MInstM
ESDU International Limited, London

SYNOPSIS. The author suggests that, with the increasing use in industry of validated data, prepared specifically for design engineers, the use of the same data in the teaching of engineering design will allow the course to match more closely real life engineering situations, enabling the newly qualified engineer to make more quickly a useful contribution to his company.

The need to introduce students to realistic and practical engineering design at an early stage in their graduate training is at last receiving the attention that it deserves.

One of the reasons for this upsurge in interest was the realisation that engineering graduates straight from university had little immediate practical expertise to offer industry. Industry was funding the transition from theory to practice.

Coupled with this, was what can only be described as culture shock for the newly qualified engineer. He found that his hard earned qualifications gave him, in the eyes of his new employer, only the skills and knowledge to select a bigger bearing from the suppliers catalogue, or specify a stronger bolt.

Consequently, not only has it been necessary to introduce a design and analysis element into the engineering course, it is also necessary to ally this course to the commercial realities of the real world. Designing a component without an awareness of the cost is unlikely to endear the designer to the company he is working for, so questions need to be answered that are far more complex than the simple engineering question "Will it work?"

Questions such as: What is the expected life of the component? Is this long enough? Equally important is it too long, bearing in mind the intended life of the whole machine? It is evident, with the cost of materials today that overdesign has become as important as underdesign. We can no longer afford the Victorian attitude that "massive is best".

What are the penalties of using a cheaper material? Could some form of surface treatment overcome these penalties? At what cost?

Could we use a stock item rather than manufacture from scratch? What are the consequences of using a standard bolt that is bigger than the ideal, or of a different material?

Should the component be more robust in an attempt to reduce down-time? Where is the balance between cost and saving?

Now we have said that to provide the ability to answer questions of this complexity is one of the objectives of the engineering design module within the engineering course. But to develop that ability successfully requires not only an appreciation on the part of the lecturer of the multiple pressures on designers in industry, in a practical sense, but appropriate data that will allow the engineering students to undertake practical design tasks within the strictly limited time of an engineering course, in which they must also gain a working understanding of engineering science.

Where is such data to be found? Is it readily available? Is it written to help design engineers? And equally importantly does this data match that found in industry?

Let's consider some of the sources of engineering data and their availability.

First of all the college library, major source of standard, straightforward engineering text books and handbooks of compiled data of various types.

Add to this the learned papers, of varying degrees of practicality, published in various journals and conference proceedings; manufacturers catalogues, and so on, and you can see that the problem is not a lack of information, rather a surfeit.

But how accessible is all this data? Naturally the college library is rapidly and easily accessible, but in many cases limited in scope. Papers and conference proceedings, provided that you know the title and source, are readily accessed by mainframe retrieval systems such as Dialog, but takes time to acquire and the cost is not inconsiderable.

However, one might think that, with such a selection of material available, the budding engineering designer, or the practising designer in industry, was well served.

I believe not.

Let's consider what students, post graduates and practising engineers need from their engineering data.

They need accuracy, but since absolute accuracy is unobtainable, they need a clear statement of the tolerances to be applied to the data. And they need applicability, a high signal to noise ratio, by which I mean data that doesn't need hours of work to extract that which is relevant to the engineering designer. They also need authoritative data so that they may be confident that the results derived from this data will be as accurate as engineering science will allow.

But what is the situation at the moment? How accurate is the data generally available? We have found serious shortcomings in a high proportion of the thousands of reports, text books and journals whose content we have needed to study and analyse in the course of our work during the fifty years of our existence. Almost every report, text book or journal we look at contains an error. Some are very obvious, typographical errors for instance; but some are the much less obvious technical errors; the authors' made a mistake, or left a term off an equation, or an equation supposed to represent a graph on the same page, simply doesn't. Let me illustrate this point. I'm sure that the non-engineers amongst us may find it difficult to believe, although readers of one of the 'quality' daily papers will realise how easy it is for typographical errors to creep in.

In an early publication, still to be found on library shelves, the Strength of Materials says on Page 309 that Perry's formula is

$$p = \tfrac{1}{2}\{f + p_e(1 + \eta)\} - \tfrac{1}{2}\sqrt{\{f + p_e(1 + \eta)^2 - 4fp_e\}}$$

This is incorrect and should read

$$p = \tfrac{1}{2}\{f + p_e(1 + \eta)\} - \tfrac{1}{2}\sqrt{\{f + p_e(1 + \eta)\}^2 - 4fp_e}$$

Note the repositioning of the final bracket.

In a recent publication Structural Analysis of Laminated Anisotropic Plates from the Wright Patterson Air Force Base there is superb confusion engendered by the use of the same symbol R for two quite different purposes culminating in the expression

$$\lambda_{2mn} = N_0 m^2 - R p_0 n^2 R^2$$

Even NASA can get it wrong! A publication from their Scientific and Technical Information Office called Vibration of Shells, on page 416 shows an expression

$$a = \lambda\{E/[\rho h(1 + \nu)(1 - 2\nu)]\}^{1/2}$$

relating to frequency intervals and the relevant Bessel functions. It should read

$$a = \lambda\{E(1 - \nu)/[\rho h(1 + \nu)(1 - 2\nu)]\}^{1/2}$$

My concern is shared by Malcolm MacCallum of Queen Mary College, London who said in the New Scientist of the 23 October 1986 "the best of the large books of tables of integrals have about 7 per cent of the formulae wrong".

So you understand my concern regarding the quality of the data used by engineers, both student and qualified engineers alike. I'm not suggesting that designers don't notice the errors, I'm sure they do, but think of the time it takes to work out what is wrong. And what of applicability? With research reports, even assuming their accuracy, the data is buried in a mass of background noise, like notes on previous research; why new research was needed; the experimental techniques and their justification; so there is still a lot of work to be done before the data can be applied to the problem in hand. And all of this takes time, a very scarce resource both in education and industry.

Earlier I mentioned authoritative data. Unfortunately authors, however eminent, are only human and usually have a biased opinion of their work, while researchers usually have an axe to grind and are more concerned with experimental methods than with the presentation of the results. For these reasons the majority of data are singularly lacking in authority and can seldom be taken at face value. And returning to applicability this means of course that very little data can be applied to the problem without a lot of work by the engineer.

Returning to the need for specified tolerances the editor of the Journal for Fluids Engineering from the American Society of Mechanical Engineers says that "authors estimates of uncertainty, generally given as + - 1%, are just not true". He suggests that the uncertainty figure is closer to + - 10% with half the information having the uncertainty of + - 30%. If he's right, and I suspect that he is very close to the truth, the designer can't even rely on the author's estimates of tolerances.

A director of the US National Bureau of Standards has said that "50% of all published data are unusable", for many of the reasons I've just mentioned.

Add to this alarming situation, the very low signal to noise ratio they provide to the practical engineering designer, because design is not the main objective of the publication, and you realise it is not just a matter of reading one article or extracting one number.

In addition to the accuracy, applicability and authority I've mentioned, the engineering student needs to know that the data and techniques he uses during his engineering education are those in use in the real world.

In the Aerospace industry design engineers automatically work to a safety margin in the region of 1.4. To work to these tolerances it is necessary to use data that has been validated. Validation is a long and very expensive programme requiring the collection of all the available information and test results from the many sources I've already mentioned, reading and assimilating the information, comparing the test results and test methods, before deriving the best data for the problem in hand.

To reduce as much as possible the costs involved in such an operation, the Aerospace industry both in the UK and overseas, especially in the USA, use engineering data supplied by my company.

In general engineering, this safety margin may be 2 or 3 times as great due to unknown loads and other factors, including of course a lack of trust in the data upon which the design is based.

I call this the "just in case" syndrome, where rather than believe the results of their calculations they make the component more robust or bigger or heavier - "just in case".

However, over the last twenty years or so, as industry has coped with the increased cost of materials and labour, the requirements of the market for higher and higher performance from the products of industry, and the need to produce these products continuously, with minimum downtime to interrupt the production process, it has realised that the key to all these requirments, the key to a succesful operation in these times of hard international competition, is accurate initial design. We have seen over the last years a sharp increase in interest in validated design data written specifically for the design engineer.

The comment has been made that different industries require different solutions. This is self-evident, and a major shortcoming of data produced in house is that its application is limited to use on similar problems.

This non-universality, this inability to transfer technology and its solutions between industries can however be overcome.

The production of data must be an end in itself, without concern for a particular industry. That is, the data must be discipline orientated and with sufficient scope to encompass the design requirements of any industry, and consequently any department of engineering design with an interest in that discipline.

Let me illustrate this point with a simple example.

It doesn't matter which industry you are in, or what product you are designing, from washing machines to aero engines, if it has rotating parts it will require some sort of bearing, among the most common are dry rubbing bearings, oil impregnated porous metal bearings, rolling bearings and hydrodynamic oil film bearings. There is a mass of information for each type of bearing from the various manufacturers.

By assembling and correlating this information by bearing type it was first possible to derive a series of curves of load against rotational speed for each type, before assembling these curves as shown here (**Fig 1**) to produce a general initial guide to the suitability of each type for the parameters shown.

A more complex example concerns the deflections of flat square plates under uniform normal pressure. Derived from more than fifteen references including unpublished information from Pilkington, City University, and the University of Birmingham, Russian and Japanese papers and many others, the information, in the form of curves (Fig 2), is now in use in aerospace engineering, structural engineering and shipbuilding, a good example of the universality of the data produced by the techniques I have described.

So Aerospace is not the only industry using such data. The atomic energy industry, the structural engineers, the motor industry and many others have found that the use of trustworthy, universal data removes one of the major areas of uncertainty in the design process, enabling the designer to approach more closely the optimum answer to his design problem, while saving the cost of both his time and overdesign - "just in cast".

Where does this leave the design engineering students? 40 Universities, Colleges and Polytechnics with engineering departments, in the UK alone, have already recognised the value of data that allows even first year students to prepare detail design, and which can then follow the student right through his course until he can deal with the problems encountered by the mature designer, problems embracing more conceptual design, with the need to check the feasibility of a number of possible approaches before proceeding to a definitive answer.

With the expanding use of ESDU data in engineering education and industry alike, one can see that engineering design education and engineering design in the real world are moving closer together. No longer does the employer need to teach his new designer the hard facts of industrial life and the new designer need no longer feel that his knowledge of engineering science has little relevance to the job he has undertaken.

By using practical, validated design data written specifically for the design engineer, and applicable across a whole range of engineering industries, the transition from student to design engineer is smoothed, to the advantage of employed and employer alike.

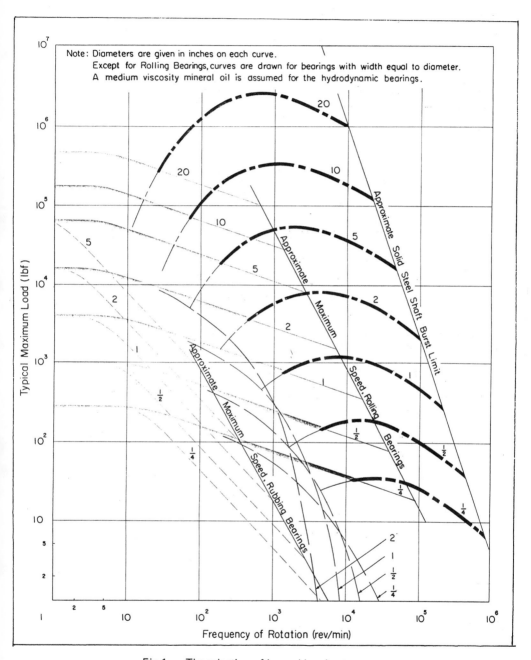

Note: Diameters are given in inches on each curve.
Except for Rolling Bearings, curves are drawn for bearings with width equal to diameter.
A medium viscosity mineral oil is assumed for the hydrodynamic bearings.

Fig 1   The selection of journal bearing type

1255

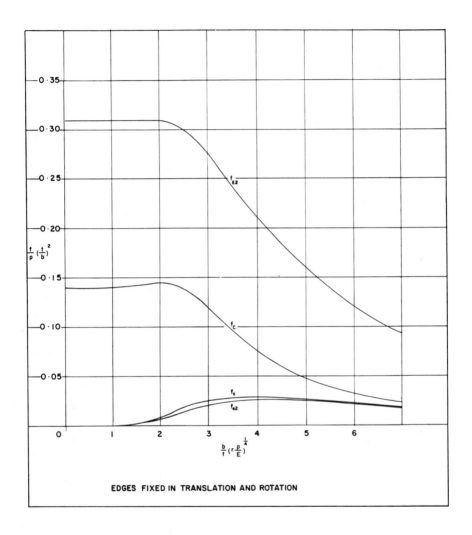

Fig 2   Stresses in a flat square plate under uniform pressure

C388/080

Market information processing—the first step towards successful product design

R G RHODES, MSc, CEng, MIM, ALA and **D G SMITH**, BSc(Eng), DLC, CEng, MIMechE
The Pilkington Library, University of Technology, Loughborough, Leicestershire

SYNOPSIS A structured approach has been developed for
the thorough processing of market information in order
to establish a definitive product design specification
so essential to product success in the market.   The
approach comprises a number of steps with the emphasis
on obtaining the full spectrum of information, the
establishment of a comprehensive information base and
its subsequent synthesis and analysis.   This approach
has been developed from the study of a wide range of
collaborative projects between education and industry
and has been adopted by a number of companies.

1 INTRODUCTION

A definitive product design specification is essential for product success in
the market.   However, the product design specification itself must be soundly
based demanding a thorough market investigation for which information is the
key.

There is evidence of increasing recognition on the part of some company
managements and others[1] of the importance of the product design
specification.   However, design teams are likely to lack an awareness of all
essential search areas and knowledge of market information processing
procedures so essential to its compilation.   Furthermore, where this total
approach is adopted the information flow profiles to design teams radically
changes.   Indeed the analysis of information flow has shown that where
projects are thoroughly executed, in a total sense, the major information
requirements are during the market investigation phase of design.   Thorough
information processing, frequently a time consuming activity, is therefore
nowhere more demanding and necessary than at this phase.

A further complication arises from the fact that the total set of
information employed is unique for every project.   This may lead to the
conclusion, that each project must be treated on its own merits.   However, the
analysis of a number of design projects, covering a wide range of products,
carried out on a collaborative basis between sponsoring companies and academia
has clearly shown that there exists a common approach to market information

processing for the majority if not all product designs.

Published literature deals with either single information areas, such as patents, or other specific aspects of information. This paper provides a proven, practical and structured approach to total market information processing. If comprehensively and thoroughly executed, it provides, in the authors' experience, a reliable basis for a sound definitive product design specification.

2 INFORMATION AND THE DESIGN ACTIVITY

It is considered essential at the outset to state briefly the premises fundamental to the approach to be discussed which are the views taken of:

* the design activity
* information
* the design activity-information relationship.

2.1 The design activity

Design is seen as 'the total activity necessary to provide an artefact to meet a market need that commences with the identification of the need and is not complete until the product is in use, providing an acceptable level of performance'[2]. It is also seen to comprise, within the total activity, identifiable core phases with information and technique inputs to enable the core phases to be operated. It is these inputs to the first two phases, market investigation and product design specification with which this paper is concerned.

2.2 Information

Problems arise with information because its value is not always clearly appreciated, it is confused with knowledge and literature for its own sake and is difficult to analyse in order to utilise to full benefit. It also requires the resources of at least either time or money in order to obtain it. In contrast to this it is in fact a valuable and vast resource in its own right along with time, money, materials, energy, manpower and machines. Furthermore, it is also re-usable.

A further problem arises from the fact that in many cases the required pieces of information are varied, scattered, in various forms and embodied with non-relevant information in a variety of locations. It is available from our memories, by discussion with other people, through publications and by observation. This will be more apparent later when information areas relating to the initial design phases are discussed.

2.3 The design activity-information relationship

Information may be regarded as the 'life-blood' of design[3]. Therefore, it must be considered as a major, not as a subsidiary, activity which must be planned, organised and controlled.

Where a project is carried out in a total and thorough sense the first major information processing task is that necessary to formulate the product design specification. It is the authors' experience that for many design projects this task is still sadly neglected or given only superficial treatment. If, in such cases, information input is considered in relation to the core phases of design a profile approximating to the dotted line in Fig 1 results. However, where greater importance is attached to the first two phases of design there is a fundamental and significant change in the profile. Indeed for a number of projects covering a wide product range carried out in the total sense, the profile has taken on a form similar to the solid line in Fig 1. Furthermore, the establishment of a comprehensive and reliable information base at the outset has the effect of reducing the amount of information processing in the conceptual and detail design phases thus aiding in increasing the efficiency with which they are carried out.

Before discussing the approach to the compilation of such an information base it is necessary to establish common ground in regard to the expression 'information processing'.

3 INFORMATION PROCESSING

Information processing is regarded as commencing with the realisation of a need and is not complete until either the required information is in a form which forwards the project or else is not obtainable. The main steps in market information processing may be defined as[4]:

* Clarify objective
* Search, Locate and Obtain
* Synthesise and analyse

Clarifying the objective is often overlooked with the subsequent waste of time and effort resulting in vague and incomplete activity. In contrast, a clear objective helps the subsequent process by providing additional viewpoints and keywords so vital to effective information gathering.

Contributory factors to a clear objective include the span of time for published material and indexes, relevant countries and associated terminology.

4 SEARCH, LOCATE AND OBTAIN MARKET INFORMATION

Factors which influence the product environment are many, varied and interactive. They include technology, politics, fashion, change in population age profile, change in income profiles, market sector and existing products to name but a few. In order to accommodate these factors and assess their influence on a particular project, it has been found essential to obtain information in the following eight areas, as depicted in Fig 2:

*   Business and statistical information
*   Competitive and analogous products
*   Patents
*   Standards, codes, regulations and legislation
*   Books, papers and reports
*   Manufacturing facilities
*   Specialists
*   Buyers and users.

The establishment of these eight categories of information and a structured approach to their use within the market investigation phase is seen as a significant factor in understanding a product market environment. Some of the relevant factors for the importance of this approach are as follows:

*   It is a move towards the ideal of 'total information' in which uncertainties are minimised.
*   It is tailored to the design process and in particular to the product design specification.
*   The categories are generic with a relationship to design.
*   It is applicable to all design projects.
*   It is efficient and effective.

Each of the areas will now be considered in turn in respect of scope, purpose and sources of information.

4.1 Business and statistical information

The category includes information on the market sectors and sizes, details of competitors' activities and any published market surveys.

The information is required in order to identify product trends such as market growth or decline, whether imports are increasing or decreasing and to relate competitors performances to the market situation.

Much information can often be obtained, either directly or indirectly, from a few relevant source publications. The main problem is that information is not available for many single products, but only for product groups.

4.2 Competitive and analogous products

Competitive and analogous products are existing products which either compete with the proposed design or provide a comparable function.

In designing a new product the aim should be to better any similar product already on the market. In order to be in a position to do this, knowledge of competitive products is vital.

The main source of information is manufacturers' and suppliers' literature but whenever possible it is essential to study the products themselves and to speak with users.

4.3 Patents

Patent specifications are documents providing details of registered inventions and their legal ownership.

It is essential for design teams to be aware of relevant patents lest infringement occur which may result in a legal action. They can also be of value in stimulating ideas in the later design phases or it may be an option to obtain a licence for use rather than designing a new product or component.

Patent specifications are obtainable from libraries holding patent collections but the use of a patent agent to carry out a search may be advisable.

4.4 Standards, codes, regulations, legislation

Standards, codes of practice, regulations and legislation are collectively the various technical and legal requirements for a product.

Their significance is usually self evident and design teams need to consider the various levels such as local, national and international requirements.

Initially it is fairly easy to locate standards from appropriate source documents such as handbooks from national standards organisations but there are pitfalls for the unwary. For example, sometimes standards and codes are set by organisations other than the recognised standards institutions. Also a range of standards and codes may be applicable to one product. Legislation can be a difficult area to search and assistance is likely to be needed.

4.5 Books, papers and reports

Books, papers and reports are the published works from experts, academics, researchers and others providing a wide variety of material including background, product surveys and latest developments.

For design teams, a variety of useful information can be gleaned from the enormous amount of published literature.

The ease of searching varies and professional help from libraries and computer based systems can be used.

4.6 Manufacturing facilities

Manufacturing facilities are the available processes and materials for making the products either in-house or by sub-contractors.

Design teams need to be aware of the scope of facilities available and also the constraints in order to execute the design for compatibility with proposed manufacturing facilities.

In-company information should be readily available and information on sub-contractors from trade directories.

4.7 Specialists

Specialists are individuals and organisations which have gained an expertise in selected areas.

They have particular value to design teams in their intellectual response to informal questions. Literature cannot cover all knowledge and sometimes information is only obtainable by word of mouth.

The design team needs social as well as technical skills to gather information from specialists but they are usually keen to discuss their work. They can be identified through such means as appropriate directories and publications.

4.8 Buyers and users

Buyers and users are the people most concerned with the products as to their value and how well they perform.

Their views are one of the most vital sources of information. The main difficulty is one of relating views to potential products rather than possibly prejudiced views of former products and brand images.

In a limited number of product areas consumer surveys have been carried out. However, the task usually amounts to the use of well-planned questionnaires directly by interview and indirectly by post.

5 METHODS OF SEARCHING

5.1 Recorded Information

The traditional method of searching for recorded or documented information is 'desk research' in which the searcher peruses a variety of indexes and directories. These lead to greater variety of documents for reading and names of possible contacts. The process normally takes place in a suitable library where business librarians and information specialists have detailed knowledge and skills. The main disadvantages of 'desk research' are the time needed for thoroughness and the varied accessibility of both an extensive range of sources and skilled help. Furthermore, information specialists do not always appreciate the needs of design teams. For example, market information is often dominated by information on potential buyers rather than the product features they require.

Traditional 'desk research' is being supplemented and replaced in part by computer based searches of large databases. These methods are popular with those who have used them because of their speed and convenience. They appear to be comprehensive but recent research[5] shows that a typical computer search provides only about 5% of the documents found in a comprehensive search. Another factor to be taken into account is the need to budget for computer usage particularly for the more expensive patent and company background searches.

The abilities needed to search for 'recorded' information is one of finding sources and of identifying terminology and concepts used by others. Individuals who like crosswords and similar competitions are likely to be good at searching for recorded information!

5.2 Non-recorded information

Information 'not on record' can be obtained by talking to people by observation and also by such means as bench testing competitors' products.

Talking to people, directories being one means of identifying potential contacts, needs different skills. Sometimes, the term 'field research' is used to describe this activity. An obvious factor is courtesy so that the individuals do not feel exploited or dissatisfied. The reason why questionnaires have been found to be convenient is that they make the best use of interview time and simplify the subsequent handling of the information.

The order of searching the eight categories of market information sources is a matter of choice. However, it has been found that it is best to search for 'recorded' before 'non-recorded' information. This enables the searcher to identify the gaps in recorded sources and phrase questions to relevant non-recorded areas with greater confidence.

It is appropriate to comment here that the value of libraries, severely under-rated by many industrial companies, is the prime source of recorded information. Some of the sources of information found useful have been noted and published[4].

6 SYNTHESISING AND ANALYSING MARKET INFORMATION

Upon receipt, information should be sifted, filed and assimilated. Synthesis and analysis should follow as soon as a representative quantity of information is to hand and should not be delayed in the hopes that all the information requested will be received.

Synthesis and analysis is another crucial stage in information processing. The purpose in so doing is to identify patterns and trends which lead to the compilation of the product design specification. Even so it is frequently not given the attention it merits. All too often we obtain information, perhaps read it, or more accurately scan it, think about it, file it, and probably forget much of it! Yet from this process we expect to extract critical data to progress a project forward. In so doing we are trying to manipulate, in our minds, perhaps hundreds or even thousands of pieces of information and fit them into patterns which will be helpful in progressing a project. Incredible as our minds are, we are unlikely by this procedure to get the maximum benefit from the information gathered. It is essential to employ systematic methods and techniques as an aid to the synthesis and analysis of information in order to ensure that the maximum benefit is being obtained and deductions made with an acceptable level of confidence. The systematic synthesis and analysis of information is becoming increasingly important in the current environment with the rapid growth in the information available to design teams.

The main steps to be taken in information synthesis and analysis, which will be realised as being interactive and iterative, are:

* organisation
* categorisation
* structuring
* deduction.

6.1 Organisation

Whatever the information area to be analysed it is necessary, as a first step, to synthesise it into a common format or information base. The use of the information matrix has been found to be a useful aid for this purpose. An example of such a matrix is shown in Fig 3 for the features of air centres supplied by a range of manufacturers. It will be noted that the matrix has the manufacturers on the horizontal axis and the features listed on the vertical axis. The chart shows only ten out of the forty features identified. The numbers adjacent to the ticks indicate footnotes which are not given.

A word of warning is necessary in regard to published ready-made matrices. Whilst they may contain much useful information, sole reliance should never be placed on them as they will almost certainly have been compiled for use by others than designers. They have always been found to be insufficiently comprehensive.

Even the organisation of data in this simple way begins to show patterns and trends. For example, in Fig 3 some manufacturers incorporate a large number of features in their products whilst others market products with minimum features. Furthermore, all manufacturers include one of the features whilst another feature is included by only one manufacturer. Observations of this nature prompt further investigations in order to clarify real customer needs.

6.2 Categorisation

Information is found in two forms each having a sub-division:

* numeric   – actual
                – estimated
* non-numeric – fact
                – opinion.

In compiling an information matrix care must be taken to differentiate actual data from estimated and fact from opinion. Care must also be taken to ensure that consistent units are used for numeric data.

Non-numeric data may be handled by the use of comments and/or by the use of the 'Yes – No' approach both of which have in effect been employed in the example given in Fig 3.

6.3 Structuring

In many cases the organisation of information itself causes some deductions to become self-evident. However, data usually yields much more benefit when parameters are considered not in isolation but in combination with information structured to allow for this such comparisons. For example, Fig 4 shows the sales of compressor units of various size categories against time. The graphs make immediately obvious the different sales patterns for reciprocating and rotary machines.

The plotting of quantifiable product parameters known as 'product-parametric analysis' has been proven to be an extremely valuable technique[6]. Fig 5, for example, shows a pattern of results related to air centres in which free air delivered is shown plotted against selling price. The plot shows an expected trend of increasing cost with increasing output. The closely packed points at the lower end of the range are indicative of the highly competitive nature of the market. Such plots if thoroughly carried out give considerable insight into the understanding of product areas.

The foregoing examples have demonstrated the value of linking parameters as opposed to reliance simply on the information matrix alone valuable as it is. However, it is when search areas are linked together that the real value of information analysis is likely to be revealed. A cross-linking which frequently proves informative is that of buyers and user perception of products compared with the manufacturers perception.

7 FORMULATION OF PRODUCT DESIGN SPECIFICATION

The product design specification is a statement of the market need arising from the results of market information processing. Fig 6 shows a comprehensive set of elements to be included in a product design specification[7] in relation to the information areas on which, from the authors' experience, they are likely to be dependent. A note about the specification elements is appropriate at this point. In speaking with companies who take the trouble to write product design specifications it has been found that they record about one-third of the elements listed, think about another one-third and neglect the remainder!

In Fig 6 the information areas have been arranged in order of the influence that they have on the specification elements. Thus, for example, patents are related to only two elements whereas the needs of buyers and users influence nearly all of the elements. Ironically, it is the views of buyers and users which are so often ignored in designing new products and for some of the elements is the sole source of information. Hence, it is impossible to complete a specification without adequate feedback from customers and information on competitors. The importance of the customer viewpoint to product success has also been established by others.[1] Further, there is little published information in this area of customer views.

Also in Fig 6 the specification elements have been arranged in order of decreasing reliance on information areas. For example, it may be necessary to consider information from as many as six areas as far as materials are concerned. However, in compiling patent details only two information areas are likely to be involved. Certain information requires analysis before being compiled into the specification, as indicated earlier, whilst other, such as that on manufacturing, is factual information which once obtained can be included directly.

8 SUMMARY AND CONCLUSIONS

The paper provides a structured approach to information processing at the market investigation phase in order to provide a reliable information base for a sound product design specification. The approach comprises the following main elements:

* The identification of eight areas as potential sources of information.
* Obtaining relevant information from each of these areas.
* Synthesising and analysing the information in each area using structured techniques.
* Relating the trends and patterns identified in each area.
* Compiling the product design specification from the information base and from the trends and patterns identified.

The thoroughness of the process outlined clearly calls for the allocation of people, time and finance and hence there must be a conviction on the part of company management that such allocations are justifiable. There can never be one hundred percent certainty in foresight that this is so. However, it is the authors' experience that in the wide variety of projects carried out between industrial companies and academia the resources expended have always been considered to be well justified. This experience is now being confirmed by a number of industrial companies which are also employing the proposed approach.

It must be borne in mind that it is the initial setting up of an information base that is the most resource consuming. Once established, and if properly organised, it can be systematically updated to suit company design policy. Thus, information processing begins to become a 'regular' rather than a 'sporadic' activity. Not all information requires updating with the same frequency. Certain types of fundamental data are likely to be always valid whilst other, for electronic products for example, changes very rapidly. It is necessary to identify these different categories and update accordingly.

An information base established at the outset of a project can be used throughout all the phases. Thus, for example, information on legislation, standards, codes of practice, ergonomics and competitors' products is required not only to compile the product design specification but also at the conceptual and detail phases. This helps to increase efficiency in the later phases as information is ready at hand; frequently it is sought only when required resulting in delays.

Information is recognised as the 'life-blood' of design and hence a vital resource which must be exploited to the full in the current highly competitive product environment. Nowhere is this more important in product design than at the market investigation and product design specification phases. Failure here jeopardises the remainder of the project.

REFERENCES

(1) COOPER, R.G. and KLEINSCHMIDT, E.J. Success Factors in Product Innovation. Industrial Marketing Management, 1987, Vol 16, p215-224 (Elsevier).

(2) Curriculum for Design - Engineering Undergraduate Courses. Proceedings of Working Party, 1985 (SEED).

(3) CORFIELD, K.G. Product Design. A report carried out for the National Economic Development Council, 1979 (NEDC).

(4) RHODES, R.G. and SMITH, D.G. Information Retrieval. In the series Curriculum for Design - Preparation Material for Design Teaching, 1987 (SEED).

(5) DAVISON, P. and MOSS, A. International Bibliographic Review on Costs and Modellingin Information Retrieval. British Library Research Paper, No. 37, 1988 (British Library).

(6) PUGH, S. Concept selection - a method that works. Proceedings WDK5 ICED '81, Rome, 1981, p 497-506 (Techniche Nuove).

(7) PUGH, S. Specification Phase. In the series Curriculum for Design - Preparation Material for Design Teaching, 1987 (SEED).

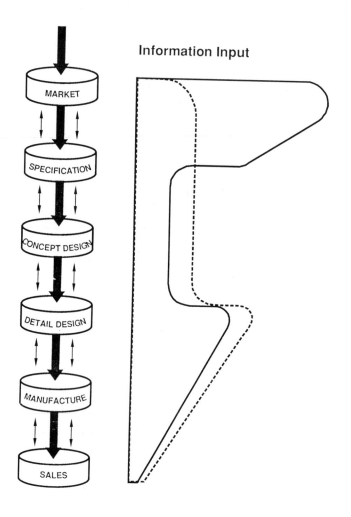

Fig 1   Variation in information input during the design activity

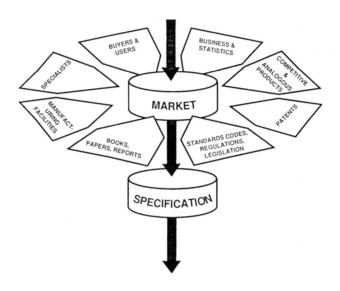

Fig 2   Information areas relevant to the market investigation phase of design

MANUFACTURER Vs FEATURES OFFERED COMPARISON CHART	A	B	C	D	E	F	G	H	I	J	K	L	M
FEATURE													
1 HOURS 'ON' COUNTER	✓	✓	✓	✓	✓	✓	✓	✓	✓	✓	✓	✓	✓
2 HOURS 'ON-LOAD' COUNTER		✓[31]					✓						
3 OUTLET AIR/OIL TEMP. LIGHT & RESET		✓		✓	✓	✓	✓	✓	✓	✓	✓	✓	✓
4 AIR/OIL TEMP. GAUGE (COMPRESSOR TEMP.)	✓		✓	✓[1]		✓		✓		✓	✓	✓	
5 POWER 'ON' LIGHT		✓			✓	✓	✓		✓	✓	✓		✓
6 RUNNING LIGHT		✓	✓	✓		✓	✓	✓	✓	✓	✓	✓	
7 KEY OPERATED ON-OFF CONTROL						✓							
8 CHECK OIL SEPERATOR		✓[2]		✓		✓	✓	✓[3]	✓			✓	✓
9 CHECK INLET FILTER		✓[7]		✓		✓	✓	✓	✓	✓		✓	✓
10 BLOCKED OIL FILTER		✓			✓		✓[5]	✓		✓		✓	

Fig 3   Partial information matrix for air centres

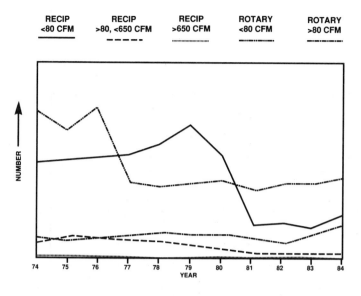

RECIP <80 CFM	RECIP >80, <650 CFM	RECIP >650 CFM	ROTARY <80 CFM	ROTARY >80 CFM

Fig 4   Graphical presentation of information on air compressor sales with time

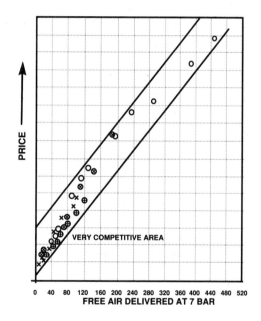

Fig 5   Parametric analysis graph for the price of air centres against free air delivered

1271

	BUYERS & USERS	COMPETITIVE & ANALOGOUS PRODUCTS	STANDARDS CODES, REGULATIONS LEGISLATION	SPECIALISTS	BOOKS, PAPERS, REPORTS	BUSINESS & STATISTICS	MANUFACTURING FACILITIES	PATENTS
MATERIALS	✔	✔	✔	✔	✔		✔	
ERGONOMICS	✔	✔	✔	✔	✔			
PERFORMANCE	✔	✔	✔	✔	✔			
SAFETY	✔	✔	✔	✔	✔			
PACKAGING	✔	✔	✔	✔	✔			
COMPETITION	✔	✔				✔		✔
AESTHETICS	✔	✔		✔				
COST	✔	✔				✔		
ENVIRONMENT	✔	✔	✔					
POLITICS	✔			✔	✔			
Q & R	✔	✔	✔					
SHIPPING	✔		✔	✔				
STANDARDS & SPECIFICATIONS	✔	✔	✔					
TESTING	✔	✔	✔					
LIFE IN SERVICE	✔	✔						
MAINTENANCE	✔	✔						
MARKET CONSTRAINTS	✔					✔		
PATENTS		✔						✔
PRODUCT LIFE SPAN	✔	✔						
QUANTITY	✔					✔		
SHELF LIFE	✔	✔				٠		
SIZE	✔	✔						
TIME SCALES	✔					✔		
WEIGHT	✔	✔						
COMPANY CONSTRAINTS							✔	
CUSTOMER REQUIREMENTS	✔							
MANUFACTURING							✔	
PROCESSES							✔	

SPECIFICATION ELEMENTS INFLUENCED BY INCREASING NUMBER OF INFORMATION AREAS

INCREASING INFLUENCE OF NUMBER OF INFORMATION AREAS ON SPECIFICATION ELEMENTS

Fig 6   Typical relation between information areas and product design specification elements

1272

C377/224

The performance of a mechanical design 'compiler'

A C WARD and **W P SEERING**
Massachusetts Institute of Technology, Cambridge, Massachusetts, USA

Abstract

A mechanical design "compiler" has been developed which, given an appropriate schematic, specifications, and utility function for a mechanical design, returns catalog numbers for an optimal implementation. The compiler has been successfully tested on a variety of mechanical and hydraulic power transmission designs, and a few temperature sensing designs. Times required have been at worst proportional to the logarithm of the number of possible combinations of catalog numbers.

1 Introduction

Among our research goals is the development of "mechanical design compilers"; that is, programs which take as input a schematic (or other high-level description) of a mechanical design, plus specifications and a cost function, and return a description of the optimal implementation of the design, sufficiently detailed to support manufacture. Such programs should decrease design time and cost, increase design quality, and allow designers to explore more alternatives in greater depth.

We have chosen to work initially in the domain of mechanical and hydraulic power transmission systems built from cataloged components. For this domain we have substantially accomplished our goal; this paper presents the evidence for that success, and discusses its limitations.

After mentioning some related work, we provide a very brief over-view of the compiler, intended only to allow the reader to understand our performance evaluation. For a more detailed introduction to the compiler, and in particular the theory on which it rests, see [1]; for full details see [2]. We then examine the capabilities of the compiler in three different respects: 1) the range of design problems it has been tested on; 2) its reliability; and 3) its efficiency or time complexity.

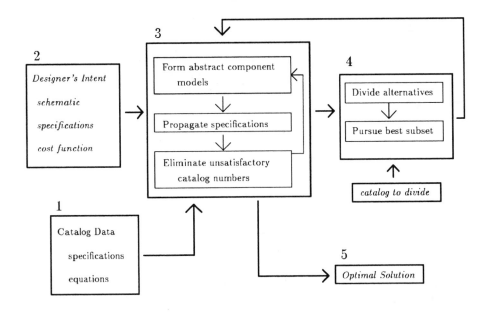

Figure 1: The Compiler Block Diagram

2 Related Work

We have found no other programs identified as "mechanical design compilers" by their creators. [3] and [4] discuss programs which offer the designer a schematic language, but which perform analysis only. We argue in [1] that the traditional "constraint propagation" methods they use are inadequate to represent essential mechanical design information.

[5] and [6] discuss programs able to find the optimal selection of a single component, given constraint and cost equations. These use "hill-climbing" optimization routines, with heuristics to modify the hill-climbing process when they get stuck. The "hill-climbers" are called by supervisory programs which can represent combinations of components. The supervisory programs supply the hill-climbers with specifications, based on "expert knowledge", and in [5] using iteration to improve the initial guesses.

The approach these programs use has some disadvantages. They appear to require substantial effort to set-up each configuration of components; a "domain" in [5] is equivalent to a single schematic for our system. They use nominal values; manufacturing tolerances, and variations in operating conditions are not explicitly represented. The "hill-climbing" search process can become stuck on local optima. Because the set of artifacts implicitly represented by the design is continually changing, all calculations must be repeated at each step, and correcting one deficiency can introduce others. For these reasons we have chosen a radically different approach.

3 Overview of the Compiler

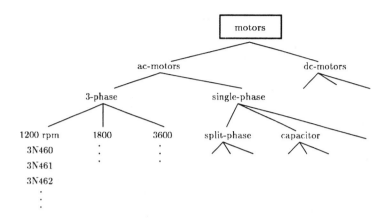

Figure 2: A hierarchy of motors

Figure 1 illustrates our approach. Our data base is built up (block 1) from "basic sets" of artifacts. Each basic set is represented by a single catalog number. The set consists of the individual artifacts one might receive by ordering that catalog number. For example, on ordering Dayton motor number 2N103, we will receive any one of an effectively infinite variety of motors, each slightly different because of manufacturing tolerances, each with its own serial number; these motors make up the basic set denoted by catalog number 2N103.

The basic sets are modeled by an engineer, using equations and specifications in a special "labeled interval" specification language. For example, the speed regulating characteristics of Dayton motors 2N103 might be represented by $\langle \mathbf{A} \overset{only}{[\ \ ]} RPM\ 1740\ 1800 \rangle$. This specification tells us that we are *assured* ($\mathbf{A}$) that for any motor we might get by ordering number 2N103, the the RPM will take on *only* ($\overset{only}{[\ \ ]}$) values between 1740 and 1800, under normal loading.

The engineer groups the catalog numbers into a hierarchical structure, and the compiler **abstracts** (block 3) the information about the basic motor sets to form descriptions for higher levels in the hierarchy. For example, the next level up might be all the 1800 rpm three-phase motors represented; these have varying degrees of speed regulation, so the set as a whole might only guarantee speed regulation between, say, 1700 and 1800 rpm: $\langle \mathbf{A} \overset{only}{[\ \ ]} RPM\ 1700\ 1800 \rangle$. Finally, a schematic symbol (Figure 2) represents the whole hierarchy of catalog numbers, and therefore the union of their basic sets. The motor symbol might initially represent all of the electric motors listed in the Dayton catalog[1].

The compiler's user, a mechanical designer, **composes** new designs by pointing at schematic symbols (block 2). The system automatically makes appropriate connections, asking for help if needed to resolve ambiguities; for example, in adding the first cylinder to the schematic of Figure 3, the compiler would have to ask which valve to attach it to. Having defined such a design schematic, the user may assign it a symbol of its own, for

---

[1]The currently implemented catalogs only include a subset of the Dayton catalog.

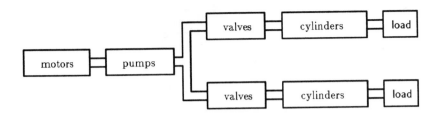

Figure 3: A Hydraulic Power Train

recall or use in more complex designs.

The compiler automatically **eliminates** catalog numbers which are incompatible with any implementation of the connected components (block 3). For example, on connecting the motor schematic to one representing a 220-volt power supply, the system automatically eliminates any 110 volt motors.

After building the schematic, the user provides specifications. These specifications describe sets of operating conditions; $\langle \mathbf{R} \stackrel{every}{\leftrightarrow} speed\ 0\ .2 \rangle$ applied to a cylinder means that the speed of the cylinder shaft is "Required" ($\mathbf{R}$) to take on every value ($\stackrel{every}{\leftrightarrow}$) in the interval from 0 to .2 feet per second.

In this example, the maximum output pressure available from any of the pumps, together with the highest of the range of forces required, sets a minimum diameter requirement on the cylinders. These, together with the speeds required, establish flow requirements. This use of equations and specifications to form new specifications is **propagation** (block 3); it can be regarded as a generalization of "the constraint propagation of intervals" [7]. More specifically, the constraint propagation of intervals corresponds to one of 21 propagation operations employed in the compiler.

The propagated specifications for flow, horsepower, torque, and so on cause further eliminations, leaving subsets of the original catalog numbers. Descriptions for these subsets are then abstracted to produce new specifications, which trigger further propagation and elimination.

When the cycle of abstraction, propagation and elimination ceases, a variety of alternative combinations of catalog numbers often remains. The user then provides a cost function, for example the weighted sum of the price and weight of the components. He also directs the compiler to split one of the catalogs in half, for example to look at 3600 and 1800 rpm motors separately (block 4)[2]. The compiler then generates two daughter designs, one for each motor set; the abstraction operators formulate new specifications describing the new, smaller motor sets. These specifications trigger another cycle of eliminations.

Repeating this splitting process generates a binary best-first search tree. The compiler always splits the leaf of the tree offering the lowest possible cost. The search continues until a single catalog number remained for each component.

The output of this compiler thus consists primarily of catalog numbers. Given these numbers and the schematic, most mechanics could probably buy the components and

---

[2]Having the user guide the search in this way improves efficiency; catalogs could be selected for splitting randomly, or by a heuristic.

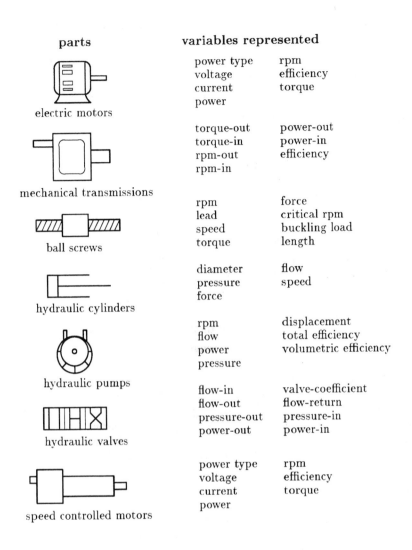

parts	variables represented	
electric motors	power type voltage current power	rpm efficiency torque
mechanical transmissions	torque-out torque-in rpm-out rpm-in	power-out power-in efficiency
ball screws	rpm lead speed torque	force critical rpm buckling load length
hydraulic cylinders	diameter pressure force	flow speed
hydraulic pumps	rpm flow power pressure	displacement total efficiency volumetric efficiency
hydraulic valves	flow-in flow-out pressure-out power-out	valve-coefficient flow-return pressure-in power-in
speed controlled motors	power type voltage current power	rpm efficiency torque

Figure 4: Some test parts

construct the system without further input from an engineer. A future, more complete compiler would provide a drawing of the base-plate. The most complete compiler possible would instruct an automated manufacturing system to build the design.

4 Some Examples

In this section I discuss in general terms my experience with the compiler. Figure 4 shows some component types, with the primary variables used to model them.

The component models now used include specifications for most of the non-geometric criteria that the vendors discuss in the "engineering" sections of their catalogs. Formulating a representation for a component type requires the engineer to extract a precise model, in our formal specification language, from the usually vague and often inconsistent catalog data. He must compromise between simplicity and completeness. For example, we have chosen to represent the efficiency of mechanical transmissions as a range of possible values, from 90 to 98 percent. We could instead have entered an equation relating efficiency to speed; manufacturing variations would then be represented by interval specifications on the coefficients of that equation.

Once the form of the precise model has been determined, a simple program can be instructed to translate the manufacturer's catalog into the labeled interval specification language. Entering further catalog numbers for components of this type is then a typing exercise. It generally takes about one day to decide the form of the specifications and equations for a new kind of component, generate the transformation procedure that converts the catalog to the desired form, and test the results.

The system has been tested on a few temperature measurement system design problems and more than a dozen different arrangements of power transmission components. Figure 5 shows some of these, with machines in which the power trains might be used.

Let us now consider in more detail the two-cylinder hydraulic system example of Figure 3. The catalogs for the components shown include the following numbers of alternatives: 7 types of electrical supply (omitted from the schematic), 36 motors, 13 pumps, 3 valves, and 12 cylinders. There are thus 4,245,696 possible combinations; of these, because our catalogs are still sparse, only 505,440 remain after the eliminations caused by connecting the components.

After composing the schematic, the user then enters load specifications, for example:

Load-1: $\langle \mathbf{R} \overset{every}{\leftrightarrow} speed\ 0\ .2 \rangle$, $\langle \mathbf{R} \overset{every}{\leftrightarrow} force\ 0\ 1000 \rangle$
Load-2: $\langle \mathbf{R} \overset{every}{\leftrightarrow} speed\ 0\ .15 \rangle$, $\langle \mathbf{R} \overset{every}{\leftrightarrow} force\ 0\ 3000 \rangle$.

For the first load, this means that the system must provide every speed from 0 to .2 feet per second, with forces from 0 to 1000 pounds.

The compiler uses these specifications, and those built into the catalogs, to eliminate unsatisfactory alternatives and to generate further specifications. For example, the linear horsepower equation is built into the "load" component, $hp - \frac{(force)(speed)}{550} = 0$. The compiler incorporates an inference rule which can be written

$$\langle \mathbf{R} \overset{every}{\leftrightarrow} x\ x_l\ x_h \rangle \& \langle \mathbf{R} \overset{every}{\leftrightarrow} y\ y_l\ y_h \rangle \& G(x, y, z) = 0$$
$$\longrightarrow \langle \mathbf{R} \overset{every}{\leftrightarrow} z\ \text{RANGE}(G, \langle x\ x_l\ x_h \rangle, \langle y\ y_l\ y_h \rangle) \rangle.$$

The left hand side of the rule matches the input data and the equation:

$$\langle \mathbf{R} \overset{every}{\leftrightarrow} speed\ 0\ .2 \rangle \quad \sim \quad \langle \mathbf{R} \overset{every}{\leftrightarrow} x\ x_l\ x_h \rangle$$
$$\langle \mathbf{R} \overset{every}{\leftrightarrow} force\ 0\ 1000 \rangle \quad \sim \quad \langle \mathbf{R} \overset{every}{\leftrightarrow} y\ y_l\ y_h \rangle$$
$$hp - \frac{(force)(speed)}{550} = 0 \quad \sim \quad g(x, y, z) = 0.$$

The RANGE function on the right side of the rule is one of three operations on equations and intervals discussed in [1]. In effect, it solves the equation for the hp, forming $hp = \frac{(force)(speed)}{550}$. It then determines the range of the horsepower subject to the constraints

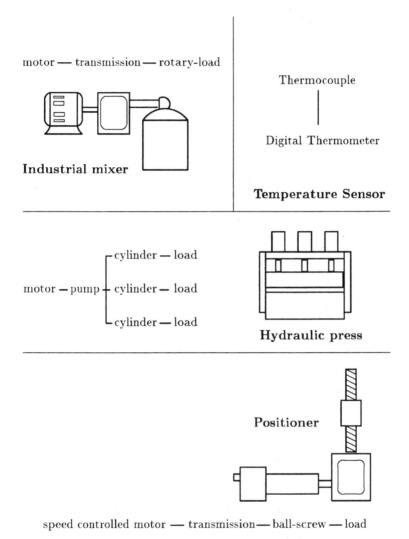

Figure 5: Some test designs

that force and speed are restricted to the intervals $[0\ 1000]$ and $[0\ .2]$. The numerical results of RANGE are thus identical to those produced by the "constraint propagation of intervals". (The other operations discussed in [1] can be thought of as inverses to RANGE.) But the new specification which would be formulated by the right hand side of the rule , $\langle \mathbf{R} \overset{every}{\leftrightarrow} hp\ 0\ .36\rangle$, is not a "constraint" in the usual sense of a limit on the values. Rather, it says that the cylinder must have available to it power flows from 0 to .36 horsepower; higher powers are acceptable as well.

These specifications eliminate many potential implementations; for example, motors unable to supply the required horsepower, adjusted by the efficiency of the pumps. The designer then splits the catalog for one of the components, for example one set of cylinders, generating daughter designs. One daughter design has only large cylinders, the other only small; this starts a new cycle of abstraction and elimination.

On the particular data given, the compiler searched 71 daughter designs, generating 15,663 new specifications in the process. The cost function used was price plus one half weight. The design run took about 20 minutes, a normal time for the program to complete a hydraulic problem of this size. Optimization of the code will speed this considerably. The output for this problem included:

```
The optimum solution, with cost 441.97, is:
For POWER-SUPPLY, US-3PH-220 with cost 0
For MOTOR, 3N593 with cost 192.72
For GEAR-PUMP, TYPE-103 with cost 133.0
For VALVE, TYPE-1 with cost 50.0
For CYLINDER, 1.25 with cost 6.25
For VALVE-2, TYPE-1 with cost 50.0
For CYLINDER-2, diameter 2.0 with cost 10.0
```

5 Assessing Program Reliability

We have used basic set theory, predicate calculus, and analysis to develop formal correctness proofs of most of the individual compiler operations; see [2]. Such proofs add greatly to the reliability of the program, and to our understanding, but they are no better than the assumptions on which they rest; the program must still be tested empirically. We have done dozens of "runs", with varying specifications, on more than a dozen different arrangements of components. We evaluate these runs by determining why particular alternatives are eliminated, and by examining the "optimal solutions" resulting.

The system appears to eliminate only invalid designs. It frequently surprises us, but we always find either a correctable bug, or that our understanding of the design problem was incomplete.

We are also quite sure that the designs selected are "optimal" with respect to the cost function, but our confidence here is based on the simplicity and clarity of the optimization process rather than on empirical results. Finding an optimum solution by hand is extremely slow on even these simple design problems, and no optimization program we know of can easily be set up for problems of this kind. Even an exhaustive check of combinations of components would still involve sets of operating conditions, hence require most of the mechanisms of the compiler and not constitute an independent check. The most we can say, as a human designers, is that the designs produced look like they could well be optimal.

A subtler question is whether the program eliminates all the implementations it should—whether its rule set is complete enough to guarantee that the designs it produces will work. It is not, in three senses. First, we know that there are propagation operations we have not yet implemented. We implement operations only as needed, because new operations slow the system and require testing. Second, as we discuss later, the compiler does not propagate every specification it could.

Third, and pragmatically most important, the selected design can always be unsatisfactory because of criteria not represented in the component models. Our formalism imposes restrictions on the criteria it can represent. In particular, equations must be algebraic, and have three variables, though intermediate variables can be used to break up complex equations. We must be able to solve for each variable, and the resulting functions must be continuous and monotonic. The equations must be "instantaneously true"; they cannot values which occur at different times. Values must be non-negative. Specifications must be stated as equations, cost expressions to be minimized, or "hard-edged" intervals. Finally, variables must be divided into only two "causal categories"—parameters, which are fixed at manufacturing, and state variables, which change during operation.

These restrictions limit expressive power. Lack of differential equations probably prevents the system from compiling servo-system designs, or detecting vibration problems. Speed controller catalogs often provide ratios between the highest and lowest controllable speeds, thus relating two different operating conditions. An attempt to model automobile seat design failed because seat-back position is neither a state variable nor a parameter.

Nonetheless, within the domain and problems we have implemented the system appears to select correct designs. It is at present probably less reliable than a very skilled designer working on familiar problems, because very skilled designers make use of information omitted from the catalogs. However, it is probably more reliable, faster, and more likely to produce an optimal design than the average designer.

6 Time Complexity

How long does it take to solve these design problems? "About 20 minutes for a problem involving half a million alternatives" is correct but not very useful, since this depends mostly on implementation and hardware. What we really want to know is how the time required to solve the problem increases as the size of the problem increases.

6.1 Theoretical Results

We will consider two measures of the size of the problem. The first is the total number of possible alternatives, where an alternative is a combination of catalog numbers without regard to feasibility. This is proportional to C^n, where n is the number of components in the design, and C the average catalog length for each component.

The program searches for an optimal solution by creating a binary search tree; the forks in the tree are generated by dividing the catalog for a single component into two parts, splitting the "artifact space". The program then pursues the "most promising" daughter design. There is no guarantee that the "most promising" decision will be correct, and unless it is correct most of the time, back-tracking may require time at least proportional to the number of alternatives.

The situation grows even worse when we consider the other measure of size, that is the number of equations involved. The compiler subsumes the conventional constraint propagation of intervals, and it can be shown [7] that the constraint propagation of intervals can run forever. For example, suppose we have two equations, $x = y$ and $x = 2y$, and we start with intervals $0 \leq x \leq 1$ and $0 \leq y \leq 1$. We first conclude from the second equation that $0 \leq y \leq .5$, then from the first that $0 \leq x \leq .5$, then $0 \leq y \leq .25$ and so on. We never arrive (barring round-off error) at the solution, $x = y = 0$.

It may be possible to avoid such pathological cases in real design problems, but even much simpler forms of constraint propagation can require time double-exponential in the number of variables involved, and therefore singly exponential in the number of equations[7].

6.2   Empirical Results

Fortunately, the system actually performs much better than the worst case theoretical projections. In figures 6 and 7, we have used the number of specifications generated by the searching compiler as the measure of time; this measure is independent of the particular hardware and software implementation. Most of the compiler's operations take time proportional to the number of specifications generated. One, the elimination of alternatives, can at worst take time proportional to the number of specifications generated times the average length of the catalogs.

Figure 6 shows a semi-logarithmic plot of the number of specifications generated against the number of alternatives. At worst, the number of specifications generated grows according to the logarithm of the number of alternatives.

Figure 7 shows a plot of the number of equations involved in the design against the number of specifications generated; growth is no worse than linear with the number of equations. This, in turn, is linear in the number of components in the design.

6.3   Explaining the difference

There seem to be five principal reasons why the empirical results are so much better than the worst case predictions.

First, note that eliminating a single catalog number eliminates many alternatives, since that catalog number is involved in a combinatorial set of alternatives.

Second, the artifact space is organized, for example by horsepower. A single specification can eliminate many catalog numbers. More importantly, the optimal solution generally involves the smallest (or nearly the smallest) of the devices meeting the horsepower requirement. Since these are clustered together in the search space, only a few branches of the search tree need be followed.

Third, the equations used to describe mechanical components establish a fairly sparse network between variables. In particular, all information passed between components is channeled into a small number of "port variables", such as rpm and torque. (These components have been selected for manufacturing and cataloging in part because they

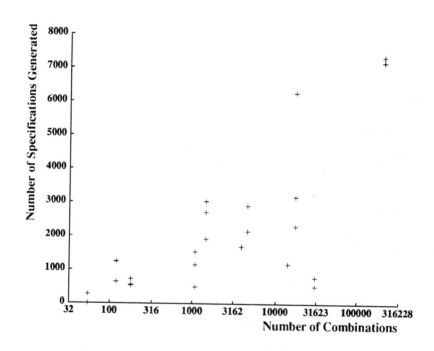

Figure 6: Specification generation vs alternatives for a variety of designs

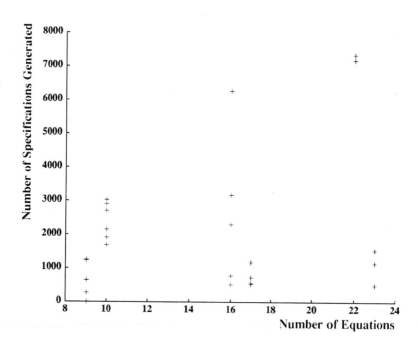

Figure 7: Specification generation vs equations for a variety of designs

have relatively simple connections with the rest of a design.) This sparseness helps limit the growth in execution time as a function of the number of equations.

Fourth, some of the propagation operations are correct only if each input specification is independent of the other variables in equation used. The compiler in fact requires independence for all propagation operations, thus preventing infinite loops of the kind discussed above.

Fifth, the compiler propagates only the "strongest" specifications, for example the tightest required limits.

These last two reasons involve restrictions on the constraint propagation process. We have not proven that these restrictions cannot cause failures to eliminate, but have not observed any such errors in practice.

7   Conclusions

To summarize, the compiler has been tested on a range of mechanical and hydraulic power transmission designs; new designs can be entered by the designer in minutes. Results have been correct and optimal for the tested problems. Time required for solution grows reasonably slowly as the problem grows. These results are evidence that the theory outlined in [1] is both essentially correct, and useful.

References

[1] Allen C. Ward and Warren Seering. Quantitative Inference in a Mechanical Design Compiler. memo 1062, MIT Artificial Intelligence Laboratory, November 1988.

[2] Allen C. Ward. *A Theory of Quantitative Inference for Artifact Sets, Applied to a Mechanical Design Compiler.* PhD thesis, Massachusetts Institute of Technology, 1989.

[3] David Serrano and David Gossard. Constraint management in conceptual design. In *Knowledge Based Expert Systems in Engineering, Planning and Design.* Computational Mechanics Publications, 1987.

[4] R. J. Popplestone. The Edinburgh Designer system as a framework for robotics: the design of behavior. *Artifical Intelligence for Engineering Design, Analysis, and Manufacturing,* 1(1), 1987.

[5] Kenneth L. Meunier and John R. Dixon. Iterative respecification: A computational model of hierarchical mechanical system design. In *ASME Computers in Engineering Conference.* ASME, 1988.

[6] Jack Mostow, Lou Steinberg, Noshir Langrana, and Chris Tong. A domain-independent model of knowledge-based design: Progress report to the National Science Foundation. Technical Report Working Paper 90-1, Rutgers University, 1988.

[7] Ernest Davis. Constraint propagation with interval labels. *Artificial Intelligence,* 32, 1987.

C377/085

The impact of intelligent information systems on designer efficiency

J VOGWELL, BSc, PhD and **S J CULLEY**, BSc
School of Mechanical Engineering, University of Bath

SYNOPSIS

Automated information systems are beginning to make inroads into the design office for aiding the selection of catalogue components. This gives the designer access to information quickly and efficiently and in a form which enables better design choices to be made. Such a package has been produced in which the first module covers selecting bearings. It has been distributed to a diverse range of industrial users for testing and evaluation. The results indicate that such a system will radically influence engineering design efficiency.

1 INTRODUCTION

A computer package has been developed for automating the selection of mechanical engineering components using data taken from manufacturers' catalogues. It is called CASOC (derived from Computer-Aided Selection Of Components) and has been produced specifically for engineers in the design office.

This software design aid has been produced by the Design Group at Bath University's School of Mechanical Engineering and the work has been sponsored by Technical Indexes Limited and by a Department of Trade and Industry grant.

The system runs on a micro computer and uses a novel mouse driven windows interface. It has been designed for both infrequent and regular use by either a novice or expert. The system performs the necessary design analysis and selection functions to identify all suitable bearings for a specified application. These may then be evaluated by the system to enable an optimum to be chosen.

The first completed module covers bearings and has recently been supplied to a range of design offices in industry for evaluation. The outcome of the site trials has confirmed earlier market research - that intelligent information systems will significantly improve design.

This paper describes the desirability of such component selection packages, gives salient details of the CASOC system and discusses the general response by users from site trials.

2   THE NEED FOR COMPONENT SELECTION SYSTEMS

With the trend in the design office towards aids for improving drafting efficiency, other essential tasks can also benefit from computerisation. It is well recognised, for example, that identifying a satisfactory catalogue component and verifying technical suitability for an application is usually a tedious task. Indeed, searching for information, particularly from catalogues, is recognised as taking up a high proportion of a design engineer's productive work time (Ref.1).

The amount of catalogue information which needs to be considered, though, is increasing due to ever increasing competition. This increasing pressure to use optimum components implies the need to consider larger numbers of alternative types, sizes and manufacturers. The problem is further compounded by the fact that :

(a)   manufacturers' product ranges, catalogue data presentation format and selection procedures vary significantly.

(b)   manufacturers are continually improving and diversifying their catalogue product ranges as new materials and designs are developed.

Since companies are finding it increasingly more difficult to attract good engineers into design so it becomes more important that available engineers are used effectively. Consequently, the efficient selection of optimum components from catalogues is essential. It has always seemed illogical to the authors that an engineer might be using an advanced CAD system for draughting, 3-D modelling or finite element analysis and yet would have to select a standard component by manually working through a catalogue.

3   CHARACTERISTICS OF COMPONENT SELECTION

Manually selecting standard components from catalogues tends to be very tedious and inefficient. This is for the following reasons:-

(a)   There are usually a large number of parameters that must be specified for an application.

(b)   Usually some analysis is necessary **prior to using the catalogue** so that parameters correspond with component data in a catalogue. Calculations may also have to be performed

using actual component data to confirm suitability in the application.

(c)   Graphical charts are frequently used for establishing design factors used in calculations. These are usually of an empirical nature and vary with manufacturer.

(d)   The design process by nature is iterative; a component is chosen then evaluated for suitability and the procedure repeated until a suitable version is found.

It is these characteristics (and the problems described in Section 2) which **distinguish catalogue component selection from conventional information retrieval applications** - such as searching library records, medical data or airline reservations data. Consequently, when the process is computerised the approach has to be significantly different.

4   BENEFITS OF COMPUTERISED SELECTION

Previous research (Refs.2,3,4) carried out at Bath University's Mechanical Design Department has shown that a well designed and implemented computer system for the selection of standard components has many advantages for the design engineer . These include :

(a)   A rapid **speeding up** of the selection procedure.

(b)   **The best solution is identified** - both from technical considerations and availability (since a readily available ex-stock version is more likely to be obtained when a range of solutions is generated).

(c)   Using a **common selection procedure** for all component types and manufacturers

(d)   Considering all component types and manufacturers **without bias** (versions which may not otherwise have been thought worth trying by manual means may prove to be the best available).

(e)   **Catalogue component use is encouraged** this is often desirable since the difference in performance between a **special** (perhaps a mathematical optimum) and a**standard** (a carefully selected catalogue version) can be very minor.

(f)   Improving **accuracy and reliability** of results since a computerised approach is used - manual selection can be complex and tedious and thus prone to human error.

(g)   Output **presentation quality is improved** and can be more comprehensive because both catalogue details and derived performance data of selected components are produced. These

are then displayed, for a range of technically viable
solutions, if required, in hard copy format.

5   PRELIMINARY RESEARCH

Over a period of some six years prior to commencing producing the
commercial version of CASOC, numerous stand-alone component
design analysis and selection programs were produced by the
authors. These dealt with gears, chain and belt transmission
components, plain and rolling element bearings, coil and disc
springs.

The authors' research covered numerous aspects of automated
component selection including :-

(a)   Evaluating commercial database packages for suitability.
      These largely proved inadequate, however, for handling the
      predominantly numeric data and carrying out pre and post
      search analysis that is necessary with engineering catalogue
      components as described in section 3).

(b)   Developing purpose designed databases for handling
      engineering catalogue components.

(c)   Establishing techniques for reorganising the irregular
      format presentation of catalogue component data into more
      manageable lists.

(d)   Developing techniques for handling the high proportion of
      graphical chart data into numeric form.

(e)   Producing graphical displays for presenting acceptable
      ranges of solutions so that users can best assess the
      virtues of alternatives available and make the optimum
      choice.

(f)   Considering whether it is preferable to consider assemblies
      of components together as a system or analyse components
      individually.

(g)   Applying optimisation and decision making techniques to
      standard component selection.

6   INVOLVEMENT WITH TECHNICAL INDEXES

Having produced a comprehensive range of prototype component
selection programs, the company Technical Indexes (**ti** as they are
commonly known) were approached, the programs demonstrated and
their potential discussed.

ti is the undoubted market leader in the supply of
information in compact form to all parts of industry -

electrical, mechanical, civil, chemical, defence, construction, and architecture. They have a turnover in excess of £7 million, a workforce of 160 people and a range of 150 products; the majority of which are produced at the company's headquarters in Bracknell.

The **ti** product range is predominantly microfiche or hardcopy based. Consequently they recognised that it was desirable to diversify into electronic delivery of information technology. The advanced structured search database concept being offered by the applicants was, therefore, timely and attractive.

To confirm the commercial viability of the concept, ti arranged presentations of the programs (demonstrated by the applicants) to numerous potential users from a diverse range of industry. They also carried out a market survey of their microfiche system users.

As a consequence of the response to the survey, **ti** decided to sponsor the production of a commercial system at Bath University. The system would include a comprehensive range of widely used manufacturers catalogues and component types. A DTI grant (under their Support for Innovation Scheme) was obtained to cover the necessary further research aspects of the work.

7 SALIENT FEATURES OF CASOC

The outcome of the ti/DTI sponsorship has been the generation of a highly sophisticated computerised design aid. Although space does not permit a detailed description of the package some of the important novel features will be discussed.

Whilst developing CASOC a primary objective has been to produce a system which is an asset to a users regardless of their experience analysing and selecting components or their level of familiarity with computer systems. Also it was considered desirable to supply CASOC either as a complete stand-alone system, or alternatively, to supply software to run on users' own equipment. To realise these aims a number of strategic decisions were taken which has resulted in the system having the following features :-

7.1 P C Based

To avoid the time delay difficulties and dialogue limitations associated with on-line mainframe database retrieval systems, CASOC operates as a stand-alone micro computer based system. For reasons of standardisation CASOC runs on an IBM PC XT or AT and compatible computers. A typical specification would be:-

Intel 80286 or 80386 processor
Memory of 640 k

Hard Disc of 20 MBytes
Crystal Frequency of 12 MHz (or greater)
Colour Graphics (EGA) and monitor
Intel 80287 or 80387 maths coprocessor
Serial or bus mouse.

CASOC will also run on the new range of PCs conforming to IBM's Micro Channel Architecture as incorporated in the PS/2 models.

7.2 Purpose Designed Windows Interface

To simplify the running procedure, CASOC operates under Microsoft Windows using specially designed windows which are accessed in an **open and non-hierarchial** way. The principal advantages are:

(a)   The system is extremely easy for the novice to use and fast and efficient when used by an expert. Indeed, it is only necessary to use the **mouse** and the **SHIFT, TAB, DEL** and **NUMERIC** keys to operate the full system.

(b)   Input data can be entered (and edited) in **any order**.

(c)   Technical information and help instructions can be **accessed at any stage** during running.

(d)   The range of satisfactory solutions is concisely and comprehensively displayed in an innovative components found matrix.

(e)   Solutions can readily be reordered into optimised lists based upon a number of performance criteria.

(f)   A common operating approach is taken with each component type and manufacturer. This is completely transparent to the user and is necessary since one major manufacturer alone has 13 different analytical techniques to be considered!

(g)   Built-in intelligence in input dialogue boxes prevents nonsensible information being specified.

7.3 Efficient Search Algorithms

In order that the response time is kept to an acceptable duration, considerable effort has been taken with coding. This has resulted in the following features being incorporated into CASOC:

(a)   Use of a multi-tasking environment on a microcomputer.

(b) A component indexing system to enable lists of catalogue component data to be reordered into ascending or descending lists to facilitate structured database search techniques.

(c) Use of binary searching techniques over a range of up to seven parameter fields.

These have been necessary because of the presentation order of much of the catalogue data. Unfortunately data (for example bearing dimensions, load ratings, limiting speeds etc.) do not appear conveniently in either exactly ascending or descending order.

7.4 Structured Data Entry Procedures

The organisation of an efficient procedure for extracting data as it appears in the different manufacturers catalogues and convert it into a consistent ASCII format for the database has been developed.

7.5 Modular Approach

CASOC has been designed so that it can be supplied to users in different forms - incorporating as many or as few manufacturers and component types as required. This has encouraged a totally modular approach to coding; which has been in the C programming language. In addition this has meant that all coding modules can be fully tested individually as well as when installed as part of a fully functioning system thus ensuring operating reliability.

7.6 Coding Structure

The coding produced for the CASOC System has used Jackson's Structured Programming (JSP) techniques throughout. The system comprises an **Input Stage** followed by a database **Search Stage** to identify suitable alternatives. An **Evaluation Stage** is then entered in which solutions can be ordered into optimised lists of ascending priority. Hardcopy details of solutions can be obtained at any stage after the search phase has been completed.

7.7 Inputs and Default Conditions

To assist the user when commencing a run, CASOC has a number of default conditions. If the user does not provide a value for an input parameter then the default value will be used at the search stage. Specifying an input value will over-write the default condition and the current state of data values is displayed in what is called the 'Client Area' window.

It is not necessary to provide values for all inputs. However, the effect of specifying various input values is to reduce the number of bearings searched thereby shortening the search time and the number of viable solutions found.

8 THE HIDDEN IMPACT

The impact of these systems on designer efficiency can be judged not only from the previous sections but also from Figures 1 to 4. Indeed, the comparison of bearing solutions for specified applications as described in the following sections would not be possible without a system such as CASOC. Undoubtedly, it is the combination of the extent of the search and its objectivity that enables solutions to be found that would otherwise not even have been considered.

8.1 Life Expectancy

The effect on the number of suitable bearings of varying size under life requirements varying from 1000 to 10 million hours is shown for the same radial load in Table 1. At greater life not only do the numbers of suitable bearings reduce but the numbers of suitable types also - but at different rates. The suitability of cylindrical roller bearings under just radial loading is well illustrated but clearly self-aligning bearings are restricted to a much lower life.

8.2 Radial Load Demand

At fixed life and increasing load conditions, a similar trend is seen in Table 2. Again it is the roller types rather than the ball bearings that are able to cope at the higher loads.

Although the general trends identified in these tables show roller bearings to be the most suitable under high life requirements with radial loads, other factors such as axial load and accommodation of misalignment will radically alter the outcome. Making an objective selection without an automatic selection system, though, is unlikely because of the difficulty of assimilating the effects of all factors.

8.3 Specific Type Performance

When considering different bearing types of identical size the performance behaviour varies significantly as shown in Table 3; ISO bearing 30210 has been chosen to illustrate this point. The difference in the load rating between ball and roller bearings is only about 1:2.5, however, when the resulting life is computed the variation becomes 20 fold (from 360 to 8500 hours). Consequently the designer is able to make a detailed comparison of bearing features and this can result in the specification of either:-

(a) a higher performance bearing within the same space envelope
or

(b) a smaller bearing (hence more compact and light) to a fixed performance requirement.

8.4 Variation in Manufacturers' Versions

When apparently similar bearings are compared from different manufacturers there can be significant variations. This effect is shown for a common ISO designation number taper roller bearing in Table 4 for five major manufacturers. There is as much as a 13% variation in dynamic load rating between the standard bearings but when the predicted life is computed and compared, under precisely the same operating conditions, this translates into a 41% variation!

The differences occur because bearing design inevitable is a compromise. Manufacturers, for example, will use different quality steels, production quality will vary and in addition, some will be more conservative with the performance figures specified in their catalogues than others.

The design engineer, though, must identify the most suitable bearing for an application using available information. Such impartial comparisons can only be carried out using a powerful selection system rather than use manual means or rely on a company representative.

9   RESPONSE FROM TEST-SITE USERS

Nine sites were chosen for the tests by Technical Indexes. The Bath University team developing the CASOC system had no say in the companies selected and this was considered preferable in the interest of achieving an unbiased response. The sites were selected because they represented a broad cross section of industry.

The findings of the site test reports was that generally everyone found the system a workmanlike and high added-value development. It was found easy to use and engineers almost unanomously seemed to like the graphical displays. Certainly the conventional aspects of bearing selection such as parameter specification and operational information definition was predictably the most successsful aspect of the system to these engineering users.

The less successful aspect of the system related to the most innovative part – the 'evaluation' and 'optimisation of acceptable solutions' stages. These are the areas which are new and so it has not been too surprising that they met with some uncertainty by some users. This clearly is an aspect of the system in which more emphasis is necessary for educating the user as to the potential benefits available so that it can be exploited.

10   CONCLUSIONS

It has been very clear from the previous research work undertaken, from the development of the CASOC system and by the response from test site trials that component selection systems will make considerable impact upon designer efficiency. The system module and examples referred to in this paper relate to bearings but the principles will apply to all types of engineering components.

Undoubtedly, for engineering information systems to be successful, as the site trials have shown CASOC to be, they must be totally acceptable to engineering designers. This implies that key features must be incorporated into the software. These include an easy to use, versatile and open interface, a means of presenting solutions in a clear, explicit and unambiguous manner, a procedure for evaluating alternatives in a logical and unprejudiced way. It is important also that the designer remains fully in control with the system presenting information enabling the user to make the final selection rather than the computer impose a version.

The effort which would be involved in manually generating alternatives is so great that it is not practical. This means that it will not be done and so this data is effectively hidden from the designer.

With available technology, it has only been possible to achieve the innovations described under a windows environment and with a system that is designed specifically to undertake this particular information retrieval activity.

11   REFERENCES

(1)   CAVE, P.R. NOBLE, C.E.I. Engineering Design Management, EMTA'86, University College, Swansea

(2)   VOGWELL, J CULLEY, S. J. Optimal Component Selection Using Engineering Databases. ICED'87

(3)   CULLEY, S.J. VOGWELL, J. HULME, G.W. The Use of Component Selection Systems for Automatic Component Selection. Effective CADCAM'87 IMechE, London

(4)   VOGWELL, J. CULLEY, S.J. Improving Design Efficiency by Automating Component Selection ICED'88 Budapest, Hungary

Radial Load = 10 kN
Speed = 2000 RPM
Number of Bearings = 5436

Bearing Type	Life (hours)							
	10^3	10^4	10^5	$5*10^5$	10^6	$5*10^6$	10^7	$5*10^7$
BALL: Deep Groove	190	64	6	0	0	0	0	0
Angular Contact	113	66	9	1	0	0	0	0
Self Aligning	95	34	0	0	0	0	0	0
ROLLER: Cylindrical	497	373	220	106	68	8	3	0
Taper	139	106	42	4	1	1	1	1
Spherical	146	124	80	35	12	0	0	0
TOTALS	1180	767	357	146	81	9	4	1

Table 1   Effect of Life Expectancy on Bearing Solutions

Speed = 2000 RPM
Life = 1000 Hours
Number of Bearings = 5436

Bearing Type	LOAD (kN)						
	1.0	5.0	7.5	10.0	15.0	20.0	25.0
BALL: Deep Groove	191	5	0	0	0	0	0
Angular Contact	113	7	1	0	0	0	0
Self Aligning	95	0	0	0	0	0	0
ROLLER: Cylindrical	523	220	135	68	16	3	1
Taper	148	42	10	1	1	1	0
Spherical	148	80	44	12	0	0	0
TOTAL	1218	354	190	81	17	3	1

Table 2   Effect of Radial Load on Bearing Solutions

```
                    bore diameter = 50 mm
                outside diameter = 90 mm
                        breadth = 20 - 23 mm
```

Bearing Type	Dynamic Load Rating, C (N)	Life (hours) (at 2000 RPM)	Relative Life
BALL:			
Deep Groove	35100	360	1
Angular Contact	37700	447	1.24
Self Aligning	23400	107	2.97
ROLLER:			
Cylindrical	62700	3788	10.5
Taper	70400	5573	15.5
Spherical	79900	8498	23.6

Table 3   Effect on Life of Different Bearing Types

```
        Bearing ISO Designation 30210 (except NSK)
                Bore Diameter = 50 mm
            Outside Diameter = 90 mm (88.9 mm for NSK)
                Thickness = 21.75 mm (20.638 mm for NSK)
```

Performance	SKF	FAG	NSK	TIMKEN
Load Rating(N) (dynamic), C	70400	75000	73085	66300
Load Variation	0	+ 7%	+ 4%	- 6%
Relative Life (10^6 Revs.)	1	1.23	1.13	0.82
Speed Limit (grease),RPM	4300	4000	4000	1364 → * 3547

* indicates only applicable with radial loads

Table 4   Comparison of Manufacturers' ISO Designation Taper Roller Bearings

C377/174

Machining centre design through component selection and specification

B R MOODY, MSIE and **R P DAVIS**, PhD, PE
Department of Industrial Engineering, Clemson University, Clemson, South Carolina, USA

A design methodology based on a dynamic programming
approach is developed as it specifically relates to
the design of machining centers. The resulting methodology includes configuration decomposition, component characterization, definition of specifications,
and a selection algorithm for configuration of a
center to met given specifications.

1 INTRODUCTION

The purpose of this research is to develop a formalized approach for the
selection of components for machining center design. This will result in an
approach which will yield a machining center configuration that will meet
given specifications. The methodology to be introduced will be based on a
dynamic programming approach to problem solving.

The methodology to be developed will not deal with the specific design
of individual components. Characteristics of components, as designed, will
be used in specifying different configurations of centers. These characteristics will not be altered as part of the methodology.

The perspective employed in this research will be that of the producer.
It is expected that such a methodology can aid producers in selecting components of a machining center to yield a purchasers desired requirements. This
methodology will be useful in that producers may use it to search for configurations from existing components to yield a design which does not require a
redesign of the entire machining center. It will not aid designers in the
actual design of individual components but in the selection of components
from those which already exist.

The foundation for this research lies fundamentally in two specific
areas of contemporary literature. They are: component selection and specification problems [5, 7] and machining center technology [2, 3, 4, 9, 10,
11].

2 DYNAMIC PROGRAMMING FRAMEWORK

The first characteristic of a dynamic programming approach is the ability to
decompose the problem into a series of smaller, interrelated problems. Each

of the smaller problems is a stage of the larger problem, and each stage will have a policy decision. This can be directly related to the design of machining centers. It has been recognized that machining centers can be described by the basic components which make up a center. Each component can be thought of as a stage of a larger problem -the problem of specifying and selecting a complete machining center configuration. The selection of a specific component from the component group (stage) is comparable to a stage decision.

The second characteristic is the definition of stage characteristics. In dynamic programming, each stage has states associated with it. Comparably, each component of a machining center has certain characteristics and features which define it. The problem of component descriptions will be discussed.

The third characteristic is stage coupling. At each stage of a dynamic programming problem the policy decision changes the current state into a state which relates to the following stage. This feature is also recognized in the designing of machining centers. Each time a component is selected, constraints are placed upon the selection of the next component. It also relates to the redefinition of the component as one which is not singular but includes the group of components which have been thus far developed (i.e., a subsystem). Due to characteristics of the previously selected component, constraints may be placed on which specific components may then be chosen in the next selection stage.

The fourth characteristic, sequential decisions, closely relates to the policy decision made in each stage. When using a sequential decision policy, the selection of an optimal component is dependent upon the component selected in previous stages. The selection of a component in a prior stage will place constraints on the selection of a component in future stages.

These characteristics of dynamic programming are most useful in developing a methodological approach to the design of machining centers. It allows for the optimization of component selection based upon a given criterion (or, criteria). It also places attention on the component characteristics and interfacing characteristics which can constrain the selection of components. Figure 1 represents the relationships discussed previously.

3 CONFIGURATION DECOMPOSITION

The first step of the methodology is to define the component breakdown of the product to be considered. Determining the level of decomposition which is required to adequately define the product must first be determined. Many levels of decomposition may be considered. The product should be decomposed first into recognizable modules which relate to the overall operation of the product.

The center configuration is decomposed into separate, recognizable components. The machining center component breakdown, which has been chosen, is given below:
1. pallet,
2. work table,
3. front base,

4. back base,
5. column,
6. automatic work changer,
7. spindle.
8. tool magazine,
9. tool changer, and,
10. control system.

This component breakdown is not extensive, but allows for an illustration of the methodology.

The main components of a product can be further decomposed into sub-assemblies, and then further into sub-sub-assemblies. This may be desired to allow for recognition of greater standardization and modularity of the product. For example, the above mentioned work table could be further decomposed by recognizing the following sub-assemblies of the work table:
1. rotary drive motor,
2. directional drive motor, and,
3. work table structure.

The question which must be answered is whether further breakdown into sub-assemblies will add any additional information to the component selection process. Another question is whether further breakdown will recognize a level of standardization of components on the lower level. If standardization cannot be recognized within the layer of decomposition, it may be advantageous to ''back-up'' one level of decomposition.

4   COMPONENT CHARACTERIZATION

After defining the product's components, it must be determined how these components will be described or characterized. When describing the components, they should be described in terms of the operational and the physical characteristics of the product, or the component if sub-assemblies are present. The methodology employed allows for defining these two types of characteristics for individual components as well as characteristics of the interfaces between the components.

Such physical characteristics as the dimensions of the component can be used to describe the component. For example, the physical dimensions of the pallet (the length and width) can be used to characterize the pallet.

In addition to physical characteristics of a component, its operational characteristics should also be considered. These operational characteristics relate to the actual operation of the component itself. For example, a motor's horsepower and its rpm rating can be considered to be operational characteristics. This information can be stored in tables for each component group. See Table 1 for a listing of the characteristics which will be used to describe the components of a machining center configuration.

5   SPECIFICATION LIST

The specification list is necessary to determine what characteristics are required to meet a given set of processing needs. When a consumer purchases a product, a given set of specifications are given which must be met for the machine to perform as required.

Some of the specifications which are given as requirements can be re-lated to specific components. For example, the dimensions required for fix-turing a work-piece would relate directly to the pallet size. The specifica-tions such as this, which can relate directly to a given component, must be mapped to each com-ponent so that individual component selections can be made.

The quantity of information which is given in the specification list can vary. The more control that purchasers are allowed to have over the characteristics of the final machine will mean that more specifications must be given. Also, as more components are recognized in the decomposition of a complete configuration, more information may be needed in the specification list.

6 SELECTION ALGORITHM

Having defined and characterized a machining center's components, an algo-rithm may be developed by which components are selected to develop complete configurations. The information which is required to describe the components may be stored in direct access files which are accessed by the selection algorithm program.

The general structure of the selection algorithm may be seen in Figure 2. The selection algorithm may be implemented through different techniques. As discussed earlier, the basic framework is that of dynamic programming. A decision tree analysis is one suitable method for solving the selection algo-rithm. The fundamental rationale employed in the selection algorithm is one of implicit enumeration with back-tracking.

7 INDIVIDUAL SCREENING OF COMPONENTS

To begin the selection algorithm, suitable component alternatives must be selected from each of the component groups. In Figure 2, it is shown that feasible component alternatives are found with the aid of component charac-teristic files. These files contain only basic information which can be compared to specifications, manipulated through equations, or compared to information which is a function of the specifications. The examples to be discussed will clarify some of the ways in which this component data and the specifications can be used to remove infeasible components from considera-tion.

Infeasible alternatives are removed from consideration by determining if the component can meet, and/or exceed, the requirements which are given. The specifications list (characteristic set 1) is compared to the component description data (characteristic set 2). The use of simple numeric compari-sons, equations, and zero-one relationships can all be used to determine feasibility of components. The order of selection is not important in this phase of the selection algorithm.

8 CONFIGURATION DEVELOPMENT

After all of the feasible components have been chosen, the task of configura-tion development must be undertaken. In this phase, those components which have been determined to be feasible will be placed together to allow for the development of a complete configuration. Figure 2 shows that configuration development is completed through the use of files which contain physical and operational interface characteristics.

After each component selection based on general physical compatibility (through characteristic set 3) is made, any additional physical or operational interface characteristics (characteristic set 4) will be considered. Much like the initial screening of components, simple comparisons of values or equations can be used to check compatibility of components as the configuration develops.

It should be noted that the selection algorithm develops configurations in much the same way as a decision tree is traversed. The decision tree shown in Figure 2 illustrates the manner in which a configuration is developed by advancing down the tree.

9 IMPLEMENTATION EXPERIENCE

A family of machining centers, produced by a U.S. machine tool builder, was employed to develop a database of component characteristics. Subsequently, in conjunction with this manufacturer, a set of typical application specifications was defined. Based upon these specifications, the methodology described in this paper was employed to generate a set of feasible machining center configurations.

In total, over 15 million possible configurations could be defined from the component database (of course, not all of these would be feasible for a particular application). Four feasible configurations were produced, using this methodology, with a microcomputer-based implementation. This output result was obtained in a matter of a few minutes total session time (i.e., from specification data input to generation of output results.)

10 BENEFITS OF RESEARCH

Several benefits can be recognized from this research. One of the most evident benefits is the ability to design a machining center for a given set of specifications based upon the charac- teristics of the individual machine components and their interfaces. A configuration can be defined, by the algorithm developed in this thesis, very quickly and easily.

Another previously mentioned benefit is the possible prevention of unnecessary redesigning of centers. Centers, and other products, are often designed from ''scratch'' when in reality a configuration which meets the specifications may be constructed from the existing components. This approach may reduce the cost of design.

Both of these benefits lead to another advantage of this approach. When using such a methodology, it is necessary to have adequate descriptions of all of the components. This in itself is an advantage since more is known about what components are available and their specific capabilities.

An important characteristic of the nature of the problem being discussed is that modularity and standardization is present among the components. The standardization of components provides an important advantage in applying this methodology.

REFERENCES

(1) Doyle, Lawrence E., Carl A. Keyser, James L. Leach, George F. Schrader, and Morse B. Singer. Manufacturing Processes and Materials for Engineers, Prentice-Hall, New Jersey, 1985.

(2) Friedmann, Axel, ''The Modular Fixturing System, A Profitable Investment!'', Proceedings of the International Conference on Advances in Manufacturing, Elsevier, North Holland, 1984, p. 165-173.

(3) Groover, Mikell P., Automation, Production Systems, and Computer-Integrated Manufacturing, Prentice-Hall, New Jersey, 1987.

(4) Jablonowski, Joseph, ''What's New in Machining Centers'', American Machinist, 1984, p. 95-114.

(5) Kilmartin, B.R. and R. Leonard, ''Selecting Advanced Tools by a Systems Approach Based on Key Machined Components'', Proceedings of the Institute of Mechanical Engineers, 1983, 197 (b) p. 261-269.

(6) Muck Rainer, ''Modular Mechanical Engineering - A Revolution in Engineering Industry'', Proceedings of the International Conference on Advances in Manufacturing , Elsevier, North Holland, 1984, p. 271-282.

(7) Nnaji, Bartholomew O., Computer-Aided Design, Selection, and Evaluation of Robots, Elsevier, New York, 1986.

(8) Reidelbach, John A., Modular Housing - 1971, Cahners Books, Massachusetts, 1971.

(9) Sanders, A.J., ''A New Concept of Machine Tool Construction.'' Proceedings of the Institute of Mechanical Engineers, 1983, 197 (b) p. 183-186.

(10) Strakeljahn, Lutz, ''Design Elements, Developmental Tendencies and Application of Modern Milling Machines, Machining Centers and Flexible Manufacturing Systems'', Proceedings of the International Conference on Advances in Manufacturing, Elsevier, North Holland, 1984.

(11) Wick, Charles, ''Advances in Machining Centers'', Manufacturing Engineering, October 1987, p. 24-32.

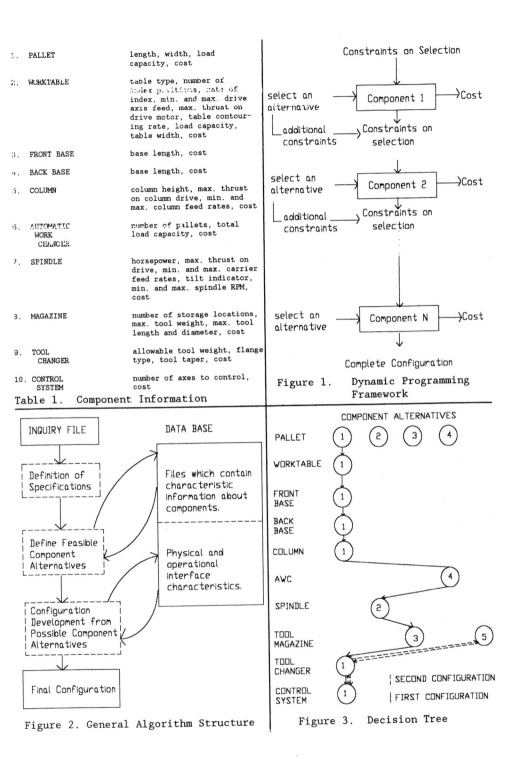

Table 1. Component Information

1.	PALLET	length, width, load capacity, cost
2.	WORKTABLE	table type, number of index positions, rate of index, min. and max. drive axis feed, max. thrust on drive motor, table contouring rate, load capacity, table width, cost
3.	FRONT BASE	base length, cost
4.	BACK BASE	base length, cost
5.	COLUMN	column height, max. thrust on column drive, min. and max. column feed rates, cost
6.	AUTOMATIC WORK CHANGER	number of pallets, total load capacity, cost
7.	SPINDLE	horsepower, max. thrust on drive, min. and max. carrier feed rates, tilt indicator, min. and max. spindle RPM, cost
8.	MAGAZINE	number of storage locations, max. tool weight, max. tool length and diameter, cost
9.	TOOL CHANGER	allowable tool weight, flange type, tool taper, cost
10.	CONTROL SYSTEM	number of axes to control, cost

Figure 1. Dynamic Programming Framework

Figure 2. General Algorithm Structure

Figure 3. Decision Tree

1305

Information systems for designers

W E EDER
Royal Military College of Canada, Kingston, Ontario, Canada

ABSTRACT

Starting from a definition and model of designing, the range of knowledge required by designers is developed. Various constituents of this knowledge are placed in the context of design science. Aspects of contents and presentation of this knowledge system for designers are discussed.

1. ASSURING PRODUCT QUALITY DURING DESIGN

One of the declared aims of designing an engineering product is to achieve "high quality". A companion paper (**1**) discusses the implications of the term "quality" for a product, and for the design process. The effects that various factors that influence the design process have on the resulting quality of the product are outlined. This paper contains an extended discussion of the nature of one of these factors, the information usable by and useful for the engineering designer, and the typical contents of a designer's information system.

2. DEFINITION OF DESIGNING

A definition of engineering design has been proposed as follows (**2**):

> Engineering design is a process performed by humans aided by technical means through which information in the form of REQUIREMENTS is converted into information in the form of descriptions of TECHNICAL SYSTEMS, such that this technical system meets the needs of mankind.

This statement about engineering design contains various implications, especially for information and knowledge. In order to be able to design effectively, designers must have:

(a) a knowledge (information and experience) about the needs to be fulfilled (the aims for designing), and about the requirements and constraints for the problem;

(b) other knowledge (information and experience) that a designer possesses, and enables that designer to propose means to satisfy those needs;

(c) externally available knowledge (including consulting) about various properties and constituents of the technical processes and systems.

Designing can be modelled as shown in figure 1 (**3**). According to this model, the design process consists of two major constituents, (a) a process, and (b) a set of operators that drive the process by exerting their effects onto the operand of the process. The quality of the product depends on the quality of each of the operators of the design process, AND on the quality of the information (inputs) to be transformed as operand, as well as on other factors that operate after designing has been completed (e.g. manufacturing, etc.).

The design process transforms the operand "information" from an input state to an output state. The inputs can be summarized as:

- information about the problem, specification by a customer, sponsor, or sales (internal contract),
- information about general and branch-specific knowledge;

and the outputs are:

- information about the (designed) technical systems to be manufactured,
- information about manufacturing, operating, maintaining, etc. for the life cycle of the technical system.

The design process is driven by the effects exerted individually or in a combination by the following operators:

- designers, individuals and teams, including design consultants for more specialized areas of knowledge,
- working means (tools, including computers),
- information about designing,
- management of designing, including goal setting and directing towards goal achieving,
- the active design environment.

3. WHAT KNOWLEDGE IS NEEDED

Knowledge for designing and about the design process can be classified under various headings (4), the resulting morphology is shown in figure 2. The most important are the two shown at levels 1 and 4. These provide the basis for a map of design-related knowledge shown in figure 3 (3,5). The four major quadrants of this figure (and the supplementary sector) may be characterized as follows:

A. Prescriptive knowledge – "know-how" – consists of two parts:

1. design knowledge related to the technical system to be designed:
   (a) knowledge about natural phenomena (as investigated by science in its narrower definition), – part of the operand in the design process;
   (b) knowledge about how to apply that science, and any other experiential knowledge to designing a technical system – another part of the operand;
2. design knowledge related to the design process:
   (c) knowledge about general strategic approach to designing, – a part of the operator "information system" of the process;
   (d) knowledge about tactics and methods for designing. – another part of the operator "information system"

B. Descriptive knowledge – theories – consisting of two parts

3. design knowledge related to the technical system to be designed:
   (e) knowledge about properties and constituents of socio-technical and technical system – a further part of the operand of the process;
   (f) theories of properties, especially mathematical models, derived by science (in its narrower definition) – a further part of the operand of the design process, but more removed than parts (b) and (a);
4. design knowledge related to the design process:

(g) knowledge about design processes (and theory) – a further part of the operator "information system".

(h) knowledge about using working means (tools for designing) – a part of the operator "technical systems" for the design process;

4. CONTEXT OF TECHNICAL SYSTEMS

A major need that is expressed within a social system is to transform some material, energy and/or information from an existing form into one that is more suited to human use within that social system. Such a transformation may be viewed as shown in figure 4 (3), as a model of the socio-technical system. This diagram obviously involves various components, including a process, humans, "machines" and tools in a conventional sense, information for operating the process, and management (including goal-setting and directing towards goal-achieving), and an environment.

The transformation process in this socio-technical system should make something (as an operand) more suitable for human use, to fulfill a need. This process is usually artificial, it does not exist naturally. It needs to be driven, which is the duty of the various operators: a human system, technical systems, information systems, and a management and goal system, working within an active environment.

It is obvious that our model of the design process (figure 1) is completely analogous to that of the socio-technical system. When designing, an appropriate process (and sequence of operations) to fulfill the recognized needs has to be selected from the various alternatives that are available (i.e. they either exist, or can be developed or invented). Tasks for driving the process must be allocated to the operators, particularly the human and technical systems, the main (executing) operators, according to their properties which are compared in figure 5 (3). For this purpose (and others), designers need a knowledge of ergonomics. And:

- IF the technical systems DO NOT EXIST – they must be designed,
- IF they DO EXIST – the technical systems must be selected from those on offer in the market.

5. CONSTITUENTS OF TECHNICAL PROCESSES

Processes have finer structures. All processes include three major constituents, figure 6, and may be sub-divided down to separate operations. Technical processes may be characterized in various ways, figure 7.

Their operands can be materials, energy, or information, or any combination of these. Materials and energy must be conserved, i.e. all inputs (transformed) must re-appear in the outputs. Information can be added to or destroyed. Processes also need various additional features that allow them to perform their main tasks, at all or more effectively. Typical constituents of technical processes may be classified as (see also figure 8):

- main stream of transforming the operands from an input state to an output state (preferably better suited to human needs);
- phenomena (technologies) that achieve the transformation, and require effects to be supplied by the operators;
- supply of signals, the regulating and controlling effects, control processes to determine, regulate, adjust and control the main process;
- supply of auxiliary materials;

- supply and conversion of energy, the propelling effects;
- connection and support processes.

6. PROPERTIES OF TECHNICAL SYSTEMS

Technical systems of many different levels of complexity exist, see figure 9. Even so, they have many features and properties in common, and this is the subject of the theory of technical systems (3).

Every technical system goes through a typical life cycle (compare figure 3 in (1)) during which the system is designed and manufactured (technical system as operand), used (technical system as operator), maintained, repaired and liquidated (technical system as operand). It must perform (or have performed on it) various operations, and must survive them in (or to reach) the desired state.

Consequently, each technical system has properties of various kinds that allow it to (or make it) perform and behave in a certain way. The properties are listed in figure 11, and the relationships between these properties are indicated in figure 10 (compare also figure 2 in (1)). The outer zone of figure 10 shows the properties that a customer or user wants and sees. The external properties belong to the technical system, and are what it visibly offers. The internal properties are only observable (and some may be measurable) by an expert who is familiar with such technical systems. The life-cycle stages are thus reflected in the properties of the TS. Each of these properties must be generated by a designer's work, by designing the whole system, i.e. by establishing the design properties, especially for each and every component in the technical system.

The knowledge about properties needed by designers to perform the tasks of designing is situated in the "western" half of figure 3. It contains, in the upper (NW) quadrant, the areas of "design for ... " related to each individual property. Such "design for ... " knowledge is backed by the knowledge of natural phenomena, labelled outside the circle in figure 3. This knowledge is closely related to the broad class of theories describing each property, which is part of the lower (SW) quadrant of this figure.

Typical internal tasks (as constituents) of technical systems may be classified as:
- output work effects that drive (a part of) the process;
- regulating and controlling functions;
- auxiliary (or secondary) functions (those not listed below);
- propelling functions (energy delivering);
- supporting and connecting functions;
- receiving functions, inward connections at the TS boundary;
- effector functions, outward connections at the TS boundary, including the ones that prepare the output work effects for delivery to the operands of the process.

The hardware of the technical system, supported by firmware and software (or their equivalents) must perform those functions.

7. DISCUSSIONS ABOUT DESIGNING

Designers establish the design properties (the innermost circle of figure 10) by an appropriate sequence of (conscious or unconscious, methodical or intuitive) steps or stages. In each step, they try to define the aims (or goals) to be achieved in that step, and then to define a range of means that may be suitable for achieving those goals. Typically, this is a divergent (one-to-many) transition, many (or at least a few) different means are available to achieve any one goal. The available means must then be criticized, evaluated, checked for possible improvements, etc., and the most promising selected for further processing, in a convergent activity. During this latter activity, some mathematical and analytical work is likely to be required.

A useful sequence of design stages (6) may consist of deciding (compare also figure 6 in (1)):

- what the future technical system must be able to do (a design specification),
- the process that must be performed to fulfill the needs,
- the effects that the human and the future technical system must deliver to drive that process,
- the technology and principles according to which the technical system could possibly do that task,
- the internal modes of action of the technical system, and major action sites for achieving those actions,
- overall concepts and arrangements,
- sketch and dimensional laying out, and
- detailing, including specifying parts lists, bought-out components, auxiliary materials, etc.

Each of these stages presents the goals for the next following stage (i.e. the next stage that is actually executed by the designer), and each stage supplies the means to achieve the goals set by its preceding stage. Only if each statement of goals at each level is of optimal quality, are the conditions for obtaining a solution of optimal quality present. A major pre-requisite for good designing is a good design specification.

Such a sequence (at least the useful parts of it, if the future TS is not a complete newly designed system) can enable designers to establish the design properties for technical systems of all levels of complexity, from large-scale systems, through machines, to assemblies, sub-assemblies, and components. There will, of course, be some differences in design processes for these different levels, but also many features in common. This is the subject of design science currently under development (2,4,5,6). The designer's tasks include imagining, creating (using "creativity"), procedural activities, intuitive leaps, analyzing according to an available theory to determine and optimize various parameters, evaluating, judging, diagnosing, deciding, and communication.

More optimal quality of the designed product is easier achieved by adopting a systematic design approach (1). A preferred approach to systematic design consistent with design science has been published (6), and has been shown to be useful as a guideline for designers' actions. Various case examples (7) exist as learning tools for this preferred approach.

8. FORMS OF PRESENTATION

Each property of the socio-technical system should be supported by knowledge about "how to design for" it, as indicated in the upper left quadrant of figure 3. This must include properties of technical systems, but also of human, social, economic, financial, ecological, political and many other forms of systems (compare also (8)). Many of these "design for ... " groups will be traceable to current scientific knowledge, which must be available in a verified and substantiated form for design use, e.g. (9). This "design for ... " knowledge will be substantially different in its form of presentation.

Designers are interested in (say) the theory of mechanisms, but only while they are analyzing a proposed linkage system, e.g. converging, for the purpose of evaluation. While they are in a divergent step, they are faced with generating various alternatives, e.g. linkages with different numbers of elements, other mechanisms using different principles and modes of action such as cams – either by literature search, or by mental experiment (an aspect of "creativity"). Currently available literature does not provide sufficiently good support for designers, it is usually "science-generic", categorized according to the phenomena and subject areas of conventional science and knowledge to fit into research fields and the scope of learned journals.

Designers typically ask questions about "how can I transmit an estimated force over an estimated distance?" They expect to find various principles (hydraulic, pneumatic, mechanical linkage, gear, wedge, electro-magnetic, etc.), each of which are capable of performing the task. Each of the principles should have adequate information about its properties, parameters, experience knowledge, mythology (conventional wisdom), technique, but also theories that describe the phenomena, so that designers can make their selections. A suitable form of presentation for much of this knowledge exists in the design catalogues (10), mostly in German language. Further collections of knowledge appear in some limited expert systems that are currently being developed. Such knowledge is "function-comparison" based, and may be labelled "design-generic".

The two forms of presentation, "science-generic" and "design-generic", complement each other.

9. CONTENTS OF INFORMATION SYSTEM

Following from the above discussions, the contents of an information system for designers must therefore consist of:

1. design knowledge related to the technical system to be designed:
   (a) knowledge about how to apply science and any other experiential knowledge to designing a class of technical systems:
   - design for all needed processes on operands
   - design for all individual properties of technical systems,
   - design to include all necessary constituents of processes and systems,
   - design for all life stages of a product,
   - state of the art in the field,
   (b) knowledge about natural phenomena, including science in its narrower definition, but also including techniques, experience-derived knowledge, a mythology (conventional wisdom) of prior applications, standards and laws, knowledge and components available in the market-place, etc.
2. design knowledge related to the design process:
   (c) knowledge about general strategic approach to designing,
   (d) knowledge about tactics and methods for designing,
   (f) knowledge about design processes (and theory),
   (g) knowledge about using working means (tools for designing).

Much of this knowledge has two major elements for designers: (A) awareness, (B) the details. There are also two main ranges of knowledge of which designers must have at least "awareness": (a) immediately useful for the class of technical systems to be designed, and (b) peripheral, trends, "nice-to-know", current awareness, state of the art in related fields, etc.

Awareness, and the main implications of the related subject-matter, must exist as a designer's learned knowledge. Whether the details of this knowledge are made available as a set of manually arranged documents (paper, journals, books, etc.), available expertise from consultants, or on a computer (as a data base) is not critical. Some of the information dates quickly, in which case a computer format would be advantageous for easy up-dating. Other information changes only slowly (including most national and international standards), the danger of having "only" an out-dated hard copy available is then less.

In any case, the information system needs adequate techniques, means and media for recording, updating, interrogating, retrieving, sorting, transmitting, processing, etc.

Making this range of knowledge available and easily accessed for designers is the main purpose of "knowledge engineering". The tasks and investigations of artificial intelligence and expert systems are a sub-set of knowledge engineering. (In parentheses, the people involved in AI and expert systems have taken this term in a narrower sense, and thus distorted it to refer only to computer applications)

10. IMPLICATIONS

The context and contents of both the designer's personal information system, and of external information systems that support the designer, has many implications in two distinct directions – industry, and engineering education.

For industry (including government), adequate information sources must be urgently developed along the guidelines discussed above. The quality of products and services on which the country depends for its future survival critically depends on these information sources, as one of the vital operators of the design process.

For education, the implications are more severe – whole curricula need to be revised to provide direct education and training for future designers. Current course structures are aimed largely at producing researchers and specialists. Designers are different, tending to be generalists with broad and comprehensive awareness about all matters pertaining to the creation of products.

11. REFERENCES

(1) Hubka, V., Design for Quality, in **WDK 18 – Proceedings of ICED 89** (in preparation)

(2) Eder, W.E., Design Science - A Survey of Some Approaches, in **Engineering Focuses on Excellence**, Proc. ASEE Annual Conference 1987, p. 668-674, Washington D.C.: ASEE, 1987

(3) Hubka, V. & Eder, W.E., **Theory of Technical Systems**, New York: Springer-Verlag, 1988

(4) Hubka, V. & Schregenberger, J.W., Paths Towards Design Science, in **WDK 13 – Proceedings of the 1987 International Conference on Engineering Design**, p. 3-14, New York: ASME, 1987

(5) Hubka, V. & Schregenberger, J.W., Eine Neue Systematik Konstruktionswissenschaftlicher Aussagen, in **Proceedings of ICED 88**, p. 103-117, Zürich: Heurista, 1988

(6) Hubka, V., **WDK 1 – Principles of Engineering Design** (translated and edited by W.E. Eder), Zürich: Heurista, 1988 (reprint, originally published by London: Butterworths, 1982)

(7) Hubka, V., Andreasen, M.M., & Eder, W.E., **Practical Studies in Systematic Design**, London: Butterworths, 1988

(8) Hubka, V. & Eder, W.E., Special Design Knowledge, in **Proc. Information for Designers 1988** (in press)

(9) ESDU Index (published annually), Engineering Sciences Data Unit, London, England

(10) Roth, K.-H., **Konstruieren mit Konstruktionskatalogen**, Berlin/Heidelberg: Springer-Verlag, 1982

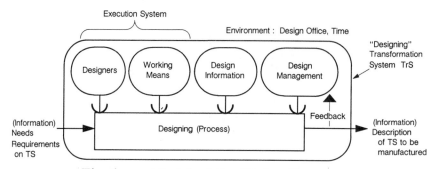

Fig. 1   Model of the Design Process

		State of Embodiment		
	Characteristic	A	B	C
1	Methodological Category of Statement	Primarily descriptive (d–statement)	Primarily prescriptive (p–statement)	Normative (n–statement)
2	Empirical Support for Statement	Pre–scientific (practice – experiences)	Scientific singular "understanding"	Scientific statistical (inductive)
3	Intended Recipient of Statement	Novice, student	Teacher, researcher	Practitioner, designer
4	Aspects of Designing (interest filter)	The technical system (in usable state)	The design process itself	
5	Range of Objects as Subject of Statement	Universal: all artificial real and process systems	Technical real systems = TS	Branch–specific objects: machine systems, building systems,etc.
6	Declared Aims of Researcher	Automation of parts of the design process	Better empirical foundation for methodology	Others
	Any given statement of design science may be characterized by combining the appropriate states of embodiment from each characteristic.			

Fig. 2   Morphology of Statements on Design Science

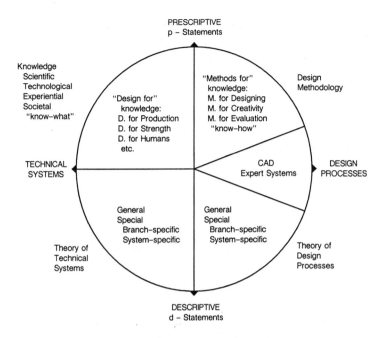

Fig. 3   Two Dimensions of Design Science

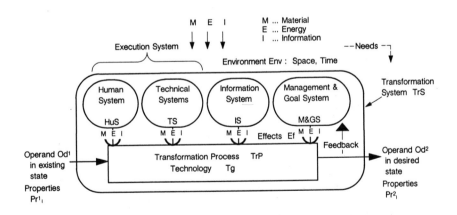

Fig. 4   Model of the Transformation System -- Elements

Human	Technical System
– is capable of reaching important decisions based on limited information	– works only according to given orders
– correct reaction even under unexpected conditions	– unexpected (by the human) conditions can lead to disaster
– can perform certain operations in various ways, important when the mechanism is damaged	– number of operations performable by a TS is limited
– is flexible in work programming	– program changes are usually difficult and costly (exception : computers)
– capacity of information reception is limited; reception rate for information is limited	– capacity of information channels can be readily increased
– limited capability of (receptor) sensing organs	– possibilities of increasing sensing parameters
– power and attention reduce with time	– almost constant power and precision
– working capability requires certain conditions : temperature, humidity, pressure, noise	– can be made for any environmental conditions
– thinking operations relatively slow; large probability of errors	– rapid execution of logical operations almost without error
– limited memory and recall capabilities	– almost unlimited information storage; limited by ready access to stored information
– information can be complex; can readily combine and interpret	– form of information basically simple; needs full instructions to combine or interpret information

Fig. 5   Characteristic Properties of Humans and Technical Systems

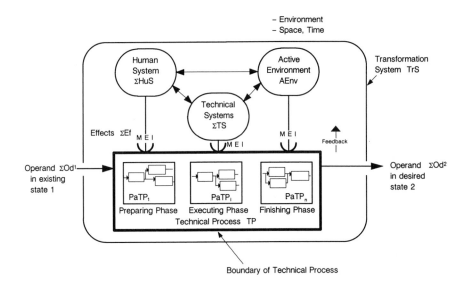

Fig. 6   TP Model -- Model of the Technical Process

Fig. 8 — Systematics of Technical Processes

Classifying Aspects	Classes of Process	Comment
Operand	– processes that treat material – processes that treat energy – processes that treat information – processes that treat biological objects	
Phenomena to achieve Working Effects	physical processes – mechanical – electrical – thermal, etc. – chemical processes – biological processes – combined processes	
Working Effects	– transport process – sorting process – comminution process – manufacturing process – assembly process	
Type of Propelling Effect	– processes with manual work – processes with animal power – mechanized processes	
Type of Regulating and Controlling Effect	– human regulated and controlled – automated – computerized (programmable) – hybrid	
Level of Complication of Process	– operation – partial process – complicated process	
Relationship between Input and Output	– combining process : number of inputs > outputs – dividing process : number of inputs < outputs	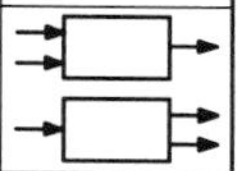
Progress of Transformation	– continuous process – discontinuous process (discrete, batchwise)	

Fig. 8   Systematics of Technical Processes

Fig. 7 — Characterization of Technical Processes

Characteristic Group	Examples of Characteristics		
	Technical	Economic	Planning
Operand	Material, dimension, form, surface finish, pressure, temperature, and other parameters	Price, costs	Production quantity, delivery deadline, supplier
Technology (Working Procedure)	Specification of operations, sequence of operations	Costs	Working time, worker, workplace
Worker	Technical knowledge, experience, personal character properties	Labor costs	Work period hours, number of workers
Technical System	Functional properties, operational properties, appearance properties, ergonomic properties, distribution properties, and others, see fig. 5.1	Price, Operational costs	Delivery deadline, quantity, supplier
Technical Information	Information index, information source	Costs	Deadlines, examiner, workplace
Regulating and Controlling Processes	Organizational system, planning system, remuneration system, administration	Costs	Deadlines, examiner, workplace
Conditions of Environment	Physical – space arrangement, – space requirements, – temperature – humidity – light – noise	Costs	Deadlines, worker, workplace
	Psychological – working climate		
	Social – political situation	Economic situation	

Fig. 7   Characterization of Technical Processes

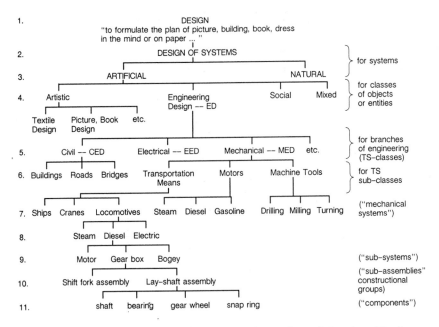

Fig. 9   The Systems Family -- Hierarchy of Design Tasks

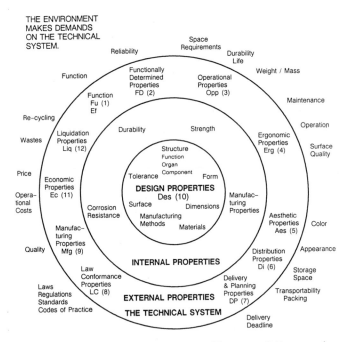

Fig. 11   Relationships between Classes of Properties

Abbreviation	Class of Properties	Questions about Class	Groups or Examples of Property Class
Fu (1) Ef	Function Effect	What does the TS do ? What capability does the TS have ?	Working function Auxiliary function Propelling function Controlling & Regulating function Connecting function
FD (2)	Functionally Determined Properties	What conditions are characteristic of the function ?	Power Speed Size Functional dimensions Load capacity
Opp (3)	Operational Properties	How suitable is the TS for the working process (operation) ?	Operational safety Reliability Life Energy consumption Space occupation Maintainability
Erg (4)	Ergonomic Properties	How is it to be operated, and what influence does the TS have on human beings ?	Operator safety Way of operating Types of secondary outputs Requirements for human attention
Aes (5)	Aesthetic Properties	What influence does the TS have on human sensory feelings ?	Form Colour Surface distribution Surface juxtaposition
Di (6)	Distribution Properties	How suitable is the TS for transport, storage, packing ?	Transportability Storage suitability Packaging suitability Suitability for commissioning
DP (7)	Delivery and Planning Properties	When can the TS be delivered ? Manufacturing quantity	Delivery capability Quantity production One–off production
LC (8)	Law Conformance Properties	Does the TS conform to laws, codes of practice, standards ?	Standardized Patent clearance Legal
Mfg (9)	Manufacturing Properties	How suitable is the TS for manufacture ?	Manufacturability Assemblability Manufacturing quality
Des (10)	Design Properties	With what are the external properties ?	Structure Form, Shape Dimensions Materials Surface quality Tolerances Type of manufacture
Ec (11)	Economic Properties	How economic is the working and manufacturing process ?	Operating costs Manufacturing costs Effectiveness Price Manufacturer
Liq (12)	Liquidation Properties	How easy is the TS to liquidate ?	Re–cycling Danger of wastes

Fig. 10   Classes of Properties of Technical Systems

C377/238

Design for quality

V HUBKA
ETH, Zurich, Switzerland

ABSTRACT

The term "quality" is discussed. Factors that influence the quality of a product during the design process are derived from the general model of the design process. These factors are explained in more detail. Design quality is contrasted to the concerns of ISO 9000 series, which relate more to production and management.

1. ASSURING PRODUCT QUALITY DURING DESIGN

One of the aims of designing is to achieve "high quality" in a product or process. In this paper we aim to investigate the ways and means by which the quality of a product can be influenced during the design phases, in order to derive a basis for a system for quality assurance in design.

2. THE TERM "QUALITY"

As first task we must reach a consensus about the meaning of the term "quality". It is used in various contexts, but always in connection with evaluation of a product or process. We will use the interpretation of a DIN-standard as basis for discussion.

DIN 55350 defines quality as the totality of the properties and characteristics of a product or activity that relate to its suitability to fulfil the given requirements. Quality thus concerns statements about the "what" and "how" of an object or process, a total judgement about a system of properties that make an object what it is, and in what ways it is different from other objects.

It is useful to also consider the expression "high quality". On closer analysis one can see that a combination of conditions that are regarded as "high quality" for one product may not be applicable to another product. This expression is therefore a relative judgement. A more appropriate expression is the "optimal quality" for a given set (or system) of conditions.

3. FACTORS THAT INFLUENCE OPTIMAL QUALITY

The factors in the phases of engineering design that influence the optimal quality of a future product may easily be derived from the model of the design process, figure 1 (**1**). This model shows the following components:
(a) Form 2 of the processed information: a full description of a technical system that completely fulfils the given requirements, as the goal (output) of the design process;
(b) Form 1 of the processed information: the given requirements (design specification) as the input to the design process;

(c) a design process, during which a transformation of information from Form 1 (requirements) to Form 2 (description of a technical system with appropriate capabilities and properties) takes place;

This transformation is realized or influenced by the following six factors (as operators of the process) which are also components of the transformation system "engineering design":

(d) the engineering designer (usually as member of a design team);

(e) the working means available to the designer (tools, etc.);

(f) the existing technical knowledge;

(g) the management of the design process;

(h) the active environment in which the process takes place; and

(i) the design process itself, the procedures and techniques employed by the designer, the technology of the process.

These influencing factors decide whether the design process can be successfully completed at all, what quality the result (output) of the process will have, and what parameters of the design process (e.g. design costs and duration, committed costs of the proposed technical system) will be attained.

We have in this discussion considered exclusively the quality of designing – we have analyzed the prospect and probability of how successfully the given list of requirements can be fulfilled by the proposed technical system (i.e. the maximum that can be achieved after manufacture of the product or implementation of the process, if no further changes are introduced during manufacture and development of the system). A further (and primary) consideration must be the quality inherent in the task that is set for designing – the quality of the list of requirements (design specification). This element of the model is also one of the influencing factors to be considered.

The questions about how an optimal quality of each of the stated factors can be assessed, and by what means it can be supported and enhanced, are the subjects of the following sections of this paper. The permitted length of this paper only allows us to show the possible range by a few examples.

4. ERROR SOURCES AND ERROR AVOIDANCE

When striving for quality, one must avoid as many faults and errors as possible – errors are usually the causes of "the wrong". We could have used the term "mistake" as synonym, but this word is used in pedagogy to reflect lack or wrong usage of knowledge about certain facts, whereas an error results when one or more of the three main functions of working life fail – attention, memory and/or thinking.

A further distinction must be drawn between **errors in the product**, for instance as indicated by Hansen (2), and **errors made during designing** which are the causes of many product errors. Our discussion concerns particularly the second of these categories, errors during designing.

The quality system under consideration must make adequate provisions for preventing errors, particularly in the following areas:

- technical (specialist) knowledge - no errors, no gaps
- attention - always "fully with it"
- memory - no tiredness, stimulated, no forgetfulness

• thought, procedure - no thought errors, thorough.

The type of means that may be applied to avoid errors will be discussed in the sections dealing with individual factors that influence the design process.

5. QUALITY OF THE LIST OF REQUIREMENTS (DESIGN SPECIFICATION)

A list of requirements must be:

• complete: all requirements (and constraints) must be explicitly formulated;
• correct: the measure (quantity, size, value) of all requirements must conform to the desires of the customer and the state of the art;
• quantified: the requirements must if possible be quantified and permitted variations (tolerances) stated;
• qualified: the importance of each requirement must be clearly visible, e.g. fixed requirement, minimum requirement, desire, etc.;
• formalized: appropriate forms of presentation, use of proforma sheets can improve overview and avoid faulty interpretation.

In particular, the completeness and qualification of the requirements are (and will continue to be) points of contention, because opinions and conceptions about them diverge. We are concerned about a few hundred statements of requirements that must be formulated at the start of a product's life, in and before the conceptual phases of designing. It would be useless to stir a controversy at this point based on some well-known starting-points. We present here a few important insights from the Theory of Technical Systems (1) as a range of beneficial knowledge towards solving this problem.

Figure 2 represents a technical system (TS) by means of its classes of properties. The blocks on the circumference show the external properties of the technical system that appear as the interface to the user and customer. The classes of properties are formed on the basis of two aspects, (a) with respect to the suitability of the technical system to withstand each separate "life phase" of existence, and (b) important considerations which can not be directly allocated to the life phases, e.g. function, law and standards conformance, environmental, quality assurance, etc.

The life phases of a technical system are illustrated in figure 3, which also shows the influencing factors (operators) of each process taking place during the life of a TS. Regarding the properties of a technical system that make it suitable for these life phases, the most important properties (see figure 2) are:

• suitability for realization, including planning and work preparation
• suitability for distribution
• suitability for operation, including repair and maintenance
• suitability for liquidation.

Within these individual classes shown in figure 2, a few characteristic properties are listed.

The inside area of figure 2 contains three classes of properties which are labelled internal properties. Most users of a product will regard it as a "black box" and will never make contact with these properties. Engineering designers have the task of establishing the individual properties in these classes during their design work, they are the causes which have the external properties as their consequences. All properties are directly or indirectly dependent on the class labelled "elementary design properties". For instance, reliability of a product depends (among others) on strength and corrosion resistance, which in turn depend directly on form, dimensions, material, surface characteristics, etc. of the individual components.

This example also demonstrates that figure 2 also shows the relationships among these classes of properties. The quality of the internal properties therefore are the cause of the external properties. Central to these classes are the elementary design properties that are the ultimate cause of all other properties.

These insights may be used as basis for generating various check-lists or catalogues that can be used by designers as part of their technical knowledge and as methodical aids for designing. They relate particularly to aspects of the quality of the list of requirements. The literature contains many other works that treat this set of problems (e.g. **3,4,5**), space does not permit further discussion here.

6. QUALITY OF DESIGNERS

A description of the quality of designers contains the same elements as their job description (profile of requirements). Such a document serves also to derive the learning objectives for education in engineering design, particularly for the areas of knowledge and ability. It is obvious that the job profile of design engineers is substantially different from that of detail designers and draughtspersons.

The profile of requirements for engineering designers must be derived from the characteristics of the design process. Results of this derivation have been reported (**6,7**), some of which are summarized in figure 4 (a list of requirements) and figure 5 (some bases for evaluation). Details may be found in the quoted references, with some further developments reported in (**8,9**).

The details of requirements placed on designers can show strong dependence on other factors (operators) of the design process, particularly in the areas of their knowledge and abilities. Such more stringent requirements relate especially to the available working means (including computers) and specialist technical information. Factors (operators) such as management and environment of the design process are usually reflected in requirements for personal properties of designers, especially attitudes.

Whether the design work has to be performed by an individual designer, or a design team, plays a large role for evaluating the qualities of designers.

7. QUALITY OF WORKING MEANS

The working means available to designers support them in their activities during the design process. Various classes of working means may be recognized:

(1) means for handling information – storing, classifying, relating, arranging for easy access, retrieving;

(2) means for the work of modelling and representing – i.e. for all possible representations of technical systems and their properties by graphical and other techniques;

(3) means for calculating and analysis;

(4) means for conventional office work – writing, communicating, dictating, storing/filing, arranging for easy access;

(5) means for reproducing – copying, printing, enlarging, reducing of drawings and documents;

(6) means for handling drawings (originals and copies);

(7) means for experimenting, technical testing, etc. – measuring instruments, test apparatus, simulation apparatus.

Working means of varying complexity can be employed in the different areas of activity, from the simplest tools, to "intelligent" computers that can perform some activities by themselves (automatically).

All of these working means belong to the family of technical systems, and must therefore fulfil requirements and possess properties that are similar to those described in a previous section of this paper (see figure 2). The class of ergonomic properties, related to the area of psychology, is of great importance for these working means.

8. QUALITY OF PROCEDURE IN DESIGNING

When considering questions about the structure of the design process, a prime discovery is that the contributing (partial) processes are extremely varied in their complexity, even down to the elemental operations which are repeated many times during designing. In addition, the type of thought processes that take place show different character from various aspects. Some belong to the difficult ones bordering on creative, others are merely routine tasks. A particularly important distinction is according to the way in which thought processes are used:
- the discursive way, using individual and distinguishable steps of thought;
- the intuitive way, using contemplative thought and leaps.

From this aspect we can distinguish two basic kinds of design process:
- intuitive: the traditional process with mainly intuitive work;
- discursive: methodical, with mainly discursive and planned working methods.

Methodical design can be characterized by:
- division of the total procedure into recognizable smaller processes (steps with clearly defined aims);
- formulation of variants;
- evaluation and optimization at various levels.

The planned and controlled way to achieve quality in a product can only be accomplished by using methodical design procedures, such as the one shown in the procedural plan in figure 6 (10). Preparing the necessary report in order to verify and confirm reasons for the various decisions is only possible with methodical procedures.

9. QUALITY OF DESIGNER'S SPECIALIST (TECHNICAL) INFORMATION

Technical information has a dominant place for the quality of design. We define the total of all knowledge that is needed for designing as the designer's specialist information. The designer's quality is determined by every subject that the designer has at his or her command, and is competent in. The relevant range of knowledge, and classification systems, etc. are beyond the scope of this paper, some indications may be found in (11,12).

The quality of designer's specialist information may be assessed with the help of the following criteria:
- completeness of the existing information system?
- are contents correct and reliable (correct vs. wrong)?
        are contents complete (not only excerpts without connective relationships)?

what is the range of validity of the information?

are the contents verifiable (are sources traceable)?

- is the form of information clearly understandable and unique (sharpness of information)?

is the form logical?

is the form lucid and usable (quick orientation, density)?

is the form short and comprehensible (without repetition, redundancy)?

- age of information (is it still valid)?
- volume and complexity of information?
- type of information carrier (medium) or document?
- compiler or producer of information (publisher)?
- availability of information (possibility of and permission for use)?

In a concrete instance the specialist technical information that has been used for designing a product is judged as a personal information system.

At present, the form of the available information is a quality criterion that has not been adequately acknowledged. This leads to a newly recognized area of formalized knowledge: design for properties, i.e. design to achieve certain properties, namely those shown in figure 2 – design for manufacture, maintenance, packaging, etc.

10. QUALITY OF MANAGEMENT (FOR DESIGN)

The ultimate aims of designing as a process may be formulated as follows:

- the designed technical system has optimal quality (with respect to all properties);
- the design process is quick (short design time);
- the design costs are low.

Quality assessment according to these criteria is basically correct, but because it tends to be *post hoc* (after the design work, and possibly manufacture, have been completed) it is too late. It is the quality of the causes for these criteria that contribute to achieving the stated aims. Such causes can be derived from the functions of management (leadership), and formulated as the following quality criteria:

- construction and constitution of the working system (designers and working means);
- establishing the aims and methods of progress, as well as controls and checks;
- acquisition, documentation and transmission (communication) of information (internal and external);
- leadership and motivation of personnel (working climate);
- organization and application of technical means;
- level of specialist knowledge in designing;
- general organization and methodology of leadership activities (capabilities for planning, deciding, executing and controlling).

The special character of the design process and its activities, which are substantially different from most other processes, demand a particular style of leadership and management. Consideration of all operators of the design process and their optimal use and application play a decisive role in managing design. One of the most important factors is the designer's specialist knowledge. Design managers must possess broad design knowledge, so that they can exert expert control and advice at the design work-station (drawing board, CAD terminal, etc.) (13,14).

11. QUALITY OF THE ACTIVE ENVIRONMENT

Design processes do not take place in a vacuum, they are directly influenced by a particular environment in which they occur. A macro-situation and a micro-situation can be recognized in this environment. The macro-situation can hardly be influenced and changed, one can only attempt to ameliorate or guard against its negative effects.

In contrast, the micro-situation is an important factor that can and must be actively formed. It significantly influences the operators of the design process discussed above, but also has a marked influence on the main working functions of designers, e.g. their attention. The major considerations are:

- position of working rooms (areas) in the business;
- size of the offices (large or small);
- equipment and its arrangement;
- illumination: intensity and type;
- climatic conditions: temperature, humidity, air movement;
- noise: frequency, loudness, type (music);
- relationships within the working group: psychological working conditions;
- colour: cold or warm tones.

With the possible exception of the last of these criteria, all of these can be summarized under the term "physical working conditions". Each one represents an important evaluation criterion for the quality of the working environment (**15**).

12. OVERVIEW AND WEIGHTING OF INFLUENCING FACTORS

In the preceding sections of this paper we have analyzed the individual operators of the design process, and shown some of the important criteria of quality for assessing these operators. Figure 7 shows a weighting of the factors that influence the quality (of design) of technical systems, and for other aims of the design process. A four-step scale is indicated which characterizes directly important, dominant, significant and insignificant influence of an operator on the states aim.

13. GENERAL ASSURANCE OF QUALITY

Consistent assurance of product quality is no new invention. Means and measures for ensuring and increasing the quality of a product have been sought and operated particularly in manufacturing. From recent investigations it appears that only a small proportion of quality problems are traceable to manufacturing – about 10 - 15 adopted the task of building up an integrated quality system. National institutions (e.g. SAQ – Swiss Working Group to Promote Quality, DGQ – German Society for Quality) have treated the range of problems to generate national standards (e.g. Swiss standard SN-029100 - Requirements for Quality Assurance Systems). The current 9000-series of ISO standards is now internationally obligatory.

The standard ISO-9000 prescribes three different forms of "functional and organizational capability" for quality systems. Only in one of these forms must the supplier provide assurance for quality in several phases of product life, "which may include design/development, production, installation and servicing".

According to point 4.4.1 (**16**), the "supplier shall establish and maintain procedures to control and verify the design of the product in order to ensure that the specified requirements are met." The supplier must therefore draw up development plans and provide appropriate documentation and testing procedures for design/development and the results from these activities.

These duties, it must be remarked, have a close relationship with law. Some countries hold the supplier or manufacturer liable not only for the capabilities of the product, but also for secondary damage if the fault occurs because of deficient design, manufacture or operating instructions.

14. COMPARISON OF QUALITY SYSTEMS

The quality system introduced in this paper contains more technical knowledge, rather than the organizational means and measures that form the contents of the ISO quality system outlined in the previous section. Our system uses a consistent application of design science (based on the theory of technical systems) by which the design process becomes transparent. It represents a new departure for designing which is characterized by a transition from empiricism to a scientific base. The scientifically based working methods carry further effects, particularly an extended and goal-oriented use of computers during designing.

All requirements of quality assurance according to ISO 9000 such as planning, documenting, providing bases for lay-out, controlling and verifying, etc., are explicitly or implicitly included in design science, and particularly in the design methodology derived from it (compare figure 6). The strength of the ISO system may be found in its formalizations, and particularly in the continuity of assurance throughout all processes. Design science can be seen as the preventive means for promoting quality, and can serve for avoidance or early recognition of quality-related problems.

Design science should thus be recognized as a forceful tool for assuring quality. The common goal of product quality can be achieved more quickly and effectively by cooperation of all participating institutions.

REFERENCES

(1) Hubka, V. & Eder, W.E., **Theory of Technical Systems**, Berlin & New York: Springer-Verlag, 1988

(2) Hansen, F., **Konstruktionssystematik**, Berlin: VEB Verlag Technik, 1965

(3) Hubka, V., "Klärung der Aufgabestellung", **Schw. Masch.** Vol. 76 (1976), Nr. 3 & 13

(4) Tjalve, E., "Formulierung der Konstruktionsziele", **Schw. Masch.** Vol. 77 (1977), Nr. 36

(5) Pahl, G., "Klären der Aufgabestellung", **Konstruktion** Vol. 24 (1972), Nr. 1

(6) Hubka, V., **Theorie der Konstruktionsprozesse**, Berlin: Springer-Verlag, 1976

(7) Hubka, V., **Konstruktionsunterricht an Technischen Hochschulen**, Konstanz: Leuchtturm Verlag, 1978

(8) Hubka, V., "A Curriculum Model Applying the Theory of Technical Systems", in W.E. Eder (ed), **Proceedings of the 1987 International Conference on Engineering Design**, New York: ASME, 1987, p. 965-976

(9) Eder, W.E., "Education for Engineering Design – Application of Design Science", **Int.J.Appl.Engng.Ed.** Vol. 4 (1988), No. 3, p. 167-184

(10) Hubka, V., **WDK 1 – Principles of Engineering Design**, Zürich: Heurista, 1988 (translated and edited by W.E. Eder, originally published by London: Butterworths,

1982)

(11) Hubka, V. & Schregenberger, J.W., "Eine Neue Systematik Konstruktionswissenschaftlicher Aussagen – ihre Struktur und Funktion", in V. Hubka, J. Bara'tossy & U. Pighini (eds), **WDK 16 – Proceedings of ICED 88**, Zürich: Heurista, 1988, p. 103-117

(12) Eder, W.E., "Information Systems for Designers", in **WDK 18 – Proceedings of ICED 89** (in preparation)

(13) Hubka, V. (ed), **WDK 11 – Führung im Konstruktionsprozess - Reading,** Zürich: Heurista, 1985

(14) Hubka, V., "Planen der Konstruktionsarbeit", **Schw. Masch.** Vol. 80 (1980), Nr. 50 & 53

(15) Hubka, V., "Arbeitsbedingungen beim Konstruieren", **Schw. Masch.** Vol. 77 (1977), Nr. 28 & 30

(16) **ISO 9001: Quality Systems – Model for quality assurance in design/development, production, installation and servicing,** ISO, 1987

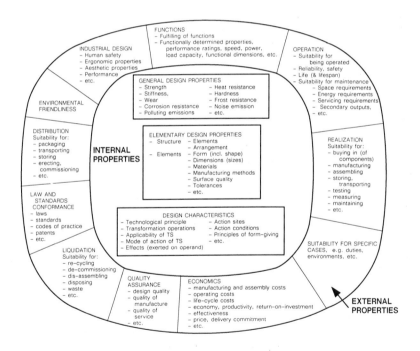

Fig. 2   Relationships Between Properties of Technical Systems

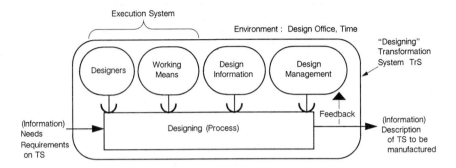

Fig. 1   Model of the Design Process

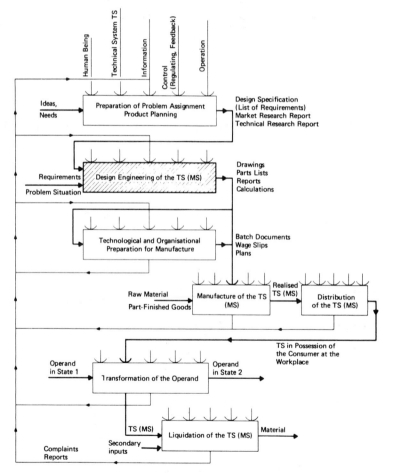

Fig. 3   Life Stages of Technical Systems

The ideal designer possesses:		
Knowledge Awareness	Abilities, Skills	Personal properties, Attitudes

General knowledge
  Languages
  Literature
  History
  Geography
  Mathematics
  Geometry
  Physics
  Chemistry
 	
Specialist technical
  knowledge
  Fundamental
    knowledge
  Specialization area
    knowledge
  Design process
    knowledge
  Manufacturing and
    technology
  Materials knowledge
 	
Economics (macro-
  and micro-)
Legal knowledge
Psychology
Technical aesthetics
Ergonomics
	Intelligence
	Memory
	Logical thinking
	Methodical working
	  mode
	Cost consciousness
	Creativity
	Mental flexibility
	Abilities for:
	  Synthesizing
	  Visualizing
	  Combination
	  Information gathering
	  Deciding
	  Representing
	  Observing
	  Concentrating
	Reliability
	Leadership
	Organization
	Personal appearance
	  and attitude
	Precise mode of
	  expression
	Persuasion
		Capability
		Persistence
		Willpower
		Honesty
		Responsibility
		Duty
		Openness
		Thoroughness
		Conscientiousness
		Care and attention
		Contact
		Broad horizon
		Objectivity
		Critical attitude, incl.
		  self-critical
		Confidence
		Enthusiasm, pleasure
		  in designing
		Cooperative
		Continuous self-
		  education
		Fairness

Fig. 4   List of Requirements as Basis for a Model of the 'Ideal' Design Engineer

Knowledge, Attitudes, Properties	L.E.	D.D.	Drm.	Abilities	L.E.	D.D.	Drm.
Knowledge				Logical thinking	3	3	2
General knowledge	2	1	1	Synthesizing	3	2	0
Basic technical	3	2	1	Cost consciousness	3	3	1
knowledge				Memory	2	3	2
Specialization	3	3	1	Visualization	3	2	2
knowledge				Creativity	3	2	0
				Mental flexibility	3	2	1
Attitudes, Personal				Methodical working	3	2	1
Properties				mode			
Capability	3	3	2	Decision ability	3	2	0
Persistence	3	3	2	Representation ability	3	3	1
Responsibility	3	2	1	Draughtsmanship	1	3	3
Thoroughness	2	3	3	Leadership	3	1	0
Confidence	3	2	0	Organization	2	1	1
Enthusiasm	3	2	1	Concentration	3	2	2
Cooperative	3	2	2	Precise mode of	3	1	0
Continuing self- education	3	1	0	expression			

Legend:		
L.E. ... Lay-out Engineer	3 ... high requirements
D.D. ... Detail Designer	2 ... medium requirements
Drm. ... Draughtsman	1 ... low requirements
	0 ... none

Fig. 5   Depth of Knowledge, and Intensity of Abilities, Personal Properties, etc. for Various Types of Designer

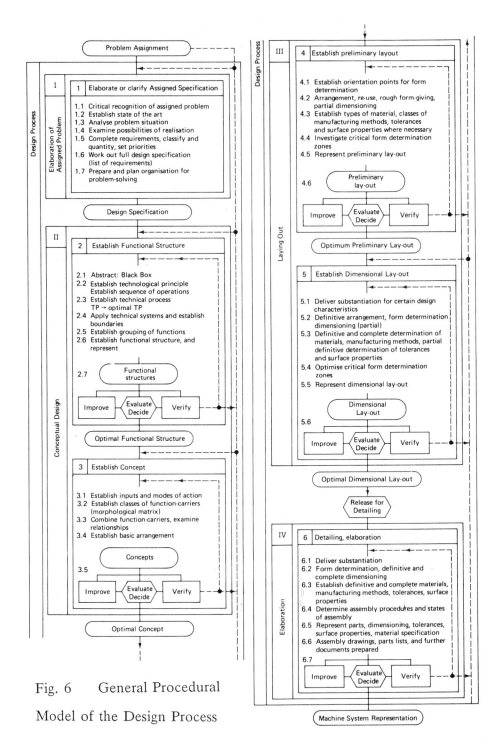

Fig. 6   General Procedural Model of the Design Process

Goals and Characterization of the Design Process	Influencing Operators of the Design Process:						
	Designer	Specialist Information	Method of Representing	Design Methods	Working Means	Design Management	Working Conditions
Quality of the TS to be designed	▼	▼	▽	V	▽	V	V
Design time (duration)	▼	▼	▼	▽	▽	▽	V
Effectiveness of the design process	▼	▼	▼	▽	▽	▽	V
Reducing the risk for the designer	▼	▼	▼	V	▽	V	-
Reducing the amount of routine work for the designer	▼	▽	V	▽	▼	▽	-
Shortening the maturation time for the designer	▼	▼	▼	-	-	▽	▽
Proportion of specialized work of the designer	▼	▽	V	-	V	▽	V
Elapsed time	▼	▼	▼	▽	▽	▽	▽
Costs of designing	▼	▽	▼	▼	▼	▼	V
Teamwork	▼	V	▼	V	-	▽	V

Legend: Influences: ▼ Direct and important ▽ Dominating V Indirect but noticable - Almost none

Fig. 7   Influence of Operators of the Design Process on some Goals and Characteristics of the Design Process